Routledge Revivals

World Mineral Exploration

Mineral exploration is an economic activity of worldwide importance. This volume, originally published in 1988, makes a substantial contribution to the understanding of mineral exploration and the major economic, political, and geologic forces that govern it. Some chapters examine the behaviour and performance of particular participants in the exploration process while others focus on specific countries. This is a valuable title for any student interested in environmental studies and global impact of economics.

W0234860

World Mineral Exploration
Trends and Economic Issues

Edited by
John E. Tilton, Roderick G. Eggert, and Hans H. Landsberg

First published in 1988
by Resources for the Future, Inc.

This edition first published in 2016 by Routledge
2 Park Square, Milton Park, Abingdon, Oxon, OX14 4RN
and by Routledge
711 Third Avenue, New York, NY 10017

Routledge is an imprint of the Taylor & Francis Group, an informa business

ISBN 13: 978-1-138-95930-9 (hbk)
ISBN 13: 978-1-315-66063-9 (ebk)
ISBN 13: 978-1-138-95939-2 (pbk)

World Mineral Exploration

Trends and Economic Issues

JOHN E. TILTON
RODERICK G. EGGERT
HANS H. LANDSBERG
editors

*A Project of
Resources for the Future and
The Colorado School of Mines
in cooperation with
The International Institute for
Applied Systems Analysis*

RESOURCES FOR THE FUTURE • *Washington, D.C.*

Printed in the United States of America

Published by Resources for the Future, Inc.
1616 P Street, N.W., Washington, D.C. 20036

Books from Resources for the Future are distributed worldwide by The Johns Hopkins University Press.

Library of Congress Cataloging-in-Publication Data

World mineral exploration: trends and economic issues / John E.
   Tilton, Roderick G. Eggert, and Hans H. Landsberg, editors.
        p.   cm.
   "Resources for the Future and the Colorado School of
Mines in cooperation with the International Institute for Applied
Systems Analysis."
   Includes bibliographies and index.
   ISBN 0-915707-28-4 (alk. paper)
   1. Mineral industries.   2. Mines and mineral resources.
3. Prospecting.   I. Tilton, John E.   II. Eggert, Roderick G.
III. Landsberg, Hans H.   IV. Resources for the Future.
V. Colorado School of Mines.   VI. International Institute for
Applied Systems Analysis.
HD9506.A2W625   1988   87-27527
338.2—dc19   CIP

This book is the product of RFF's Energy and Materials Division, Joel Darmstadter, director. It was edited by Dorothy Sawicki and Nancy Lammers and designed by Debra Naylor. The figures were drawn by Arts and Words. The index was prepared by Julie Phillips.

Resources for the Future (RFF) is an independent nonprofit organization that advances research and public education in the development, conservation, and use of natural resources and in the quality of the environment. Established in 1952 with the cooperation of the Ford Foundation, it is supported by an endowment and by grants from foundations, government agencies, and corporations. Grants are accepted on the condition that RFF is solely responsible for the conduct of its research and the dissemination of its work to the public. The organization does not perform proprietary research.

RFF research is primarily social scientific, especially economic. It is concerned with the relationship of people to the natural environmental resources of land, water, and air; with the products and services derived from these basic resources; and with the effects of production and consumption on environmental quality and on human health and well-being. Grouped into five units—the Energy and Materials Division, the Quality of the Environment Division, the Renewable Resources Division, the National Center for Food and Agricultural Policy, and the Center for Risk Management—staff members pursue a wide variety of interests, including forest economics, natural gas policy, multiple use of public lands, mineral economics, air and water pollution, energy and national security, hazardous wastes, the economics of outer space, and climate resources. Resident staff members conduct most of the organization's work; a few others carry out research elsewhere under grants from RFF.

Resources for the Future takes responsibility for the selection of subjects for study and for the appointment of fellows, as well as for their freedom of inquiry. The views of RFF staff members and the interpretations and conclusions of RFF publications should not be attributed to Resources for the Future, its directors, or its officers. As an organization, RFF does not take positions on laws, policies, or events, nor does it lobby.

Contributors

Alexander S. Astakhov
Head of Department, Academy of National Economics of the USSR, Moscow, USSR

Theo E. Beukes
Professor of Mineral Economics, Department of Mineral Economics, Rand Afrikaans University, Johannesburg, Republic of South Africa

Willy Chazan
Consulting Mineral Geologist and Retired Chief Engineer, French Ministry for Industry; General Rapporteur of the Committee for the French Minerals Plan, Saint Germain-en-Laye, France

Allen L. Clark
Research Associate and Program Director, Minerals Policy Program, Resource Systems Institute, East-West Center, Honolulu, Hawaii

Melinda Crane-Engel
Research Fellow, Natural Resources Project, Institute for Foreign and International Trade Law, Frankfurt am Main, Federal Republic of Germany

Donald A. Cranstone
Mineral Economist, Resource Evaluation Division; Mineral Policy Sector; Energy, Mines and Resources Canada; Ottawa, Ontario, Canada

Phillip C. F. Crowson
Economic Adviser, The RTZ Corporation PLC, London, England

Michail N. Denisov
Head of Department, All-Union Institute for Mineral Economics, USSR Ministry of Geology, Moscow, USSR

Roderick G. Eggert
Assistant Professor, Mineral Economics Department, Colorado School of Mines, Golden, Colorado

Lo-Sun Jen
Mineral Economist, Resource Evaluation Division; Mineral Policy Sector; Energy, Mines and Resources Canada; Ottawa, Ontario, Canada

Charles J. Johnson
Research Associate, Minerals Policy Program, Resource Systems Institute, East-West Center, Honolulu, Hawaii

Hans H. Landsberg
Senior Fellow Emeritus and Consultant-in-Residence, Energy and Materials Division, Resources for the Future, Washington, D.C.

Brian Mackenzie
Senior Research Associate, Centre for Resource Studies, and Professor, Department of Geological Sciences, Queen's University, Kingston, Ontario, Canada

Henry L. Martin
Acting Director, Information Systems Division; Mineral Policy Sector; Energy, Mines and Resources Canada; Ottawa, Ontario, Canada

Vladimir K. Pavlov
Head of Sector, All-Union Institute for Mineral Economics, USSR Ministry of Geology, Moscow, USSR

Arthur W. Rose
Professor of Geochemistry, Department of Geosciences, Pennsylvania State University, University Park, Pennsylvania

Erich Schanze
Professor of Law, University of Heidelberg, and Coordinator, Natural Resources Project, Institute for Foreign and International Trade Law, Frankfurt am Main, Federal Republic of Germany

John E. Tilton
Coulter Professor of Mineral Economics, Mineral Economics Department, Colorado School of Mines, Golden, Colorado

Roy Woodall
Director of Exploration, Western Mining Corporation Limited, Adelaide, Australia

Contents

Figures

Tables

Foreword

During Resources for the Future's first twenty-five years, mineral economics was a lively topic of research. The late Orris Herfindahl's path-
breaking work on copper costs and prices, Sterling Brubaker's accounts
of the economics of aluminum, the popular *Minerals and Men* by McDivitt
and Manners, and Fischman's *World Mineral Trends and U.S. Supply Prob-*
lems are examples of past achievements.

After a considerable hiatus, this volume marks a notable revival of
activity in the mineral economics area. At the same time, it attests to the
effectiveness of RFF's collaboration with other institutions. At its inception
the Mineral Economics and Policy Program was undertaken as a joint
enterprise by RFF and the Pennsylvania State University. Early on, the
International Institute for Applied Systems Analysis joined the effort and,
in the final phase, as the volume's senior editor shifted location, the Colorado School of Mines became involved. All of this was made possible
by a research grant from the Alfred P. Sloan Foundation, supplemented
by funds from RFF to carry the venture to publication.

The multiplicity of sponsors is paralleled by that of researchers, for it
was clear from the start that mineral exploration was an activity that, on
the one hand, crosses national boundaries and, on the other, lends itself
to comparative analysis. The results—comprising contributions from
American, Australian, British, Canadian, French, German, South African,
and Soviet experts—have fully borne out this expectation. With its publication, the book easily finds its place in the tradition of its predecessors
in the RFF catalog. Together with Roderick G. Eggert's *Metallic Mineral*
Exploration: An Economic Analysis, which conveniently summarizes much
of the material in this book, this volume should appeal to both the expert
and the general reader.

Joel Darmstadter
Director
Energy and Materials Division
Resources for the Future

Preface and Acknowledgments

Mineral exploration and research and development (R&D) have much in common. Both generate new information and knowledge. Both are high-risk investments where the rare successes pay for the many failures. With long gestation periods, even the successes may take years, or decades, to pay back their investors. Both have experienced certain similar trends—over the long run, for example, the apparent waxing of the role of the large corporation and the waning of the role of the lone prospector or individual inventor. Both facilitate the rise in living standards over time and loosen the constraints on economic development imposed by limited resources and capital. In fact, the same word, *recherche*, is used in French for both of these economic activities.

The analogy, of course, should not be overdrawn. Exploration and R&D differ in important respects. It is quite possible, for example, for a private firm or individual engaged in exploration to capture all, or at least the lion's share, of the benefits flowing from its endeavors. This is rarely the case for R&D, despite the best efforts of firms to do so through patents or secrecy.

The original stimulus for this volume arose from a discrepancy of another kind between exploration and R&D: namely, the tremendous difference in attention that these two important activities have received from the economics profession. Over the last several decades R&D and the economics of technological change have claimed the interest of literally hundreds of economic scholars. So many studies have been published in this area that the *Journal of Economic Literature* now includes this topic in its classification system for books and articles. As a result, a great deal is known about the important factors affecting the rate and direction of R&D, although the field continues to be rife with uncertainty and controversy.

By contrast, there is very little literature on the economics of mineral exploration. The resulting void is filled by "plausible assumptions." For example, it is often assumed that exploration productivity is falling over time as the deposits that are easier to find are discovered first; yet the trends in exploration efficiency and the important factors driving these trends are far from clear. We also have much to learn about trends in exploration activity: how much is spent on exploration, where it is spent, how it is spent, and for what types of mineral targets. The research that produced this volume was originally undertaken in the hope of providing insights into such issues.

It was Simon D. Strauss, chairman of the Advisory Committee for the Mineral Economics and Policy Program—a joint venture of Resources for the Future, the Pennsylvania State University (and later, the Colorado School of Mines), and the International Institute for Applied Systems Analysis (IIASA)—who first drew attention to this topic. From the beginning of the project to its end, John H. DeYoung, Jr., a mineral economist at the U.S. Geological Survey and one of the few scholars with a long-term interest in the economics of mineral exploration, offered encouragement, guidance, and particularly thorough reviews of the manuscript. The authors of the individual chapters and their home institutions donated their time and efforts. IIASA staff members Nöel Blackwell and Anna John provided secretarial and research assistance. At RFF, Angela Blake typed major parts of the manuscript, and Helen Marie Streich provided administrative assistance throughout the course of the project. The volume itself also benefited from the careful editing of Dorothy Sawicki and Nancy Lammers.

Earlier versions of the chapters in this volume served as background papers for an IIASA conference on the Economics of Mineral Exploration held in Laxenburg, Austria, in December 1983. The helpful suggestions and insights of the conference participants—mineral economists, geologists, exploration directors, and government officials from Canada, the Soviet Union, Czechoslovakia, the United States, France, Australia, and many other countries—enhanced the analyses that follow.

We would also like to acknowledge the financial assistance provided by a grant from the Alfred P. Sloan Foundation, without which the project never would have been started. IIASA and RFF also contributed significant funding of their own to assure its successful completion.

Research support, such as that provided by the individuals and organizations just identified as well as the many others whose kind assistance has not been explicitly acknowledged, carries with it a responsibility: the recipient, in turn, has an obligation to make a contribution. We like to think that the works contained in this volume substantially enhance our understanding of mineral exploration and the major economic, political, and geologic forces that govern this important activity. Yet we know that what has been done is at best only a beginning. Much work remains before our knowledge of exploration approaches our understanding of R&D and other similarly important economic activities, and we hope that others will continue where we have left off.

John E. Tilton
Roderick G. Eggert
Hans H. Landsberg

November 1987

1

Introduction

Over the past fifteen years, prevailing perceptions of worldwide mineral supply in general and mineral exploration in particular have experienced a complete turnabout. During the 1970s, concerns about resource scarcity and the specter of "running out," in part the result of insufficient exploration, dominated the debate. The following quotation illustrates this view:

The geologic, economic and political conditions in the exploration and mining of minerals are sufficiently similar to the exploration for hydrocarbons as to indicate a potentially more serious problem of rapidly escalating finding costs than for oil and gas. Indeed . . . it would appear that the next crisis will be with respect to *nonenergy* deposits [Ramsey, 1981, p. 336].

As to the causes of the predicted shortages, it was widely believed that mining companies had virtually stopped exploring the mineral-rich developing countries because of political risk, and that in the industrialized nations exploration was being severely hampered by excessively onerous government policies in otherwise attractive areas such as Australia, Canada, and the United States.

Today, less than a decade later, the fear of mineral shortages has been replaced by a sense of resource abundance and of a potentially chronic oversupply of many minerals. It is now widely believed that little, if any, additional exploration will be required for the foreseeable future. Mineral reserves and the stock of known, undeveloped deposits, it is argued, are more than sufficient to meet expected demands.

How is it that within so few years the conventional wisdom about relative mineral scarcity and the importance of mineral exploration has swung from one end of the spectrum to the other—from alarms about depletion and the need for much more exploration to confidence that there is an oversupply of minerals and little need for exploration?

1

One explanation is that what was initially viewed as largely a cyclical downturn in international mineral markets was actually a more fundamental secular or structural change. A complementary explanation is that too much of the debate about mineral exploration—and thus also about mineral supply, of which exploration is a critical component—relies on anecdotal evidence and fragmentary data, often taken out of context and quoted so often as to become "unassailable fact." Compounding the difficulties raised by inadequate data, emotional and deeply held ideological positions often block any true exchange of information on these issues.

This volume proposes to provide a more factual and dispassionate basis for future discussions about mineral exploration than now exists. To this end, the eleven studies that follow address one or more of these sets of questions:

1. How have the overall level and direction of exploration worldwide evolved in recent years? Have the real resources devoted to this activity fallen with the decline in mineral markets since the early 1970s? Have companies conducting exploration shied away from the Third World, as many believe? Have their priorities shifted in terms of the mineral commodities they are pursuing?

2. What are the forces driving recent changes in the level and direction of exploration? How important, relative to one another, are mineral prices, government policies and political risk, technologies of exploration and extraction, and other factors in explaining these changes?

3. How has exploration productivity changed over time? What factors have caused the changes? In particular, has exploration become more costly because the deposits that are easier to find are generally discovered first?

After a brief discussion of the nature of mineral exploration, the sections that follow present an overview of the chapters and draw together some of their major findings and implications.

The Economic Nature of Exploration

Mineral deposits are eventually depleted by mining. If depletion continues unchecked while other economic and technological conditions remain the same, resource scarcity and the costs of mineral production increase, causing real mineral prices to rise. A number of factors, however, are constantly at work counteracting the cost-increasing effects of depletion. Material substitution replaces scarcer materials with more abundant ones—for example, the replacement of wood in many construction applications over the past century with iron and steel, aluminum, and other materials. Technological progress reduces mining and mineral processing costs—the flotation process for concentrating sulfide minerals and the continuous casting of steel products being two of many such developments. Mineral exploration, by leading to the discovery of previously unknown resources, is also among the important forces helping to fend off depletion.

Mineral exploration is primarily an economic activity—specifically, a particular type of investment whose purpose is to find lower-cost deposits than those that are either available or expected to become available. Compared with most investments, exploration is very risky in the sense that the distribution of future payoffs is highly skewed. Most exploration projects fail to discover any economic deposits and hence do not even recover their initial capital investment, let alone earn a competitive return on that investment. Most of the earnings from exploration come from a small number of very lucrative finds. In this regard, mineral exploration bears greater similarity to research and development and the generation of new technology than to typical investments in new manufacturing facilities, roads and airports, or even education.

Nevertheless, exploration is driven by the desire for economic gain realized by finding lower-cost sources of mineral supplies, regardless of who is doing the exploring. Private corporations seek profits over the longer term; some developing countries seek to promote overall growth, using minerals and exploration as part of a larger development plan; and centrally planned countries such as the Soviet Union seek the lowest-cost mineral production for projected consumption.

Exploration is one of several interdependent stages in the mineral supply process—the other components being mine development, extraction, processing, fabrication, and transportation. The role of exploration in this process is to help identify and then evaluate the economic potential of mineral deposits that will augment reserves and provide the ore needed for future mineral production.

Mineral exploration is itself a multistage process—a search and a sifting process—that begins with library research and general reconnaissance of large regions and gradually narrows its focus to a detailed evaluation of small targets, thereby seeking to increase the odds of discovering a deposit that can be mined economically. It is an information-gathering process in which data from geologic mapping, geophysical and geochemical surveying, drilling, and other activities are used to decide whether to advance to the next, more detailed, and usually more expensive stage of exploration; to conduct additional tests at the same scale of activity; or to walk away from a prospect or target. Finally, mineral exploration is a dynamic process that responds to changes in mineral prices and consumer demand, mining and processing technologies, and government policies as well as to new exploration techniques and geologic knowledge.

While successful exploration reduces the upward pressure on the costs of mineral production caused by the exhaustion of operating mines, society has at its disposal other means of abating the adverse effects of depletion. It can, for example, develop known deposits that are not yet working mines. Such deposits are usually of poorer quality than operating mines in the sense that their production costs per unit of output are higher, or at least were higher at the time the operating mines first came on stream. Eventually, however, depletion causes a decline in the ore quality of operating mines and at some point the development of known deposits becomes attractive.

Alternatively, new technology may be developed to extract minerals from known types of deposits that had been technically or economically infeasible to mine. The economic recovery of copper from porphyry deposits, iron from taconite deposits, and nickel from laterite deposits exemplifies the alleviation of upward pressure on mineral costs through technological advances. Innovations that reduce the costs of processing and transporting minerals, and new developments in polymers, composites, ceramics, and other substitute materials that reduce the growth in world demand for the traditional mineral commodities likewise slow or even reverse the rise in material prices over time.

Exploration must compete for funding and resources with these other investments in future mineral supplies. When exploration appears to be the most promising option, it experiences a boom; when other options appear more attractive, exploration declines. Of course, society is usually pursuing all of these possibilities, constantly probing for means of offsetting the adverse effects of depletion. Moreover, since the attractiveness of exploration relative to the other means of enhancing mineral supplies is likely to rise and fall independently for different types of mineral deposits, the waxing and waning of exploration will normally be much more pronounced for individual mineral commodities than for exploration as a whole.

While exploration is fundamentally an economic activity, technological and political developments can greatly influence the overall level and direction of exploration, as is the case with other economic activities. Scientific, cultural, and psychological factors can also be important, as illustrated in subsequent chapters which highlight the impact of new geologic theories, traditional attitudes toward land ownership and mineral rights in developing countries, and the bandwagon effects of major new discoveries. These noneconomic factors influence exploration by altering its expected returns and risks. The revenues that are expected from gold exploration, for example, have been substantially enhanced by advances in heap leaching and the carbon-in-pulp technology of processing gold ores, along with the discovery of low-grade, submicroscopic gold deposits amenable to open pit mining.

Overview

Each of the following chapters is the result of an independent study conducted by its author or authors, whose backgrounds and experience in the minerals field range from that of exploration geologist to academic economist. Their subjects and methodologies vary from the presentation of a heretofore unavailable collection of data on worldwide mineral exploration to case studies of mineral exploration in the developing countries of Botswana and Papua New Guinea to a study of the economic productivity of base metal exploration in Australia and Canada. Some authors concentrate on particular actors or participants in the exploration process, such as major mining companies, while others focus on a particular country such as the Soviet Union, France, or South Africa. Most chapters deal with exploration for nonfuel minerals, and particularly metals, although some take in uranium and coal exploration; oil and gas exploration is specifically excluded.

Because of the diversity of subject matter and treatment, comparisons among chapters should be made with care. Generally, exploration and other expenditures are reported in constant (1982) U.S. dollars, although the authors of three chapters found it preferable for their respective purposes to use constant French francs, Australian dollars, and South African rand. In addition, the various studies employ different time periods, as determined by their particular objectives and the available data.

As noted above, the current state of the data on mineral exploration presents major difficulties for researchers and other would-be users. Thus it is that the authors of these chapters rely on their own sources of information, which are not always comparable or consistent with those used in other chapters. Nevertheless, the volume serves, among other purposes, to present an array of information, much of which was either previously unknown or not readily accessible even to the mineral specialist. As Phillip C. F. Crowson, author of the opening chapter in this set of eleven studies, points out, "If the figures were regularly accessible and of uniform quality, they would have been collected long ago."

Presenting an extensive collection of data in the appendix to chapter 2, Crowson provides an overview of worldwide mineral exploration. After warning of the many traps that threaten the unwary user of exploration data, he discusses the general trends in overall exploration expenditures from the early 1970s to the early 1980s, as well as the pronounced expenditure cycles that lie within the longer-term trends. The data demonstrate that expenditures are very responsive to changing short-term economic conditions. To assess the importance of government policies toward exploration, Crowson compares expenditures in Canada and Australia during the 1970s and poses the question of whether Australia experienced a longer and deeper downturn in expenditures because its policies were more onerous than those of Canada during the same period.

Crowson finds little evidence either to support or to dispute the oft-repeated claim that rising political risk has virtually killed exploration in developing countries. The available data indicate a drop in exploration by European companies in developing countries during the 1970s, but they fail to account for the likely increase in activity among other explorers—particularly state-owned mining enterprises and U.S. oil companies—in these countries. Crowson also suggests that too much attention is often paid to annual exploration activity and not enough to the resulting stocks of projects or undeveloped reserves. Since exploration responds quickly to changing economic conditions, a downturn in expenditures may merely reflect an adequate supply of reserves for a given economic outlook.

The six chapters following Crowson's study examine mineral exploration in particular countries or by particular participants in the world of exploration. Major private corporations are the focus of the study by Roderick G. Eggert in chapter 3. The author identifies and assesses the driving forces behind changes in the level of corporate exploration as well as its distribution by geographic region and by types of mineral targets, using data for the past fifteen to twenty years from a number of large firms, primarily North American and European. Eggert finds that changing mineral prices account for many of the similarities among trends in expenditures by individual companies. Changing prices alter expectations about future prices and revenues from exploration and mining and also affect the available internal funds on which many firms rely for financing exploration. Changing prices also explain some of the differences among firm expenditures, because not all mineral prices rise and fall simultaneously and not all companies depend on the same set of minerals for their incomes. Other factors that account for differences in expenditures among companies include different levels of exploration success or discovery rates and varying corporate goals.

Changing *relative* mineral prices appear to be most important among the factors explaining exploration expenditures devoted to particular types of deposits, although different degrees of exploration success, developments in exploration and production technologies, and new geologic concepts of ore occurrence can be of equal or greater significance in specific instances.

Finally, Eggert considers changes in the location of exploration, concluding that government policies and political risk, though important in particular instances, have been less significant than is widely presumed. Instead, geologic criteria are found to be of overwhelming importance. An initial discovery in an area often leads to a reappraisal of its geologic favorableness for mineral resources and a flurry of subsequent exploration.

In chapter 4, Charles J. Johnson and Allen L. Clark examine exploration in two developing countries, Botswana and Papua New Guinea, reviewing recent trends in exploration expenditures, prospecting licenses, mineral policy and legislation, and discoveries. Although the two countries have markedly different policies, each has successfully sustained a sizable exploration program involving multinational companies during a period when many developing countries have had great difficulty attracting private investment. Botswana issues prospecting licenses quickly, usually within two months of application, but its approach to fiscal terms of mining agreements is ad hoc, and special terms must be negotiated for each discovery. In Papua New Guinea, on the other hand, it may take a year or more for a prospecting license application to be processed, and applications are rejected more frequently than in Botswana. However, fiscal terms and environmental conditions for mining are fixed by legislation in Papua New Guinea, which reduces uncertainty for a foreign explorer in the event of a discovery.

Johnson and Clark conclude that Botswana and Papua New Guinea have been successful for four reasons: (1) Each has a government perceived by the international mining community to be honest, stable, and fair. (2) Both countries have established records of successful mineral development by the private sector. (3) Geologic conditions in both countries are highly favorable for certain minerals. (4) Both are adjacent to mineral-rich developed countries (Papua New Guinea being near Australia and Botswana bordering on South Africa) that provide a substantial spillover of exploration and mining expertise.

In chapter 5, Theo E. Beukes examines exploration in one of these developed countries—South Africa. After describing the structure of the mining industry there, he reviews trends in overall exploration expenditures from 1960 to 1983 and the allocation of these expenditures by mineral type. He then assesses the important factors underlying these trends. Beukes identifies four distinct periods of mineral exploration in South Africa since 1960: (1) During the 1960s expenditures remained relatively constant and were focused on precious metals, primarily gold. (2) The period from 1969 to 1974, during which expenditures rose significantly, was marked by increased interest in nonprecious metals. (3) Over the next five years, 1974 to 1979, expenditures for nonprecious metals fell by over 50 percent, while expenditures for precious metals, uranium, and coal increased. (4) Finally, from 1979 to 1983, real expenditures doubled for precious metals and increased somewhat less for uranium, coal, and nonprecious metals.

As for the determinants of these trends, Beukes concludes that the changing political climate in southern Africa brought about by the independence of former European colonies has, on balance, encouraged exploration in South Africa. Even though many of the new states are openly hostile toward that country, exploration has fled much of the rest of Africa because the risks are perceived to be much greater than in South Africa.

Beukes points to the importance of changing mineral prices in explaining the rise and fall of base metal exploration in South Africa in the 1970s, the changes in coal and uranium exploration since 1973, and the renewed interest in gold exploration since 1972. Other factors that help explain exploration trends in that country are the entry of foreign multinational companies between 1968 and 1977, favorable mineral policies, and new geologic approaches.

Exploration in the Soviet Union is discussed in chapter 6 by Alexander S. Astakhov, Michail N. Denisov, and Vladimir K. Pavlov. Their study demonstrates that, despite many similarities between the sequence of exploration activities in the Soviet Union and the West, central planning and state ownership of mineral resources have a profound influence on the search for mineral resources.

The analysis in chapter 6 begins with a review of the role of exploration in the Soviet mineral supply process and a description of the stages and organization of exploration there. It then traces the historical evolution of exploration goals since 1917, the development of mapping and other efforts to improve exploration efficiency, and recent trends in the level and distribution of prospecting and exploration expenditures. Between 1961 and 1980, these expenditures increased at an annual average rate of between 6 and 7 percent, reflecting not only increased geologic activity but also rising costs associated with exploring in more remote areas and in more difficult geologic environments. The share of total expenditures devoted to the early stages of exploration has grown over the past ten years or so as activity has increased in less developed and previously inaccessible regions of the country.

Finally, the authors discuss planning and decision making in Soviet exploration. They explain that scenarios of future economic development and associated mineral requirements lead to goals for mineral production and, in turn, to goals for mineral discovery. The level of exploration expenditure and its distribution by geographic area and mineral deposit type are based on these discovery goals.

In chapter 7, Melinda Crane-Engel and Erich Schanze examine the world's two largest programs offering multilateral mineral exploration assistance: the United Nations Development Programme (UNDP) and the United Nations Revolving Fund for Natural Resources Exploration. Multilateral exploration aid has been motivated (1) by concern over potential mineral supply problems, including higher prices and physical shortages, which it is feared will result from inadequate exploration in geologically attractive developing countries because of political risk, and (2) by the desire to promote economic development in potentially mineral-rich areas that lack the necessary capital or technical expertise.

As Crane-Engel and Schanze point out, mineral exploration and development projects are a small part of UNDP's universe of activities (recently amounting to about $18 million per year, or 3 percent of UNDP expenditures) and are carried out by other UN agencies, about 85 percent by the Department of Technical Cooperation for Development. More than two-thirds of UNDP-funded projects have involved direct exploration, while the remaining projects have focused on improving institutions involved with minerals, such as geological surveys and university geology departments. UNDP-assisted exploration has resulted in the discovery or contributed to the detailed delineation of fourteen deposits that are being mined now or that could be mined in the near future.

The Revolving Fund, established in 1973, had spent some $35 million by the end of 1982 on sixteen projects, about half of this amount in Africa. The Fund is intended to be self-supporting, with contributions from the revenues of successful projects supporting all exploration. Projects are chosen on the basis of their potential commercial viability, and therefore the Fund conducts no broad geologic surveys, training programs, or institution-building activities. It generally takes exploration only up to the point of detailed evaluation of a prospect, stopping just short of the feasibility study, and then attempts to make information available to potential investors. Although four Revolving Fund projects are considered successful (having located mineral deposits in Argentina, Benin, Cyprus, and Ecuador), they all need further work to determine if development and mining are justified.

French mineral exploration between 1973 and 1982 is the subject of the study in chapter 8 by Willy Chazan. The analysis takes in (1) all exploration on French territory regardless of the nationality of the explorers and (2) exploration elsewhere by French organizations, including those with minority French interests. French exploration is of particular interest because it includes large-scale activity by both private and public corporations.

Based on data collected annually by the Department of Mines of the French Ministry of Industry and Foreign Trade, Chazan shows that real exploration expenditures for metals nearly doubled over the ten-year study period. Much of this increase was due to uranium exploration, which consumed a growing share of exploration funds following France's decision in the 1970s to rely increasingly on nuclear power. As for French exploration productivity, Chazan estimates that the ratio of the gross value of discoveries to exploration expenditures was 22 over the 1973–82 period, which compares favorably with the ratio for most other countries.

The next three chapters examine exploration productivity in other parts of the world. In chapter 9, Donald A. Cranstone analyzes mineral discoveries in Canada since World War II. He finds that the rate of metal discovery (measured as the gross value of discovered metal) was highest in the mid-1950s, as a result of the first application of modern geophysical prospecting techniques. Since then, the discovery rate has been considerably lower but has exhibited a slightly rising trend. Another measure of exploration success, gross metal value per exploration dollar, was also highest in Canada during the mid-1950s, remaining lower and fairly steady since then.

Cranstone demonstrates the importance of large deposits in any calculation of exploration success: roughly half of the gross metal value of some 900 deposits discovered in Canada from 1946 through 1982 comes from the 30 largest and 80 percent from the 120 largest deposits. Finally, he shows that discovery rates for individual metals and specific types of deposits tend to be extremely cyclical over time for a number of reasons, including discovery of large individual deposits or districts and widespread application of new exploration techniques.

Exploration success in the United States during the postwar period and exploration planning within large U.S. corporations are examined in chapter 10 by Arthur W. Rose and Roderick G. Eggert. They find that, compared with the data for Canada, far fewer usable data are available on U.S. exploration expenditures and metallic discoveries because companies tend not to release this information and the government's figures are less complete. Nevertheless, rough estimates of the gross value of metallic discoveries show large year-to-year fluctuations but no clear upward or downward trend between 1955 and 1980. Even though success ratios (gross discovery values/exploration expenditures) are difficult to calculate— especially for recent years because of uncertainty over which deposits will come into production—they apparently have declined over time.

Based on Rose and Eggert's analysis of discoveries by types of firm making the discoveries, large oil companies appear to have been less successful than major mining companies and small mining and oil companies. The authors also demonstrate that exploration for and discovery of particular types of deposits are episodic or cyclical, even though aggregate exploration expenditures show more moderate short-term fluctuations. Finally, Rose and Eggert describe two general styles of exploration management and corporate planning—the one involving team management and the other characterized by a strong manager who dominates exploration planning.

In chapter 11, which compares the economic productivity of base metal exploration in Australia and Canada, Brian Mackenzie and Roy Woodall assess the net rather than the gross returns from exploration. They incorporate development and mining costs, the time value of money, and other factors, as well as exploration expenditures, into their productivity calculations. The authors develop and employ a set of criteria for choosing economic deposits from among the many postwar discoveries of mineralized rock. The value of these discoveries is then compared with exploration expenditures not just on these deposits but on all exploration projects.

The Mackenzie–Woodall calculations show that the expected value of base metal exploration is positive in Canada but negative in Australia. Compared with those in Canada, economic deposits in Australia cost four times as much and take four times as long to discover and evaluate. The Australian deposits are, however, on average three times as large as those in Canada.

The authors of chapter 11 suggest that base metal exploration has been less productive in Australia due to that country's deeply weathered soil and rock, rendering nearly useless many geophysical and geochemical techniques that are important in Canada. They examine exploration productivity in various geologic environments within Australia and Canada, changes in productivity over time, and the sensitivity of their findings to different assumptions about metal prices.

Chapter 12 considers a subject closely related to mineral exploration and discovery—ore grade trends over time. Henry L. Martin and Lo-Sun Jen examine Canadian ore grades for seven nonferrous metals from 1939 to 1979 and offer educated guesses for 1989. They test the widely held notion that ore grades inevitably decline in a country because of the inexorable forces of physical depletion. Their results indicate no clear and consistent decline in ore grades for copper, molybdenum, zinc, nickel, and silver. Lead and gold grades, on the other hand, have declined over the period of study. Martin and Jen conclude that the decline in the quality of ore deposits, as measured by ore grade, is proceeding much more slowly than is widely assumed.

Chapter 12 further recognizes that depletion is only one of several reasons for changes in ore grade over time. A decline may result from cost-reducing improvements in mining technology or the recognition of a new type of mineral deposit; a rise may result from new discoveries, as is expected to happen with gold because of the recent discoveries in the Hemlo area of Ontario.

Findings and Implications

This section draws together a number of points from the chapters in this volume that provide insights into the level and direction of exploration in recent years, important forces driving these developments, and changes in the efficiency or productivity of exploration over time. Some of the implications of the findings for both firms and governments are also assessed.

Recent Trends

Shortcomings in the data make trends in the overall level and direction of exploration difficult to measure accurately. Not only are official statistics not collected in many countries, but firms and other organizations conducting exploration consider their data to be proprietary. Moreover, differences in methods of collecting the available statistics and in procedures determining which expenditures to include under exploration create problems of comparability. Moving from the national to the global level to assess exploration trends compounds these problems.

Nevertheless, the available information does reveal several pertinent characteristics of world mineral exploration. First, as Crowson points out, real expenditures on exploration outside the centrally planned economies increased by roughly 50 percent between the early 1970s and the early 1980s. In light of the depressed conditions that plagued the copper, steel, and many other metal industries after 1974, this rise in real expenditures seems surprising. The increase, however, was concentrated on coal and uranium exploration; nonfuel mineral exploration over this period appears to have suffered a decline. Moreover, the available data indicate that exploration for metals, coal, and uranium has dropped off sharply since the early 1980s.

Among the market economy countries, exploration is concentrated in the developed countries, particularly the United States, Australia, Canada, and the European states. It is widely assumed that this high concentration is largely the result of political instability and unfavorable public policies in the developing countries, that these factors in recent years have caused a geographic shift in exploration away from the Third World toward Australia and North America. The evidence, however, does not reflect such a shift, at least in recent years: between one-fifth and one-third of all exploration outside the centrally planned states took place in developing countries both in the early 1970s and in the early 1980s. Still, the developing countries' share of total exploration seems relatively small considering their geographic area and geologic potential for mineral resources. Presumably, more favorable policies and greater political stability would help rectify this imbalance.

Analysis of trends in exploration activity encompasses not only shifts in location but also changes in the types of mineral commodities sought. As mentioned, the 1970s and early 1980s saw exploration activity increasingly directed toward discoveries of uranium and coal deposits. In addition, rather dramatic changes occurred in the search for nonfuel minerals. Interest in molybdenum and cobalt rose and fell with their prices. The search for large, low-grade, porphyry copper deposits all but ceased as firms redirected their efforts toward smaller, high-grade, massive sulfide deposits in which copper and other base metals are accompanied by gold and silver. Perhaps the most spectacular development during the period under discussion was the surge in exploration for gold, spurred on by technological advances in gold processing and by much higher real prices following the demonetization of gold in the early 1970s.

Within the secular trends noted, exploration has displayed its cyclical or episodic nature, often rising or falling sharply from year to year. Expenditures are highly sensitive to changing mineral prices and to mining company profitability. At first blush this seems irrational, since the expected returns from exploration depend not on current mineral prices but on prices years in the future. Expectations about prices and profitability, however, are strongly influenced by the present and the recent past. Moreover, many exploration groups rely almost solely on internal funds to finance exploration, implicitly assigning a high cost of capital to funds obtained from external sources for this very high risk activity.

This cyclical volatility is often more apparent in individual companies than in the aggregate. Lumpiness can result from large expenditures on one project at advanced stages of exploration—stages that account for a sizable portion of total expenditures. Once detailed exploration is completed, exploration expenditures may drop sharply as a project moves to the development stage or is abandoned.

Expenditures for specific minerals, deposit types, and, in particular, geographic regions also tend to be cyclical. A surge in exploration begins with some sort of stimulus—a rise in mineral prices, the identification of a new deposit type, or a discovery in a region previously thought to lack mineral resources. A flurry of exploration follows, often leading to many discoveries. The cycle is completed as exploration and discovery decline because the metal from the new discoveries appears to be in oversupply or because the easy-to-find deposits of a new type or in a new region have been discovered.

Exploration Productivity

Shifts in the overall level and direction of exploration activity are important as an indication of where new discoveries are likely to take place and hence of the location of future mineral supplies. They also reflect changes in societal priorities for particular mineral commodities.

Ultimately, however, it is the output, not the input, of the exploration process that tells the story: if exploration productivity—that is, output per unit of input—is falling, future mineral supplies could be in jeopardy even with rising exploration expenditures. Indeed, one index of resource scarcity is the cost of discovering new deposits, which is simply the inverse of one measure of exploration productivity. Growing scarcity is implied if, for example, a larger real expenditure is, on average, required over time to find an additional million tons of copper reserves of a given quality.

Tracking trends in the productivity of exploration is even more complicated and difficult than determining trends in expenditures, for it requires information on the latter (a proxy for the inputs consumed in the search for new mineral wealth) plus some measure of output or success. Not only are the data requirements more demanding, but some conceptual issues are raised as well, since an assessment of exploration output depends on how success is defined.

For a firm, success implies a reasonable rate of return on its investment in exploration. Thus, exploration productivity can be assessed by the expected financial return from exploration divided by its costs, properly discounted for time and risk. Expected financial return can be assessed by the market value of the expected discoveries if a reasonably well-functioning market for undeveloped deposits exists, or, alternatively, by the estimated net present value of the expected discoveries prior to development.

Measured in this way, exploration productivity provides an indication to a firm of the relative attractiveness of exploration as a type of investment. It also allows exploration to be compared with the other means (e.g., technological developments) available to society for offsetting the cost-increasing effects of mineral depletion. It does not, however, necessarily reflect the overall success of society in keeping the cost-increasing effects of depletion at bay. Indeed, with a rapid rise in mineral prices that was expected to continue, exploration productivity so measured could be increasing even as resource scarcity was increasing.

To serve as a valid measure of scarcity, then, the success or output of exploration must be measured in terms of reserves discovered, somehow standardized for differences in quality. One way of doing this is by assessing the value of discoveries made over time using a common set of cost and price conditions. The chapter by Mackenzie and Woodall and earlier works by the same authors are among a very small group of studies that attempt to measure exploration output in such a manner. In this volume, Mackenzie and Woodall identify the discoveries of base metal deposits resulting from exploration over the 1946–77 period in Canada and over the 1955–78 period in Australia. They then estimate the net present value of these discoveries assuming development at the present time under current cost and market conditions. By averaging the results they determine an expected value per discovery, which, when compared with the average exploration expenditure required per discovery, provides a measure of exploration productivity.

Using this approach, Mackenzie and Woodall find the tremendous difference between exploration productivity in Canada and Australia noted earlier. Their results also show some deterioration in exploration productivity in both countries over time. The authors suggest, however, that this latter finding could simply reflect a bias in the data base arising from the tendency to undervalue the most recent discoveries rather than from an actual increase in the scarcity of base metal resources.

Despite its conceptual integrity, the methodology used by Mackenzie and Woodall has not been adopted by other researchers. It is simply too complex and data-intensive. In addition, it rests on a number of fairly critical assumptions regarding the size of an efficient exploration budget and other matters. As an alternative, the chapters by Chazan, Cranstone, and Rose and Eggert assess exploration productivity by comparing the gross value of discovered metal with exploration expenditures. This procedure ignores the fact that production costs can vary greatly from one deposit to another, and so it provides a valid measure of mineral scarcity only if the production costs associated with deposits discovered over various historical periods have on average remained the same—clearly a very critical assumption.

Chazan examines exploration in France and by French companies abroad between 1973 and 1982 without attempting to calculate trends over time for this relatively brief period. He does, however, compare the gross discovery value per unit of expenditure for French mineral exploration with that for Canada and the United States. His finding that the three are roughly comparable is not what one would expect if depletion were a serious problem; one would expect exploration productivity to be lower in a country such as France where firms have been exploring for centuries and even millennia. This somewhat unexpected finding can be explained in part by the fact that French exploration productivity includes discoveries made outside of metropolitan France—in former colonies and elsewhere— by French companies. In addition, despite the long history of French mineral exploration, it is only in recent decades, of course, that substantial exploration interest has developed in uranium and alloy and specialty metals. Exploration productivity by French companies is as high in France as abroad for these mineral products. It is somewhat lower in France than abroad for precious metals, and very much lower for base metals.

Cranstone's study of exploration in Canada, covering a much longer period than that of the Chazan study, considers in some detail trends in gross discovery value per exploration dollar for the years 1946 through 1982. In the early postwar period of 1948 through 1956, exploration in Canada was extremely productive, largely because of the successful application of new geologic models and new geophysical prospecting methods. Since the mid-1950s, exploration productivity as measured by gross discovery value per exploration dollar has been lower, but it has displayed no tendency to decline over time.

In the United States and the Soviet Union, depletion may be more seriously affecting exploration productivity, although the available data for both of these countries are much weaker than for Canada, Australia, and France. Rose and Eggert suggest that the gross discovery value per exploration dollar declined in the United States over the period 1955 through 1982. The extent of this decline, however, depends on how the available data are adjusted for the tendency to underestimate the value of recent discoveries. Under certain assumptions the decline is quite modest. Though comparable data are not available for the Soviet Union, Astakhov and his coauthors discuss the prevailing view of Soviet experts as reported in a recent survey—that the discovery cost per ton of metal would rise in the future by at least 30 percent.

A third, and quite different, approach for assessing the effects of depletion on mineral scarcity is taken by Martin and Jen, who study trends in the grade of ore mined and expected to be mined in Canada over the 1939–89 period. Here the focus is on the nature of the deposits actually being exploited rather than on the nature of new discoveries, and no account is taken of the amount spent on exploration. Ore grade, of course, is only one of the factors determining the quality or production costs of mineral deposits; location, depth, deposit size, and type of mineralization are also important. But to the extent that these various determinants move together, ore grade can serve as their proxy in indicating a deposit's overall quality. When ore grades are stable or declining very slowly, it would appear that exploration has been sufficient to keep the cost-increasing effects of depletion at bay; rapid declines in ore grades, on the other hand, would suggest that depletion poses a more pressing threat to the welfare of mankind. As noted earlier, Martin and Jen find no clear and consistent decline in ore grade for most of the mineral commodities that they examine.

Overall, then, the evidence on exploration productivity and the adverse effects of depletion on the availability of mineral resources is mixed. In Canada, where exploration activity has been analyzed most thoroughly and where the available data are relatively good, exploration productivity appears to have enjoyed a golden age in the early postwar period as a result of new geophysical techniques and scientific advances in geology. And, in the past twenty years, Canadian exploration productivity, though lower, has been stable. Whether productivity in other countries would display the same stability if the available data and analyses were of comparable quality, we simply do not know. While that is possible, it would not be surprising to find exploration productivity declining in the United States, Europe, the Soviet Union, and other industrialized regions that have heavily exploited their mineral resources for domestic purposes. What *is* clear, however, is that if exploration productivity is declining, it is not declining at a rapid rate, at least in the countries examined in this volume. If it were, the available evidence would much more clearly and consistently reflect this trend.

Public Enterprises and the Role of Government

Exploration, as noted earlier, is a multistage activity that includes the development of geologic models of ore occurrence and ore genesis; the collection of basic geologic information over large regions from aerial photography, satellite imagery, airborne geophysical surveys, geochemical sampling, and geologic fieldwork and mapping; the identification of specific targets using geological, geophysical, and geochemical techniques; and the estimation of ore reserves in specific deposits by drilling and other methods. Various types of firms and other organizations around the world participate in exploration at one or more of these stages—large multinational mining companies, small exploration companies, universities, major international oil companies, lone prospectors, government agencies, state mining companies, and the United Nations and other international organizations.

Because most of the world's outcropping ore bodies and other easy-to-find deposits have already been discovered, the role of the lone prospector has declined, forcing exploration to rely on more costly and sophisticated techniques to locate deposits buried beneath thick layers of overburden.

In the early years after World War II, the traditional multinational mining companies dominated nonfuel-mineral exploration outside the centrally planned countries, but in recent years their influence has also waned, largely as a result of the growth of state mineral enterprises in the Third World and elsewhere, and, more recently, because of the growth of small exploration companies and partnerships.

Not only are there many different types of organizations participating in exploration, but among countries there are interesting differences in the roles of participants. In the United States, for example, the Geological Survey (an agency of the U.S. Department of the Interior), university geoscience departments, some mining companies, and several state geological surveys are actively involved in the creation of geologic models and the collection of regional geologic information. The expensive, downstream stage of the exploration process—identifying specific targets and appraising their size and grade—is left to private enterprise, even though about one-third of the land area of the country is federally owned. In the Soviet Union, by contrast, all phases of exploration are carried out by state institutions. Between these two extremes are France, Canada, and a host of other developed nations as well as many Third World countries, where public and private firms work side by side in the identification and economic assessment of new ore deposits.

One reason often given as justification for government involvement in exploration, particularly at the early stages, is that some of the benefits are difficult for private firms to capture or internalize. This is because exploration, viewed as a production process, generates two goods—the first being the discovery of economic deposits; the second, new information. The new information may take various forms: that of maps or aerial surveys revealing geologic characteristics of a region, of data supporting a particular theory of ore genesis, or simply of experience—for example, that a particular exploration effort did or did not result in the discovery of an economic mineral deposit within a given geologic region. The benefits associated with the discovery of a new ore deposit can normally be captured fully or almost fully by the firm conducting successful exploration. However, only a portion of the benefits flowing from new information are internalized by the firm; the rest of these benefits fall within the domain of "external economies," that is, they are benefits that other members of society enjoy but which are not captured by the firm doing the exploring. Wherever external economies exist, the private sector, without government support, tends to underinvest in an activity from the point of view of society as a whole—in this specific case by not spending enough on exploration.

A second rationale for government involvement in exploration relates to the high-risk, long-term nature of exploration investment, which, even if successful, often takes years or perhaps decades before it pays off. This argument rests on the assumption that the private sector is generally more risk averse and has a higher rate of time discount (that is, a preference for a faster payoff) than society in general, and so tends to discriminate against investments such as exploration. Government intervention may also at times be needed to ensure the security of material supplies, particularly materials coming from overseas, for critical domestic uses.

None of these considerations, however, completely justifies actual government participation in the exploration process, since any tendency by the private sector to engage in too little exploration could be corrected by government grants or other support to private enterprises. Ultimately the case for or against active government involvement in exploration must rest on the effectiveness or productivity of government in this activity as compared with other active or potential participants, although there is little empirical evidence available on the relative productivity of the various firms and organizations engaged in exploration.

Over the past several decades, research on technological change has found that the creation and adoption of new technology are stimulated by low entry barriers and a variety of types of enterprises (Scherer, 1980, chap. 15). Perhaps the same is true for exploration. Moreover, the relative effectiveness of different types of participants may vary with the stage or phase of exploration. Universities and other academic institutions may enjoy a comparative advantage in generating new geologic knowledge and models of ore genesis; geological surveys and similar government agencies in collecting and disseminating general geologic information; small, specialized exploration companies in identifying specific targets; and large, multinational mining companies in marshaling the funds and expertise required to appraise and develop the best of these targets.

The foregoing, however, is only one of the possible hypotheses regarding the relative effectiveness of different participants at various phases of exploration. Much more research is needed to resolve this issue. Nor is it merely idle curiosity to call for its resolution. The following chapters point up considerable differences among countries in organizing and carrying out exploration. It is important to discover just how these differences affect exploration productivity.

There is much more that we need to learn about mineral exploration; indeed, in many respects our understanding of the activity is still embryonic. What we do know is that exploration is fundamentally an investment in the future for the principal purpose of discovering low-cost mineral deposits. We know that it is one of several types of investment that enhance the availability of mineral supplies and offset the inevitable cost-increasing effects associated with the depletion of existing mines. Furthermore, the resources devoted to exploration vary greatly over time in response to the perceived threat of depletion and the relative effectiveness of exploration in discovering new ore deposits to cope with this threat. The perceived threat and relative effectiveness of exploration, which are influenced by technological change, political conditions, and other considerations, determine the economic benefits or returns from exploration and provide the motivation for exploration by both private firms and public institutions.

We also know that the available data, though far from being complete and consistent, show no sharp drop in exploration productivity over the last several decades. Indeed, for a number of the countries examined here, exploration productivity may not have dropped at all. Thus, advances in science and exploration technology seem largely to have offset the tendency for exploration productivity to decline as the remaining number of undiscovered, relatively easy-to-find deposits declines. And finally, as a reflection of trends in resource scarcity, exploration productivity suggests that depletion has not in recent years posed a particularly serious threat to the availability of mineral supplies.

References

Ramsey, James B., ed. 1981. *The Economics of Exploration for Energy Resources* (Greenwich, Connecticut, JAI Press).

Scherer, F. M. 1980. *Industrial Market Structure and Economic Performance*, 2d ed. (Chicago, Rand McNally).

2

A Perspective on Worldwide Exploration for Minerals

PHILLIP C. F. CROWSON

The objective of this chapter is to provide as firm a factual basis as possible for debate about patterns of mineral exploration spending throughout the world, both geographically and over time. The data compiled for this purpose are presented in the chapter's statistical appendix.[1] After describing some of the major problems in gathering, arranging, and utilizing these data, the chapter focuses on a few of the many possible conclusions that can be drawn from them. The sections that follow (1) summarize statistics on mineral exploration expenditures, (2) discuss trends in these expenditures, (3) consider aspects of exploration costs such as the relative effectiveness of expenditures in different parts of the world, and (4) describe the roles of public and private organizations that are active in the various stages of exploration.

Difficulties in Compiling Exploration Statistics

If they are published at all, mineral exploration data usually appear in national sources, and they are rarely, if ever, drawn together in a consistent fashion. In many instances even the national statistics are not widely known

---

[1]The figures and tables in the appendix are arranged in thirty-five parts, hereafter referred to simply by number—e.g., part 2-A-1. The table of contents at the beginning of the appendix (see page 40) lists the parts by title.

or quoted. This relative inaccessibility of data gives rise to generalizations or to imprecise qualitative impressions—which, of course, are invariably rejected in favor of any available quantitative information, however partial or ill-founded, especially when the latter supports a user's case. Thus, discussions of trends in exploration nearly always rely on a few well-thumbed references that may lack any firm statistical foundation, but which are quoted so often as to become unchallengeable "facts."

Tracing such "facts" back to their initial source can become a fascinating, if unproductive, game of detection. Myths have emerged about exploration patterns and, like all good myths, they may have had some factual basis, but they have also developed from special pleading by particular interest groups. The most potent of these is that exploration for minerals has virtually dried up outside Australia, Canada, South Africa, and the United States; another is that a lack of exploration for a few years will lead to acute supply problems in some ten to fifteen years' time.

Thus, the work of tracking down and putting together statistics on exploration expenditures has proved to be something of an exercise in frustration for several reasons. Far fewer data have proved accessible than presumably exist. Different countries collect data on different bases if they collect statistics at all, and the figures published by companies are even less uniform. Where statistics are collected regularly, their coverage and definitions have tended to change over the years so that long runs of data are rarely available. The whole field is littered with elephant traps for the unwary. There is, however, little to be gained from bewailing the inadequacies of statistics; if the figures were regularly accessible and of uniform quality, they would have been collected long ago.

The Stages of Exploration: A Problem of Definitions

One of the main problems in compiling statistics in this field is that there is no clear-cut and widely accepted definition of exploration. As shown in figure 2-1, several stages are typically involved in the development of a new mine. (Although the figure is based on the development of a uranium mine, the stages listed are broadly representative for most minerals.) In essence, each stage represents a closer focus on a small target area that will, with luck, prove suitable for development as a mine.

The first three stages shown in figure 2-1—reconnaissance, prospecting, and focusing on the ore body—may overlap or even be telescoped into one. Exploration clearly embraces these first three stages as well as stage 4, delineation of the ore body. It also includes at least part of stage 5—feasibility study—because the decision to mine is usually made only when such a study has been completed. The filing cabinets of mining companies and mining ministries are stuffed with feasibility studies on which no action has been taken. Hence, it is usual to include all the costs of the feasibility study in exploration.

Two stages in the exploration and development of a mine are not shown in figure 2-1. The first of these is the basic geologic surveying and mapping of an entire region or country, which sets the backdrop for subsequent exploration. This stage is seldom included in statistics of exploration spending.

When mines have been operating for some time they may need to prove additional ore reserves in order to prolong the working life of the processing equipment or to expand. In such cases there is a hazy dividing line between exploration and development expenditure. When drilling is within the mine itself it may be classified as development, but "step-out" drilling to delineate adjacent ore bodies—the second stage not shown on the figure—should properly be classed as exploration. In practice it is doubtful whether many statistics include this stage as exploration, although the expansion of existing mines and processing facilities often makes a greater contribution to increased total output than do completely new mines. (In Australia a distinction is made between exploration on production leases by operating or developing mines—but excluding advance development work in underground mines—and all other exploration. In Fiscal Year (FY) 1981/82 the former accounted for nearly 8 percent of total exploration, compared with an average 11 percent annually for the preceding decade [see statistical appendix, part 2-A-2]).

The Statistics on Exploration

The greater part of this chapter is contained in its statistical appendix, which assembles details of the trends and patterns in mineral exploration for as many countries as possible, taking the data as far back as possible.

With regard to the data themselves: they include expenditures on coal and uranium—exploration for these minerals is analogous to that for the other hard minerals—but they do not include spending on oil, gas, and other hydrocarbons. (The latter greatly dwarfs other exploration spending, as the following figures from the *Census of Mineral Industries 1977* demonstrate: in 1977, U.S. exploration expenditures for oil and gas were $8 billion, versus $0.5 billion for all other minerals.) In some instances it is not possible to separate exploration expenditures for the hard fuel minerals from the rest. Since all the hard minerals—fuels and nonfuel—compete for similar resources of both manpower and hardware, the only reason to distinguish is to show the likely impact of exploration on the availability of individual minerals. Nonetheless, an effort is made in this chapter to give an approximate breakdown between the fuel minerals and the rest.

The mineral exploration statistics that are available for many countries cover construction materials such as sand and gravel, even though such materials are not usually categorized as minerals. Including these costs does not distort the numbers, however, because little expenditure is needed

Year number	Stage	Size of concession maximum (in 1,000 km²)	Typical ranges of expenditure (millions of U.S.$)
−2	1. Reconnaissance	1,000	Up to 0.3
	• Literature studies		
	• Selection of favorable areas		
	• Initial field reconnaissance		
−1	• Land acquisition		
0	2. Prospecting	100	0.4—1.2
	Regional geologic, airborne geophysical, and geochemical surveys		
1	3. Focus on Ore Body	10	2.5—50
	• Detailed surveys, e.g., geologic mapping, radiometry, emanometry, geochemistry		
	• Trenching		
	• Scout drilling		
2	• Laboratory work, e.g., mineralogy and metallurgical tests		
3			
4	4. Delineation of Ore Body	1	2.5—50
	• Pattern drilling		
	• Bulk sampling		
	• Metallurgical work		
5	• Geologic ore reserve calculation		

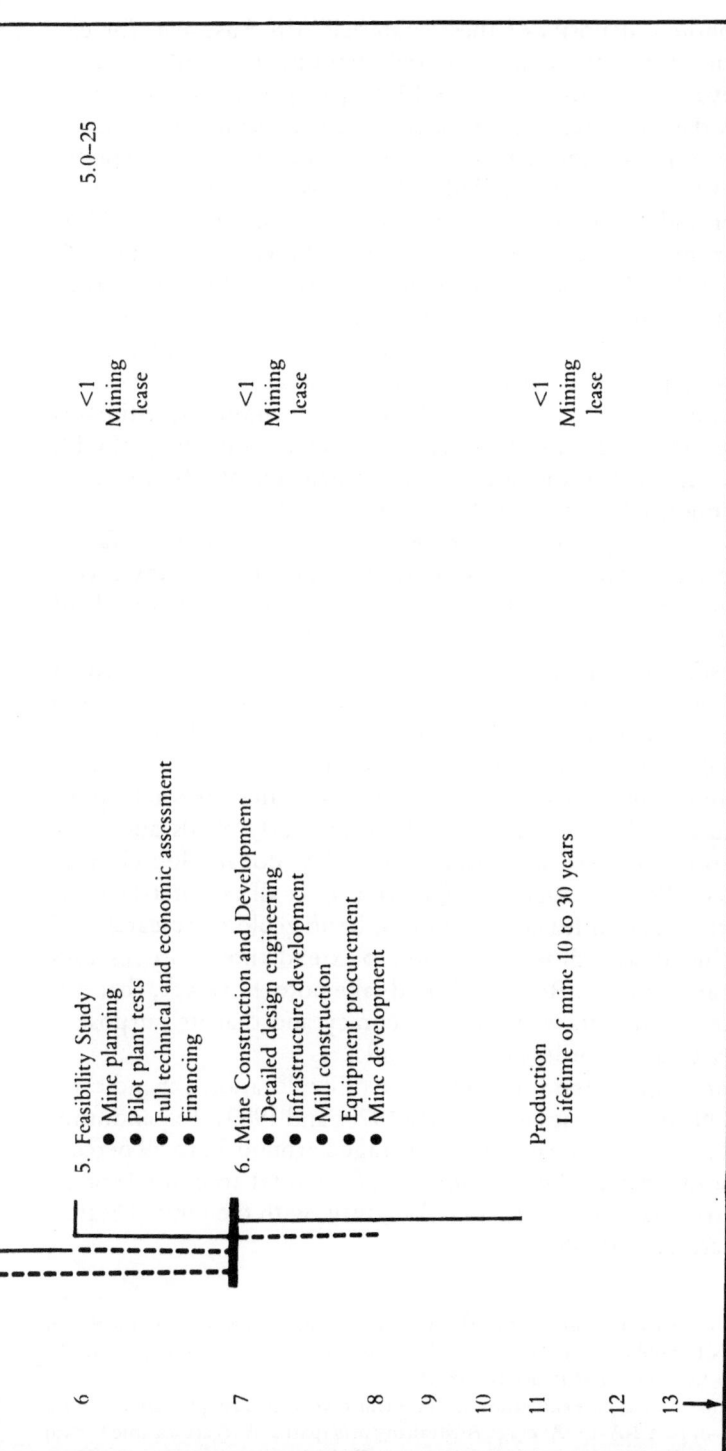

Figure 2-1. Typical stages in the exploration and development of a mine. *Source:* Based on chart in Uranium Institute, "Comparative Lead Times for Uranium Mine and Nuclear Plant Construction," unpublished paper (London, March 1982).

to delineate economic deposits of these materials. (In Australia, for example, exploration for construction materials accounted for only 0.2 percent of total private exploration in FY 1981/82 [see part 2-A-3].)

The appendix data are presented in as consistent a manner as could be managed. Time series are given in money terms in national currencies, and, where relevant, the totals are deflated to constant 1982 terms with national inflation indices. In addition, the totals are converted into U.S. funds at the appropriate average exchange rates, and then to constant 1982 U.S. dollars with the U.S. index of producer prices for all commodities.

It is important that the appendix tables be used in conjunction with their explanatory notes and sources, since in some instances different sources give different estimates for what is purportedly the same item. (For example, in the case of U.S. uranium exploration expenditures, estimates of the Organisation for Economic Cooperation and Development [OECD] are substantially higher than the numbers published annually by the U.S. Department of Energy [see parts 2-A-30 and 2-A-34].)

The statistics in the appendix are taken back in the 1960s as far as possible. Where series start later, it is because of the lack of ready access to earlier material—for example, to annual reports that provide details of corporate spending.

In a sense the statistical appendix is a mine of data from which to extract information on and insights into the driving forces behind exploration spending. The remainder of this chapter brings out a few, but by no means all, of the possible conclusions that can be drawn.

Table 2-1 draws together from the appendix estimates for total exploration expenditures in the Western world[2] in the early 1970s and early 1980s. The figures, expressed in constant 1982 U.S. dollars, loosely represent 1969–71 and 1979–81 averages. The data cover all exploration costs incurred up to the stage of feasibility studies; basic geologic research and mapping are included, as well as exploration to extend known ore reserves adjacent to operating mines. Readers should take note, however, that this table is concocted from data of varying accuracy and that its purpose is to give appropriate broad orders of magnitude.

The table clearly shows the preponderance of exploration expenditures in Australia and North America in the early 1970s and 1980s. Expenditures in all developed countries appear to have averaged around 72 to 79 percent of the total in both periods. The proportion of the total accounted for by individual countries has, however, altered slightly, with the United States, for example, increasing its share.[3]

---

[2]The term *Western world* as it is used in this chapter refers to all countries except the Soviet Union, the People's Republic of China, and the East European countries belonging to the Council for Mutual Economic Assistance (CMEA).

[3]The overall estimates of U.S. exploration in the main text of this chapter are based on the drilling statistics of part 2-A-15. Average Australian costs (part 2-A-4) are assumed, with drilling accounting for the same percentage of spending (27½ percent) as in Australia. These

Table 2-1. Estimated Mineral Exploration Expenditures in the Western World, by Developed and Less Developed Countries, Early 1970s and Early 1980s
(millions of constant 1982 U.S. dollars)

Expenditures in:	Early 1970s	Early 1980s	Early 1980s in 1980 terms
Developed countries			
Australia	420	560	500
Canada	415	530	475
Europe	200	410	370
South Africa	80	150	135
United States	500	850	765
Japan	45	30	30
Total, developed countries	1,660	2,530	2,275
Less developed countries			
Brazil	100	180	160
By European Community companies	40	55	50
By Japanese companies	30	25	25
For uranium (excluding Brazil)	30	40	35
Other companies	300–500	500–700	450–630
Total, less developed countries	500–700	800–1,000	720–900
OVERALL TOTAL	2,200–2,400	3,330–3,530	2,995–3,175

Sources: Statistical appendix in this chapter: DEVELOPED COUNTRIES—*Australia*: Part 2-A-2; the coverage of this expenditure is given in part 2-A-3. *Canada*: Part 2-A-5 totals, increased 20 percent to allow for all overhead expenses not included. *Europe*: Parts 2-A-10, 2-A-13, and 2-A-30, and approximations for exclusions from the tables (i.e., non-European companies in the European Community, national agencies not included, Norway, Portugal, and Yugoslavia). *South Africa*: Part 2-A-21, with some allowance for spending by other companies, both domestic and foreign. Non-South African spending by the companies covered is deducted. The figures are very approximate. *United States*: The uranium figures are derived from parts 2-A-30 and 2-A-34; those for coal are rough approximations, with sketchy justification from part 2-A-14 (approximately one-quarter of the total of capitalized and expensed exploration and development spending before conversion to 1982 terms). The remaining data are from part 2-A-15. *Japan*: Part 2-A-16. LESS DEVELOPED COUNTRIES—*Brazil*: Part 2-A-22. *European Community companies*: Parts 2-A-10 through 2-A-12 for expenditures on non-U_3O_8 in less developed countries. *Japanese companies*: Part 2-A-19. *Uranium*: Part 2-A-30. Other less developed countries: Author's estimate for spending by Australian, Canadian, non-European Community European, South African, and U.S. companies; by domestic companies and national agencies; and by international organizations. Parts 2-A-23 through 2-A-28 provide some support for the totals.

calculations give $310 million for the early 1970s (excluding coal and uranium) and $200 million for the early 1980s (in 1982 terms). They compare with the independent estimates provided in chapter 10 of this volume for metallic minerals of $285 million for the early 1970s and $270 million for the early 1980s (these figures, though cited in constant 1981 dollars in chapter 10, are expressed here in constant 1982 dollars). When allowance is made for spending not covered by the drilling statistics and for work on nonmetallic minerals, totals of about $350 million for the early 1970s and $300 million for the early 1980s (both in constant 1982 dollars) appear reasonable.

Before discussing the composition of and trends in exploration expenditures, additional consideration should be given to some of the statistical problems involved. Australian and Canadian figures are taken from annual surveys of companies known to be involved in mineral exploration. Australian figures take in all expenses, including all associated overhead, but the Canadian data cover only field exploration and surveys. The costs of head and regional offices, of laboratories, and of desk research of all types account for a significant percentage of total exploration budgets. So, too, does the cost of land acquisition, however temporarily it is held. A reasoned estimate is that overhead costs of all types might account for some 30 percent of total exploration spending. Part 2-A-34 shows the breakdown of U.S. exploration expenditures for uranium among land and mineral rights acquisition, drilling, and other expenses. In table 2-1, Canadian overhead costs are conservatively estimated at only 20 percent of total field expenses.

The estimates of spending prepared by companies based in the European Community, when these firms are involved in joint exploration programs with other companies, theoretically cover only their shares of an exploration budget rather than the total cost. They are meant to include expenditures regardless of the method of finance and hence should be consistent both between companies and over time. That cannot be said of any figures taken from annual reports, nor of the data on North American and South African companies. The numbers shown in company accounts are often merely a by-product of taxation and accounting policies. Some tax authorities allow the full writing off of expenditures in the year in which they are incurred, whereas others do not. Some companies may write off only unproductive expenditures in the year they are incurred and capitalize spending on projects that are thought likely to become productive mines. Only in very few cases is such capitalized expenditure noted separately in the accounts. It is not uncommon for policy to change several times with regard to a specific project, so part, but not all, of the spending may be written off. Finally, the sums charged to exploration may not include all associated costs, particularly in head offices. Although these accounting problems reduce the value of figures obtained from annual accounts, any information, however imperfect, is often better than none.

Table 2-1 covers spending on exploration only in Western countries. There are also substantial exploration programs in China, the Soviet Union, and other CMEA countries, but there are few firm, published data on their magnitude. Even if data were available, definitions of exploration might not be entirely consistent. The exclusion of data for the Eastern countries is a major gap that cannot easily be filled, although some information on exploration in the Soviet Union is provided in chapter 6 of this volume.

Table 2-2 shows estimated exploration expenditures in Western countries in the early 1980s for coal, uranium, and other minerals. If anything, the estimates for coal may be on the high side. Spending on both coal and

Table 2-2. Estimated Exploration Expenditures for Coal, Uranium, and Other Minerals in the Western World in the Early 1980s
(millions of constant 1982 U.S. dollars)

Mineral	Developed countries	Less developed countries	Total Western world
Coal	600	200	800
Uranium	600	80	680
Other	1,330	520–720	1,850–2,050
Total	2,530	800–1,000	3,330–3,530

Source: Statistical appendix in this chapter.

uranium exploration has fallen substantially since the peak of the early 1980s as interest in energy materials has subsided.

Trends in Expenditure

The estimates of total exploration expenditure show a significant increase in real terms over the past decade. However, this finding needs considerable qualification For mining companies, exploration can be a mixture of research and advertising that may have little, if any, immediate impact on profitability in the short term. Exploration in such cases is an overhead cost that can prove highly unwelcome in periods of weak demand, low prices, and poor profitability. Hence, spending tends to move cyclically with the general health of the mining industry, as the charts (parts 2-A-1, 2-A-6, 2-A-11, and 2-A-18) in the appendix clearly show. There was a peak in exploration activity in the early 1970s, a pronounced trough in the mid-1970s associated with the 1975 recession, then a very strong recovery. According to the available data, there was a sharp downturn once more in 1982. This cyclical pattern means that the totals in table 2-1 are for two roughly comparable peak periods.

The reaction of total exploration spending to mining industry activity is fairly instantaneous. Although specific programs may not be cut back in periods of decreased activity, new ones may not be sanctioned, and speculative exploration may be halted.

Changes in the target minerals of exploration bring out the cyclical patterns more clearly. It is such changes that have, perhaps, distorted the overall totals given in table 2-1. Part 2-A-11 shows the divergent trends since 1966 in European Community companies' spending on exploration for uranium and for other minerals. As part 2-A-33 demonstrates, these companies' expenditures on uranium moved very closely with the pressure of demand as reflected by spot market indicators. Part 2-A-34 shows a very similar temporal pattern for the United States. Expenditures on uranium exploration, both in total and in each major area, moved countercyclically to most other exploration during the 1970s. They rose markedly from a 1972 low to peak in the late 1970s.

Exploration for coal followed a pattern rather similar to that for uranium. In Australia, for example, coal's share of total private exploration increased from barely 4 percent in FY 1973/74 to 16 percent in FY 1979/80 and 19 percent in FY 1981/82 (see part 2-A-3). The few U.S. statistics that are available show a similar trend, as the footnote to the U.S. data in table 2-1 indicates. Coal and uranium, respectively, accounted for 14 percent and 16 percent of total U.S. exploration in the early 1970s and for 25 percent and 44 percent in the early 1980s (see parts 2-A-14 and 2-A-15).

Deducting exploration expenditures for uranium and coal from total expenditures indicates a decline in the volume of spending. That is not necessarily cause for alarm, however. It merely reflects the general weakness of most mineral markets and the varying degrees of oversupply. As is pointed out in chapter 3 in this volume, exploration is triggered by commercial need, and the lack of such need tends to cause a sharp cutback in much peripheral spending. The examples of coal and uranium demonstrate, however, that exploration activity can increase both quickly and strongly in response to perceived need.

The data in the appendix for Australia, Canada, and the United States give other examples of this responsiveness of exploration. The Canadian and U.S. drilling statistics (parts 2-A-9 and 2-A-15) show a strong variability in drilling for gold, in lagged response to price movements. Neither table is sufficiently up to date to portray the most recent boom. Very often an initial promising discovery can lead a flood of explorers to the same area. There are myriad examples of this gold-rush mentality in most minerals and countries. For example, Saskatchewan experienced a boom in uranium exploration in the late 1970s, when its share of total Canadian expenditure outside specific mining operations rose from an average 6 percent in the 1970–75 period to 17 percent in the two years 1979 and 1980 (see part 2-A-7). Another instance was the discovery of diamonds in Western Australia that brought on an increase in exploration spending there from insignificant amounts up to 1977 to $63 million (Australian dollars) in FY 1981/82 (see part 2-A-3). Going back further, the Western Australian nickel boom was triggered by Western Mining's announcement of its Kambalda deposit in 1966. The *Mining Journal*'s annual reviews of 1969 and 1970 reported that prospecting had reached almost frenzied proportions in the area and had spread from nickel to other base metals.

So far a predominantly economic explanation has been advanced for trends in exploration, and economics is probably the major influence on mineral exploration activity throughout the world. Much was written in the 1970s about a decline in exploration in less developed countries allegedly in direct response to adverse changes in fiscal regimes and increased political risks. In actuality this decline was part of a worldwide weakening of exploration for minerals other than fuel minerals because of poor market conditions. To the extent that the major known ore deposits of the less developed countries are of metallic minerals, they would automatically have suffered differentially. Explorers invariably begin searching in areas where the target minerals have been discovered in commercial abundance, and the developed countries were known to have coal and uranium. While the political-risk argument cannot be dismissed that lightly, it is difficult to find valid evidence to support it.

The interesting feature of the 1970s was the different experience of Australia and Canada, both developed countries with well-established exploration programs. In Australia the volume of total spending on exploration fell 62 percent between FY 1970/71 and 1975/76, whereas in Canada the fall from peak to trough was only 43 percent (see parts 2-A-3 and 2-A-5). Furthermore, the Canadian trough came in 1972—much earlier than the Australian trough. Was Australia's greater and more prolonged decline perhaps connected with its change in government in late 1972 and with the incoming Labor government's effective ban on uranium mining?

The Canadian statistics (see part 2-A-7) provide further possible evidence of the sensitivity of exploration to political and fiscal change. In the five years from 1968 through 1972, British Columbia accounted for 27 percent of total Canadian spending, but in 1973 through 1977 for only 20 percent (and for only 19 percent in 1974). These changes in its share no doubt partly reflect changes in the relative attractiveness of the province's resource base, but they were also a response to a large increase in mineral taxes in 1973.

Exploration Costs

Although few direct measures of changes in the various components of total exploration costs are available, there is some information on drilling expenses. No matter how thorough the initial stages of an exploration program, some form of drilling is essential sooner or later to ascertain the size and quality of a suspected ore body. As a rough rule of thumb, drilling of all forms accounts for 25 to 30 percent of a typical exploration program. This percentage, provided by exploration managers, is borne out by the breakdown of Australian expenditures in part 2-A-2 and by the information on expenditures of a Swedish mining company in part 2-A-13. In part 2-A-34, the data on U.S. uranium exploration imply higher shares for drilling—an average 50 percent in the five years from 1978 through 1982—but these do not appear to be typical.

As with all comparisons, almost anything can be proved with a judicious choice of starting points. As shown in part 2-A-4, average Australian drilling costs per meter (for core and noncore drilling combined) fell in the 1965–72 period but rose rapidly thereafter. From FY 1972/73, drilling costs per meter increased at an average annual rate of 15.9 percent, compared with an increase in the Australian implicit gross domestic product (GDP) price deflator of 11.6 percent per year. The respective increases for the period since FY 1975/76, the low point of Australian exploration

activity, were 15 percent per year for drilling and 9.5 percent per year for
the deflator. The U.S. drilling statistics for the uranium industry (part 2-
A-34) are for surface drilling costs. Between 1972 and 1982 the cost per
foot of exploration holes rose by an annual 14.4 percent, whereas the
producer price index for all commodities increased by 9.6 percent per year.
These figures suggest that the cost of exploration has risen faster than the
general rate of inflation. As a consequence, table 2-1 and the data in the
statistical appendix may overstate the real growth of exploration, at least
over the past decade, even though they report exploration in constant
dollars. More information is needed to confirm this impression because
of the burden of other costs in total exploration spending.

Another highly relevant question, examined further in chapters 8 through
12 of this volume, is the relative effectiveness of expenditures in different
parts of the world. International comparisons such as those in table 2-1
perforce use official exchange rates rather than specific exploration conversion rates. Again, some comparative figures are available on drilling
costs. Comparison of the Australian and U.S. statistics (from parts 2-A-
4 and 2-A-34) is not very helpful here—averages disguise the varying costs
of different types of drilling and take no account of diverse ground conditions. Fairly shallow drilling for uranium in sandstone deposits in the
United States is quite different from deep drilling for other metalliferous
ores in hard rocks elsewhere. Nevertheless, for what the figures are worth,
average exploration drilling costs in the United States for uranium were
some $16 per meter in 1982. In Australia in FY 1981/82 the average cost
of core drilling was U.S.$81 per meter, and the cost of other forms of
drilling averaged U.S.$21 per meter. Some rough indications are that
drilling costs in the Arabian Gulf region (for minerals, not for oil) are
approximately five times the Australian average.

More accurate comparisons can be made by comparing like with like.
One major mining company (Rio Tinto-Zinc Corporation, PLC) has made
the following comparison for late 1982 of the costs of diamond drilling
to a depth of 300 meters in a reasonably accessible area with rocks that
are relatively easy to core (the data are based on internal corporate information):

Location	Index numbers based on Australia = 100
Canada	45–70
Germany, United Kingdom, Portugal	145
Ireland	100
United States (Missouri and Georgia)	45
United States (Montana and Idaho)	115

Canada's costs are relatively lower than those in other countries partly because more drilling tends to be done. Fixed costs can be spread over a greater volume of work, and there is likely to be greater competition among drilling companies. In addition, drilling has been encouraged by the Canadian methods of calculating the assessment work necessary to keep claims in good standing.

The main point to be drawn from the comparison is that official exchange rates are indeed a poor basis of converting local exploration costs to dollars. In effect, Europe's relative exploration effort is lower than its comparative dollar expenditure might suggest. Conversely, Canada's is somewhat higher.

In addition to the relative costs of working in different areas, another type of comparison is the relative productivity of exploration spending, an example being that a dollar spent in one area is merely a method of offsetting taxes, whereas elsewhere it leads to the discovery of a rich ore body. Unfortunately, however, the available data on mineral reserves are too uneven to draw many useful broad conclusions along these lines. The basis of estimates of reserves has changed markedly even over the past decade, and sufficiently detailed statistics on exploration by target mineral are rarely available. The exception is uranium.

Part 2-A-32 highlights the approximate doubling of estimated uranium reserves and resources in response to the intensive exploration of the 1970s. The "productivity" of expenditures varied widely among countries, as table 2-3 indicates.

Table 2-3. The "Productivity" of Exploration Expenditures for Uranium in Selected Countries, 1972–80

Country	Total expenditure 1972–80[a] (millions of constant 1982 U.S.$)	Estimated resources (thousands of metric tons of recoverable uranium) Apr. 1970	Jan. 1981	Spending per metric ton of increased resources[b] (in constant 1982 U.S.$)
Australia	300	34	602	530
Canada	625	585	1,018	1,440
France	340	73	121	6,960
United States	2,730	920	1,702	3,490
Brazil	210	2	200	1,060
India	60	3	57	1,150
Mexico	30	2	9	4,000

Sources: Parts 2-A-30 and 2-A-32 in the appendix to this chapter.
[a]Annual spending for 1972–80 inclusive is deflated to 1982 terms with U.S. producer price index for all commodities and then totaled.
[b]Total expenditure divided by increase in resources over the decade.

Although it is a somewhat crude calculation, for all its manifest weaknesses table 2-3 suggests that—political and strategic considerations aside—there was little economic justification for such heavy exploration in France, Mexico, and possibly also in the United States. Admittedly that conclusion partly ignores the differences in exploration costs in those three countries; but, far more important, it begs the question: political and strategic considerations cannot be discarded so lightly, especially with regard to uranium.

Much exploration for uranium was funded precisely for political and strategic reasons. These reasons also explain the rapid growth of exploration abroad during the 1970s, shown in part 2-A-31, and, more dramatically, by the growth in exploration activities among the European companies (see part 2-A-10). The French companies in the latter category include Cogema, which is state-owned, and some of the German companies are partly owned by power utilities. The German, Japanese, and French governments provide incentives for uranium exploration, both domestically and overseas, refunding significant percentages of the cost of unproductive expenditures. The Commission of the European Communities has also supported a number of programs within the Community at a rate of between 30 percent and 70 percent of cost since the late 1970s.

This government support of exploration helps explain the growth of spending on uranium in the mid-1970s; the exploration might have been relatively unproductive, but from the companies' point of view it was cheap. Whether, with the benefit of hindsight, government funds were well spent in this fashion is another question entirely.

The Participants in Exploration

There are few solid data on the mineral exploration expenditures of different types of organizations. Much of the debate of the past decade has centered on the role of the international mining companies and on their apparent retreat into "politically safe" Anglo-Saxon areas. One facet of this situation has already been highlighted: the mineral deposits of less developed countries have, by and large, not been in demand. The same point can, however, be made with equal strength about the minerals produced by many international mining companies—they have not, by and large, been in demand. Part 2-A-20 shows clearly the much slower growth of exploration spending by a predominantly copper-producing company such as Phelps Dodge as compared with that of Homestake or Amax, which have growing interests in precious metals or coal. In 1971 Amax spent 54 percent of its total exploration budget in the United States, according to its annual report, and 9 percent of the total on energy minerals.

In the two years 1979 and 1980, its U.S. expenditure had risen slightly to 58 percent of the total and its spending on energy minerals to 23 percent.

The role of the major mining companies in exploration has never been as strong as it has often been portrayed. Some of the exploration expenditure of the twelve Canadian companies listed in part 2-A-20 was outside Canada. For example, Noranda's spending in Canada averaged 45 to 53 percent of its total exploration activity during the 1970s according to the company's annual reports. Ignoring such external expenditure (which merely emphasizes the point), the twelve Canadian companies apparently accounted for 47 percent of total Canadian exploration in the 1971–72 period (allowing a 20 percent increase in the totals of part 2-A-5) and for 34 percent in 1980–81. This is admittedly sketchy evidence, but it suggests that a large proportion of Canadian exploration is carried out by the so-called junior mining companies. Such junior companies are largely confined to Australia and Canada, with some limited manifestations in the United States and some Latin American countries.

The main statistical justification for arguments about the changing geographical pattern of exploration has been the European Community companies' spending, summarized in part 2-A-12. Leaving uranium aside as a special case, the less developed countries took an average 39 percent of the companies' exploration expenditures in the 1966–68 period and 17 percent in 1980–82. These figures do indicate a relative decline, which can only partly be explained away by the lumpy nature of spending on large projects. In money terms, exploration expenditures in the less developed countries temporarily rose strongly in 1969 and again in 1975. Throughout the period 1966 to 1982 covered in part 2-A-12, the statistics include exploration by the European-based oil companies and by the Bureau de recherches géologiques et minières (BRGM), funded by the French government. That said, the observations about the effects of market trends made earlier in this chapter should be recollected before definite conclusions are drawn. The other relevant point is that the gap in exploration caused by a decline in the mining companies' total exploration may have been partially filled by new actors on the scene. The possible star players for this role are the American oil companies, nationally owned and even private mining companies in the Third World, and international agencies. The American oil companies clearly have expanded their mineral interests both in the United States and overseas. Until the 1982 recession they were perhaps more sanguine about the outlook for mineral markets and had more ready access to funds than the mining companies did. Their broader international spread enabled them more easily to offset risks in any one country, and they had not suffered the same traumas as the traditional mining companies in the late 1960s and early 1970s. Their enthusiasm has since been tempered, however, by recession; throwing good money after bad appeals to the oil companies as little as to other companies. Published data are not available for most oil companies (e.g., Exxon), but part 2-A-20 shows a sizable growth in Atlantic Richfield's mineral exploration spending between 1977 and 1982. A significant share was overseas expenditure.

The Brazilian statistics in part 2-A-22 demonstrate the importance of nationally owned companies and agencies in just one mineral-producing country. Domestic companies have also contributed to the extent that the international mining companies (including the oil companies in this context) account for only one-third of total exploration. Their ardor may have been greatly dampened by recession and by Brazil's debt problems, but these figures nonetheless show that other sources of exploration funds are important. The Brazilian experience can be repeated throughout Latin America and Asia and even to a modest extent in Africa.

Typically governments or national agencies have carried out basic geologic research and mapping even in the developed countries. In Finland, for example, state-funded exploration makes up 20 percent of the total (with the Outokumpu Oy company not counted here as state-owned). In Spain also, national agencies play an important role even outside the uranium business, whether through direct spending or assistance to private companies. The Australian statistics in part 2-A-2 show that government expenditure has been less volatile than privately financed exploration, presumably because it is more in the nature of basic research and is less market-oriented. In FY 1980/81 and 1981/82 it averaged about 5 percent of total exploration expenditures, a somewhat lower proportion than in the late 1960s and early 1970s. Japanese government agencies have played a much greater role in mineral exploration, as part 2-A-16 demonstrates. In that country the government finances regional geologic reconnaissance, 80 percent of more detailed geologic surveys, and significant percentages of exploration. In total, the government financed 75 percent of Japanese domestic exploration in 1979. Germany and France have schemes similar to that of Japan, and the United Kingdom had a less-well-utilized system of grants. Shortage of funds is, however, usually only a minor barrier to exploration, especially in the United Kingdom.

In the less developed countries where lack of funds is reputedly a constraint on governmental activity, some international agencies have stepped into the breach. Basic geologic surveying work and mapping are often a favored form of aid. For example, BRGM, the French mining and geological research office, spent 26 percent of its total exploration outlay in less developed countries in the 1980–82 period. Part 2-A-27 comments on international assistance for Bolivian tin exploration, and part 2-A-28 on such aid in Botswana. The World Bank and United Nations agencies have both provided funds for domestic programs and also have carried out direct exploration, usually through contractors. Chapter 7 in this volume, which discusses the role of the UN agencies in detail, points out that the

United Nations Development Program (UNDP) had spent $165 million up to 1979 on more than 250 projects in its mineral programs, and governments had contributed an additional $115 million in local currencies. Since 1979 the UNDP has spent or budgeted $64 million more for such programs, with governments adding another $35 million to $40 million. The shares of these programs spent on exploration per se up to 1982 amount to $130 million by the UN and $70 million by recipient countries' governments. Much of the balance goes on overhead expenses for the UN's Department of Technical Cooperation and Development. Somewhat more than half the UN's funds spent on exploration have been in African countries, about one-third have been in Asia, and the balance in Latin America.

In 1974 the United Nations established its Revolving Fund to assist with minerals exploration in the Third World. Up to the end of 1982, as discussed in chapter 7, it had spent almost $6 million on administration, its completed projects had cost nearly $9 million, and an additional $20 million had been or were to be spent on projects then under way. Against the total exploration spending of the Western world these sums have been insignificant, although they may have proved very valuable in local contexts.

In summary, the participants in mineral exploration tend to change as exploration moves closer to the development stage (see section on The Stages of Exploration, above). Basic geologic mapping and survey work are and always have been the responsibility of governments or their agencies in most countries. In many instances contractors are employed to carry out the detailed work, but under government supervision. As noted, government activities have been supplemented by bilateral and multilateral aid. In many less developed countries the multilateral agencies have also moved into regional reconnaissance and grass-roots exploration. In Australia and Canada and to a lesser extent in the United States, individual prospectors and small exploration companies have done and still do much grass-roots exploration. Their activities were somewhat curtailed in the 1970s by tightened Australian and Canadian stock exchange rulings designed to protect small investors, but many have survived. These small concerns usually dilute their interests significantly or sell out when they find viable ore deposits; larger, more-established mining companies with technical, financial, and marketing skills then move in to carry out detailed exploration and feasibility studies as a prelude to mining.

The pattern of small concerns succeeded by larger ones also predominates in some less developed countries, but with government agencies or companies often taking the role of mining companies. In many less developed countries the national government reserves either all or some exploration for itself, which has tended to inhibit exploration. In order to encourage exploration, governments of both producing and consuming countries have, over the years, introduced various incentives. European assistance to uranium exploration has already been noted, and there are similar European programs for other minerals. Some less developed countries, such as Mexico, have attempted to encourage extra exploration effort, but their incentives may not be sufficiently broad-based. It is the total environment rather than specific exploration incentives that is of concern to mineral developers. Even well-established mining areas, however, are tempted to assist exploration. In Canada, for example, the Ontario Mineral Exploration Program (OMEP) provides grant aid and tax credits for eligible projects in that province.

The wave of nationalizations of the late 1960s and 1970s forced many foreign-owned mining companies to withdraw from less developed countries. They did not leave a complete exploration vacuum, however, as the new state-owned companies took on some exploration activities. Furthermore, the state-owned companies often coexisted with a small or medium-sized mining sector that continued to operate. Particularly in Latin America and Asia domestic companies have predominated for decades. The most obvious gap has been in Africa where political instability and lack of funds has discouraged mining companies of all types. It is here that the multilateral and bilateral agencies have become most active.

To conclude, the number and type of participants in mineral exploration have greatly expanded over the past two decades, and there is now rich diversity among them as they fill their different, or sometimes overlapping, roles. The cast of characters has always been much larger and rather more varied, however, than some of the principal actors in this field have been prepared to acknowledge. Even the data for Australia and Canada (which are available in far more detail than data for other countries) are unfortunately far too sparse to permit reliable estimates of aggregate spending by each of the main actors, even in the early 1980s let alone a decade earlier. Since any estimates are far more likely to mislead than to inform, no attempt has been made to calculate them for this chapter. The statistical appendix contains the clues for others who might attempt the task.

Stocks Versus Flows

The nearer a project moves to commercial viability the less likely are companies or agencies engaged only in exploration to be exclusively involved and the greater the role of the mining companies, whether domestic or foreign. As has been said, it is often impossible to draw clear-cut distinctions between exploration and development; the operating company's capitalizing of expenses may be one mark of the beginning of the development stage, but there are no absolute indications by which to identify the end of exploration and the beginning of development.

In the on-going debate about whether and why there is too little or too much mineral exploration, perhaps too much attention has been devoted to annual exploration activity—the flow—and not enough to the stock of projects or undeveloped ore reserves that results from the activity. Parts 2-A-29 and 2-A-32 show that total estimated resources of uranium doubled during the decade of the 1970s. The data that are available do show that there is a substantial stock of known but unexploited reserves proved from past exploration. This in itself is a strong reason for low exploration activity in recent years in those minerals, especially where prospective growth rates of demand have fallen. Copper and nickel are probably both in this category, and the data on copper projects in part 2-A-35 indeed support this suggestion. Ample undeveloped resources are, however, not in themselves a sufficient reason for reduced exploration. Past exploration presumably was guided by relative factor prices that no longer prevail. The costs of energy and capital have possibly increased relative to the cost of labor. Deposits found in the late 1960s and early 1970s may never become profitable mines because of this change in the prospective economics—a change that is not always recognized by government planners, many of whom regard any metallized rock as a prospective mine.

On a slightly different tack, much exploration expenditure will go to proving up ore bodies that have already been discovered. Development plans will be prepared and refined as projects approach the stage of feasibility studies, and little of the preparatory work will be wasted. As metal markets move over the course of the business cycle, new players will be tempted to take up some of the balls already lying on the field and run with them. Not all of these players will reach the goal before the whistle of the next downturn blows, and some may even run off the field or in the wrong direction. Some projects will, however, have been taken closer to commercial operation. In that sense a substantial portion of exploration expenditure goes to improving the economic and technical potential of the existing stock of resources rather than to the undoubtedly more glamorous discovery of hitherto unknown ore bodies.

Appendix 2-A
Exploration Statistics

Contents

Note: Charts are indicated by the word "(Figure)" after part numbers below; tables or text with tables comprise all other parts.

Part No.

Part 2-A-1

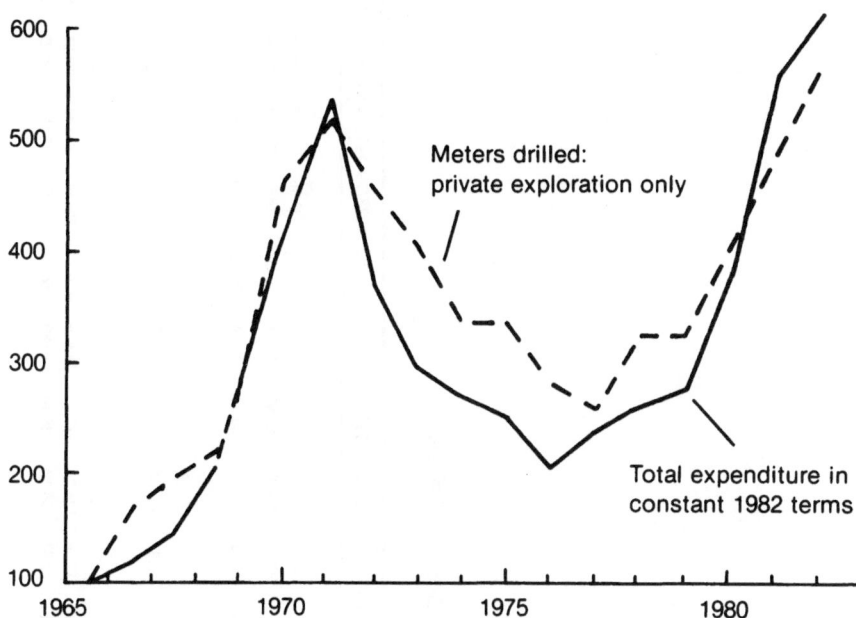

Indices of mineral exploration in Australia, 1965–83 (1965 = 100). Years 1965 through 1968 are calendar years; 1968/69 through 1983/84 are Australian fiscal years ending June 30. Oil, gas, and oil shale exploration are excluded. Money expenditures were deflated to 1982 terms with the implicit price deflator for gross domestic product (GDP). *Sources:* Parts 2-A-2 and 2-A-4 in this appendix.

Part 2-A-2

Private and Public Mineral Exploration Expenditures in Australia, 1965–FY 1981/82

(millions of Australian dollars, unless otherwise noted)

Expenditures	1965	1966	1967	1968	1968/69	1969/70	1970/71	1971/72	1972/73	1973/74	1974/75	1975/76	1976/77	1977/78	1978/79	1979/80	1980/81	1981/82
Private																		
Production leases[a]	8.39	9.44	9.33	7.65	10.18	21.09	27.55	21.18	13.70	14.36	16.76	15.13	23.93	22.82	23.96	30.53	45.25	44.59
Other[b]	13.97	18.67	25.50	44.82	62.38	97.02	133.51	95.88	86.03	87.27	93.07	84.85	110.04	135.57	158.55	255.60	425.24	530.98
(Total private drilling)	9.40	13.01	14.38	18.92	24.37	32.06	43.15	30.94	26.80	25.93	32.88	30.53	36.14	50.02	50.73	72.41	126.09	141.87
Government[c]																		
Production leases	3.22	3.57	4.57	5.86	6.53	6.70	7.31	7.41	8.39	10.91	13.37	17.16	0.63	0.73	1.26	1.29	2.52	2.20
Other													14.89	17.21	18.22	23.01	26.01	25.54
TOTAL private and public	25.58	31.69	39.39	58.32	79.09	124.82	168.38	124.48	108.13	112.54	124.20	117.14	149.49	176.34	201.98	310.43	499.01	603.31
Of which: drilling	10.51	13.99	15.49	20.45	26.20	33.52	45.11	32.86	29.04	28.82	36.17	35.10	40.89	56.28	57.91	78.84	134.22	153.57

TOTAL in constant 1982

terms	104.8	125.7	151.7	218.2	290.7	437.4	561.1	389.0	307.7	280.8	261.6	214.7	247.1	271.1	288.0	398.9	582.9	639.0
Index: 1965 = 100	100.0	120.0	145.0	208.0	277.0	417.0	535.0	371.0	294.0	268.0	250.0	205.0	236.0	259.0	275.0	381.0	556.0	610.0
TOTAL U.S.$ equivalent	28.6	35.5	44.1	65.3	88.6	139.8	188.6	145.6	139.0	165.6	170.1	147.7	172.2	198.9	229.5	346.0	579.4	667.1

TOTAL in constant 1982

U.S.$ (millions)	88.6	106.5	132.0	191.6	254.3	384.9	503.6	375.0	331.8	340.4	299.1	246.7	273.0	299.2	311.2	410.3	611.9	671.8
Index: 1965 = 100	100.0	120.2	149.0	216.3	287.0	434.4	568.4	423.3	374.5	337.6	384.2	278.4	308.1	337.7	351.2	463.1	690.6	758.2

Notes: Data for expenditures on oil shale and petroleum are excluded; data for coal are included. Fiscal years (1968/69–1981/82) end June 30.

Sources: Australian Bureau of Statistics, *Mineral Exploration, Australia 1981–82,* Catalogue No. 8407.0 (Canberra, A.C.T., February 1983 and earlier editions).

[a]Exploration carried out on the production lease by privately operated mines currently producing or under development for the production of minerals.

[b]Exploration carried out by private enterprises on areas covered by exploration licenses, authorities to enter, authorities to prospect, and similar licenses and authorities issued by national governments for exploration of minerals. Also included is exploration by private enterprises that is not directly connected with areas under lease, license, etc.

[c]Exploration by Commonwealth, national, and local government departments, commissions, and authorities.

Part 2-A-3

Private Expenditures for Mineral Exploration in Australia, by Mineral Commodity and by State, FY 1973/74–1981/82
(millions of Australian dollars)

Expenditure	1973/74	1974/75	1975/76	1976/77	1977/78	1978/79	1979/80	1980/81	1981/82
Private, by type of mineral									
Metallic minerals									
Copper, lead, zinc, silver, nickel, cobalt	*	*	*	*	*	*	91.79	152.96	166.67
Gold	*	*	*	*	*	*	29.93	69.24	94.83
Iron ore	*	*	*	14.09	10.66	8.13	10.50	14.42	15.64
Mineral sands	4.75	3.32	3.89	2.86	1.29	0.93	1.34	1.99	2.01
Tin, tungsten, scheelite, wolfram	*	*	*	*	*	*	23.09	36.93	41.88
Uranium	11.42	8.86	9.18	16.71	23.09	32.96	38.26	48.50	56.37
Other metallic	75.96*	86.54*	73.60*	78.31*	90.26*	96.00*	5.92	12.29	19.65
Total, metallic	92.13	98.72	86.67	111.97	125.30	138.02	200.83	340.0	397.04
Nonmetallic minerals									
Coal	3.81	5.63	6.30	12.83	24.47	23.07	46.71	74.94	108.73
Construction materials	0.99	0.70	0.31	0.36	0.44	0.48	1.10	1.60	1.12
Diamonds ⎱	4.69	5.78	6.71	8.82	8.18	20.07	32.59	51.39	63.11
Other nonmetallic ⎰							4.91	6.24	5.57
Total, nonmetallic	9.49	12.11	13.32	22.01	33.09	43.62	85.30	166.48	178.53

Private, by state

New South Wales	11.54	11.91	13.06	23.72	25.49	27.60	44.26	69.16	89.75
Northern Territory	9.43	6.60	6.89	11.90	13.73	17.60	21.96	29.49	32.02
Queensland	15.40	21.74	18.23	21.32	35.99	37.80	62.49	96.85	124.98
South Australia	4.49	5.62	5.12	6.09	9.34	10.51	18.88	54.67	64.77
Tasmania	4.19	5.57	4.12	4.32	7.41	9.31	12.70	19.73	22.84
Victoria	2.52	2.05	2.43	2.06	1.65	2.92	7.27	13.91	25.14
Western Australia	54.06	57.34	50.12	64.57	64.78	75.90	118.57	186.68	216.07

Notes: Exploration consists of the search for and/or appraisal of new ore occurrences and known deposits of minerals (including extensions to deposits being worked) by geological, geophysical, geochemical, drilling, and other methods. Exploration for water is excluded. The construction of shafts and adits is included if done primarily for exploration purposes. Excluded are mine development activities carried out primarily for the purpose of commencing or extending mining or quarrying operations in underground mines, and the preparation of quarrying sites, including overburden removal, for open-cut extraction.

Exploration expenditure (capitalized expenditure as well as working expenses) comprises all expenditure on exploration activity in Australia. It includes expenditures for aerial surveys (including Landsat photographs), general surveys, report writing, map preparation, and other activities indirectly attributable to exploration.

Data for expenditures on oil shale and petroleum are excluded; data for coal are included. The fiscal-year period ends June 30.

Sources: Australian Bureau of Statistics, Mineral Exploration, Australia 1981–82, Catalogue 8407.0 (Canberra, A.C.T., February 1983 and earlier editions).

*Prior to FY 1979/80 less detail was available for separate commodities.

Part 2-A-4

Private Exploration Drilling in Australia, 1965–FY 1981/82

Type of drilling	1965	1966	1967	1968	1968/69	1969/70	1970/71	1971/72	1972/73	1973/74	1974/75	1975/76	1976/77	1977/78	1978/79	1979/80	1980/81	1981/82
Core																		
Meters (000)	293	447	460	548	734	854	889	746	711	588	662	456	469	557	565	862.0	1,065	120
Expenditure (in million Australian $)[a]										15.95	22.09	18.42	23.85	30.94	30.68	45.82	74.27	88.04
Cost per meter (in Australian $)										27.12	33.37	40.39	50.86	55.55	54.31	53.16	69.74	73.31
Noncore																		
Meters (000)	422	717	933	1,022	1,163	2,448	2,785	2,489	2,169	1,815	1,739	1,553	1,364	1,772	1,763	2,055.0	2,718.0	2,824
Expenditure (in million Australian $)[a]										9.98	10.79	12.11	12.29	19.08	19.82	26.59	51.81	53.83
Cost per meter (in Australian $)										5.50	6.21	7.80	9.01	10.77	11.24	12.94	19.06	19.06
TOTALS																		
Meters (000)	715	1,164	1,393	1,570	1,896	3,303	3,674	3,235	2,880	2,403	2,401	2,010	1,834	2,329	2,328	2,917	3,983.0	4,025
Expenditure (in million Australian $)	9.40	13.01	14.38	18.92	24.37	32.06	43.15	30.94	26.80	25.93	32.88	30.53	36.14	50.02	50.73	72.41	126.09	141.87
Cost per meter (in Australian $)	13.15	11.18	10.32	12.05	12.85	9.71	11.74	9.56	9.31	10.79	13.69	15.19	19.71	21.48	21.79	24.82	31.66	35.25
Index of cost per meter: 1965 = 100	100	85.0	78.0	92.0	98.0	74.0	89.0	73.0	71.0	82.0	104.0	116.0	150.0	163.0	166.0	189.0	241.0	268.0

Notes: Core drilling is diamond drilling or any kind of drilling in which cores are taken. *Noncore drilling* is alluvial percussion and other drilling in which cores are not taken. Data for expenditures on oil shale and petroleum are excluded; data for coal are included. Fiscal years (1968/69–1981/82) end June 30.

Sources: Australian Bureau of Statistics, *Mineral Exploration, Australia 1981–82*, Catalogue No. 8407.0 (Canberra, A.C.T., February 1983 and earlier editions).

[a] A breakdown of expenditures between core and noncore drilling is not available before 1973/74.

Part 2-A-5

Mineral Exploration Expenditures in Canada, 1968–81

Total expenditures in:	1968	1969	1970	1971	1972	1973	1974	1975	1976	1977	1978	1979	1980	1981
Millions of current Canadian $	105.5	136.7	144.5	118.5	89.1	110.3	136.0	154.8	166.5	243.8	249.0	298.1	489.0	626.3
Millions of constant 1982 Canadian $	326.0	404.8	408.9	324.9	232.7	263.9	282.3	290.1	284.6	388.0	371.5	403.3	595.4	689.4
Index (1968 = 100) in constant Canadian $	100.0	124.0	125.0	100.0	71.0	81.0	87.0	89.0	87.0	119.0	114.0	124.0	183.0	211.0
Millions of current U.S.$	97.6	126.9	137.9	117.3	90.0	110.3	139.1	152.2	168.9	229.2	218.3	254.5	418.2	522.4
Millions of constant 1982 U.S.$	285.0	355.5	373.9	308.2	226.2	365.5	260.0	260.5	276.2	353.2	312.2	323.3	465.7	532.9
Index (1968 = 100) in constant U.S.$	100.0	125.0	131.1	108.0	79.0	128.0	91.0	91.0	97.0	124.0	110.0	113.0	163.0	187.0

Notes: Data for expenditures on oil and natural gas are excluded.
The totals include only field expenditures on physical work and surveys, and not applied administration, general overhead, and lease rental costs. They are obtained through an annual census, carried out since 1969.

Source: Statistics Canada, *Investment Statistics. Exploration and Development 1981,* Ref. 61-216 (Ottawa, 1981 and earlier editions).

Part 2-A-6

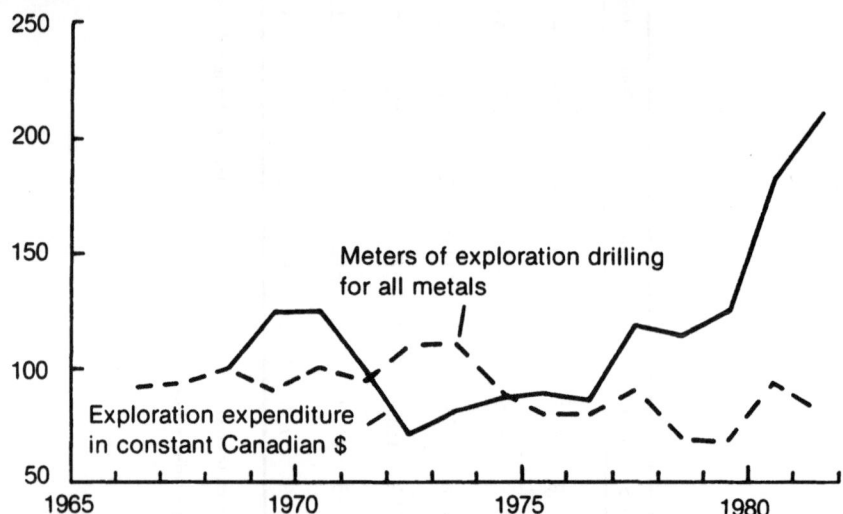

Indices of mineral exploration in Canada, 1966–81 (1968 = 100). *Sources:* Parts 2-A-5 and 2-A-9 in this appendix.

Part 2-A-7

Mineral Exploration Expenditures in Canada, "on Property" and "Other," by Province, 1968–81
(millions of current Canadian dollars)

Province or territory	Category	1968	1969	1970	1971	1972	1973	1974	1975	1976	1977	1978	1979	1980	1981
Atlantic provinces	On property	1.7	2.7	2.0	N.A.	1.6	2.0	1.7	1.9	1.9	1.7	1.2	2.2	2.7	6.3
	Other	3.7	4.6	6.4	2.9	2.7	4.5	10.1	13.4	15.0	20.3	17.6	21.1	35.5	50.8
Quebec	On property	5.2	12.9	3.2	6.4	3.2	3.1	6.3	4.1	7.3	N.A.	16.1	7.5	15.6	28.0
	Other	12.2	14.2	8.8	11.0	9.9	13.4	21.4	25.0	21.1	29.3	39.9	39.5	58.5	81.7
Ontario	On property	7.5	9.6	10.7	8.7	5.9	6.9	7.1	8.3	10.0	10.2	7.8	6.4	12.1	17.9
	Other	18.0	22.6	32.2	21.3	15.3	18.5	19.6	23.3	24.6	30.1	21.0	18.4	58.5	79.5
Manitoba	On property	2.9	3.5	4.6	4.1	N.A.	N.A.	N.A.	N.A.	N.A.	N.A.	N.A.	N.A.	N.A.	8.3
	Other	5.6	7.9	9.3	9.4	5.6	5.9	6.9	6.0	5.6	8.2	11.8	11.8	21.2	20.6
Saskatchewan	On property	1.4	0.2	0.1	0.0	N.A.	N.A.	N.A.	N.A.	N.A.	N.A.	N.A.	4.9	7.0	20.2
	Other	6.2	8.3	6.6	5.6	3.7	6.3	6.3	9.7	11.9	15.1	27.9	52.6	56.4	45.4
Alberta	On property	1.5	1.7	N.A.	N.A.	N.A.	N.A.	N.A.	N.A.	0.7	N.A.	0.5	N.A.	N.A.	2.6
	Other	0.8	0.5	2.5	5.1	1.8	2.5	4.1	3.5	4.3	3.0	7.0	8.5	14.2	23.9
British Columbia	On property	8.4	5.2	3.1	3.8	2.2	5.6	3.8	4.7	8.1	21.2	16.2	17.8	31.1	34.9
	Other	19.1	28.4	37.4	27.7	27.2	27.0	22.5	24.5	22.9	26.0	32.5	48.3	91.0	111.7

(Continued)

Part 2-A-7 (continued)

Province or territory	Category	1968	1969	1970	1971	1972	1973	1974	1975	1976	1977	1978	1979	1980	1981
Yukon and Northwest Territories	On property	3.0	2.1	2.0	1.7	2.8	2.0	2.2	2.8	3.7	3.8	5.3	5.6	8.6	16.3
	Other	8.3	12.3	15.1	8.3	6.3	9.0	19.9	23.7	25.7	38.0	37.2	48.7	68.3	78.2
TOTAL	On property	31.6	37.9	25.7	27.2	16.6	23.2	25.2	25.7	35.4	73.8	54.1	49.2	85.4	134.5
	Other	73.9	98.8	118.8	91.3	72.5	87.1	110.8	129.1	131.1	170.0	194.9	248.9	403.6	491.8
OVERALL TOTAL		105.5	136.7	144.5	118.5	89.1	110.3	136.0	154.8	166.5	243.8	249.0	298.1	489.0	626.3

Notes: On property exploration covers expenditures in searching for new ore bodies on properties that are in production or that are being prepared for production. It includes surface, underground, and airborne exploration, including costs of staking, assessment, and lease rental; diamond drilling; geological, geophysical, and geochemical work; trenching and other surface work; exploration shafts; services such as hoisting and ventilating; and general supervision.

Other represents expenditures in the search for all other new ore bodies. It covers airborne, surface, and underground exploration, including costs of staking, aerial surveys, assessment, and lease rental; diamond drilling; geological, geophysical, and geochemical work; trenching and other surface work; exploration shafts and other exploration work.

N.A. = not available.

Sources: Department of Energy, Mines, and Resources, *Canadian Minerals Yearbook* (Ottawa, 1968–80 editions); Statistics Canada, *Investment Statistics: Exploration and Development 1981*, Ref. 61-216 (Ottawa, 1981).

Part 2-A-8

Mineral Exploration Expenditures in Canada, "on Property" and "Other," by Mineral Commodity, 1968–81 (millions of current Canadian dollars)

Mineral commodity/companies	Category	1968	1969	1970	1971	1972	1973	1974	1975	1976	1977	1978	1979	1980	1981
Metals															
Gold	On property	1.6	2.5	0.5	0.4	0.6	1.0	1.9	1.7	1.6	2.8	4.2	4.1	22.6	21.7
	Other	1.1	1.0	3.7	1.2	1.0	1.0	1.9	1.6	N.A.	0.4	1.1	5.6	20.0	40.1
Iron	On property	0.4	0.3	0.4	N.A.	N.A.	N.A.	N.A.	N.A.	N.A.	N.A.	N.A.	N.A.	N.A.	N.A.
	Other	0.5	1.0	0.5	0.6	0.7	1.5	1.8	1.5	1.0	N.A.	N.A.	N.A.	N.A.	N.A.
Copper–gold–silver	On property	7.1	7.5	4.7	4.9	4.0	6.2	5.0	6.4	6.5	10.2	8.7	10.2	24.1	28.2
	Other	2.4	2.0	7.5	3.2	7.2	6.2	3.3	5.3	4.8	5.2	5.0	8.5	8.4	13.5
Silver–lead–zinc	On property	1.7	3.2	2.9	2.8	3.4	2.9	3.2	4.0	3.2	6.6	7.2	8.1	9.4	21.5
	Other	1.1	1.5	3.5	2.2	1.3	1.3	3.8	3.5	2.6	6.2	3.1	6.2	10.7	15.4
Other	On property	15.4[a]	14.4	14.0	13.6	N.A.	N.A.	N.A.	N.A.	12.7	26.0	13.6	11.5	14.7	37.3
	Other	68.1[a]	5.0	9.9	8.5	7.4	5.7	8.0	9.9	14.2	9.7	14.5	N.A.	N.A.	N.A.
Total, metals	On property	26.2[a]	27.9	22.5	21.7	14.1	17.9	20.0	21.3	24.0	45.6	33.7	36.3	75.1	(112.5)[b]
	Other	73.2[a]	10.5	25.1	15.7	17.6	15.7	18.8	21.8	22.6	(22.3)[b]	(24.5)[b]	29.8	54.4	97.0
Overall subtotal, metals		99.4[a]	38.4	47.6	37.4	31.7	33.6	38.8	43.1	46.6	(67.9)[b]	(58.2)[b]	66.1	129.5	(209.5)[b]
Nonmetals															
Asbestos	On property	0.2	1.3	1.1	2.7	0.3	0.2	0.4	0.5	N.A.	2.0	1.5	0.5	0.7	N.A.
	Other	0.3	0.4	0.3	0.3	0.2	0.1	0.2	0.1	0.2	N.A.	N.A.	0.4	N.A.	N.A.
Other[c]	On property	5.2	3.1	N.A.	1.7	0.5	4.0	2.1	2.6	N.A.	18.7	11.6	12.4	9.6	21.3
	Other	0.4	0.8	2.8	4.0	0.8	1.5	2.7	7.8	8.9	5.8	9.5	10.2	N.A.	N.A.

(Continued)

Part 2-A-8 (continued)

Mineral commodity/companies	Category	1968	1969	1970	1971	1972	1973	1974	1975	1976	1977	1978	1979	1980	1981
Nonmetals (cont.)															
Total, nonmetals	On property	5.4	4.4	(2.1)[b]	4.4	0.8	4.2	2.5	3.1	8.0	20.7	13.1	12.9	10.3	(22.0)[b]
	Other	0.7	1.2	3.1	4.3	1.0	1.6	2.9	7.9	9.1	(6.2)[b]	(9.9)[b]	10.6	18.4	38.5
Overall subtotal, nonmetals		6.1	5.6	(5.2)[b]	8.7	1.8	5.8	5.4	11.0	17.1	(26.9)[b]	(23.0)[b]	23.5	28.7	(60.5)[b]
General exploration companies[d]		[a]	92.7	91.7	72.4	55.6	70.9	91.8	100.7	102.8	149.0	167.8	208.5	330.8	356.3
OVERALL TOTAL		105.5	136.7	144.5	118.5	89.1	110.3	136.0	154.8	166.5	243.8	249.0	298.1	489.0	626.3

Notes: On property exploration covers expenditures in searching for new ore bodies on properties that are in production or that are being prepared for production. It includes surface, underground, and airborne exploration, including costs of staking, assessment, and lease rental; diamond drilling; geological, geophysical, and geochemical work; trenching and other surface work; exploration shafts; services such as hoisting and ventilating; and general supervision.

Other represents expenditures in the search for all other new ore bodies. It covers airborne, surface, and underground exploration, including costs of staking, aerial surveys, assessment, and lease rental; diamond drilling; geological, geophysical, and geochemical work; trenching and other surface work; exploration shafts and other exploration work.

N.A. = not available.

Sources: Department of Energy, Mines, and Resources, *Canadian Minerals Yearbook* (Ottawa, 1968–80 editions); Statistics Canada, *Investment Statistics: Exploration and Development 1981*, Ref. 61-216 (Ottawa, 1981).

[a] In 1968 the totals for general exploration companies were allocated to other metals.

[b] Figures in parentheses are estimates.

[c] Coal, salt, gypsum, quarries, sand, gravel, and other nonmetal mines.

[d] General exploration companies include government agencies, the exploration subsidiaries of established mining companies, nonmining companies that carry out exploration, and "junior" exploration companies.

Part 2-A-9

Exploration Drilling in Canada, by Type of Deposit, 1966–81

(thousands of meters)

Type of deposit	1966	1967	1968	1969	1970	1971	1972	1973	1974	1975	1976	1977	1978	1979	1980	1981[a]
Gold quartz	8,475.8	7,535.0	7,192.0	6,663.1	5,504.2	4,684.3	4,512.1	4,755.7	4,755.7	4,271.0	1,460.9	3,627.1	3,795.4	4,125.3	3,877.8	4,400.8
Nickel–copper–gold–silver	19,207.3	20,992.5	22,841.0	20,388.6	24,447.0	25,375.1	33,129.1	34,107.6	24,880.2	21,030.9	22,383.5	21,509.5	14,978.4	13,372.4	22,227.0	19,388.1
Silver–lead–zinc	3,962.2	3,926.6	4,742.3	3,650.0	4,922.3	3,943.8	3,626.6	3,175.3	3,885.3	3,859.1	5,007.6	4,593.3	2,501.7	2,454.5	4,129.5	4,068.6
Iron														1,549.2	1,176.6	
Uranium	5,968.4	6,176.8	6,288.7	6,201.1	6,054.8	4,933.9	3,853.8	3,360.1	3,479.0	3,582.5	3,941.0	7,765.8	7,050.7	5,489.4	6,523.1	} 6,025.6
Other metal														746.9	774.7	
TOTAL, METALS	37,613.7	38,630.9	41,064.0	36,902.8	40,928.3	38,937.1	45,121.6	45,398.7	37,000.2	32,743.5	32,793.0	37,495.7	28,326.2	27,737.7	38,708.7	33,883.1
Asbestos	2,194.1	2,129.1	2,230.6	1,816.9	1,712.9	1,763.3	1,700.8	1,526.8	1,427.2	997.2	1,627.7	1,558.8	1,242.4	N.A.	N.A.	848.9

Notes: Data cover both drilling by contractors and drilling by mining companies with their own personnel and equipment. All forms of drilling are included, whether surface or underground, with diamond or other drills of all types.

Only data on drilling for metallic deposits and asbestos are included (i.e., coal, etc., are excluded).

Sources: Statistics Canada, *General Review of the Mineral Industries*, Catalogue 26-201 Annual (Ottawa, various issues); earlier data from Statistics Canada, *Contract Drilling for Petroleum and Other Contract Drilling*, Catalogue 26-207 Annual (Ottawa, various issues).

[a]The 1981 figures are incomplete because no data were received from some firms.

Part 2-A-10

Worldwide Mineral Exploration Expenditures of European Community Companies for Uranium (U_3O_8) and Other Minerals, 1966–83

(millions of current and constant U.S. dollars)

Expenditures	1966	1967	1968	1969	1970	1971	1972	1973	1974	1975	1976	1976	1977	1978	1979	1980	1981	1982	1983
Total in current U.S.$																			
U_3O_8	0.56	2.10	11.90	16.55	20.34	15.91	14.84	28.61	45.41	61.12	65.29	160.56	208.57	207.58	224.22	251.60	176.73	139.70	111.10
Other minerals	29.10	39.34	40.54	61.18	61.50	83.73	89.56	94.39	100.94	123.12	137.91	149.35	173.97	154.55	195.82	310.94	338.75	252.40	246.90
TOTAL, ALL MINERALS	29.66	41.44	52.44	77.73	81.84	99.63	104.70	123.00	146.35	184.24	203.20	309.91	382.54	362.13	420.04	562.54	515.48	392.10	358.00
Total in constant 1982 U.S.$[a]																			
U_3O_8	1.7	6.3	29.1	46.5	55.1	41.8	37.3	63.6	84.9	104.6	106.8	262.6	321.5	296.8	284.8	280.2	180.3	139.7	109.7
Other minerals	87.3	117.7	99.0	171.9	166.7	220.0	225.8	209.7	188.7	210.7	225.6	244.3	268.1	221.0	248.8	346.2	345.6	252.4	243.8
TOTAL, ALL MINERALS	89.0	124.0	128.1	218.4	221.8	261.8	263.1	273.3	273.6	315.3	332.4	506.9	589.6	517.8	533.6	626.4	525.9	392.1	353.5

Notes: The coverage widened as of 1976; figures are given for that year on both bases. The companies included up to 1976 were BP Minerals, Charter Consolidated, Consolidated Gold Fields, Rio Tinto–Zinc, Selection Trust, Metallgesellschaft, Preussag, Uranerzbergbau, Urangesellschaft, Union Minière, Billiton, BRGM, Imetal, SNEA, Pechiney, Pertusola, SAMI. Since 1976 the coverage has included Gewerkschaft Brunhilde, Saarberg–Interplan Uran GMBH, Greenex, Imperial Chemical Industries, Greek Mining Federation, and those members of the French Mining Federation not previously included.

Source: Based on an annual survey of European companies. Until 1979 the survey was carried out by the Group of European Mining Companies; subsequent data have been collected by the Comité de Liaison des Industries de Metaux Non Ferreux, Brussels, Belgium. The statistics are not published, but are available upon request.

[a] Current expenditures are deflated by the U.S. producer price index for all commodities to obtain constant 1982 dollar expenditures.

Part 2-A-11

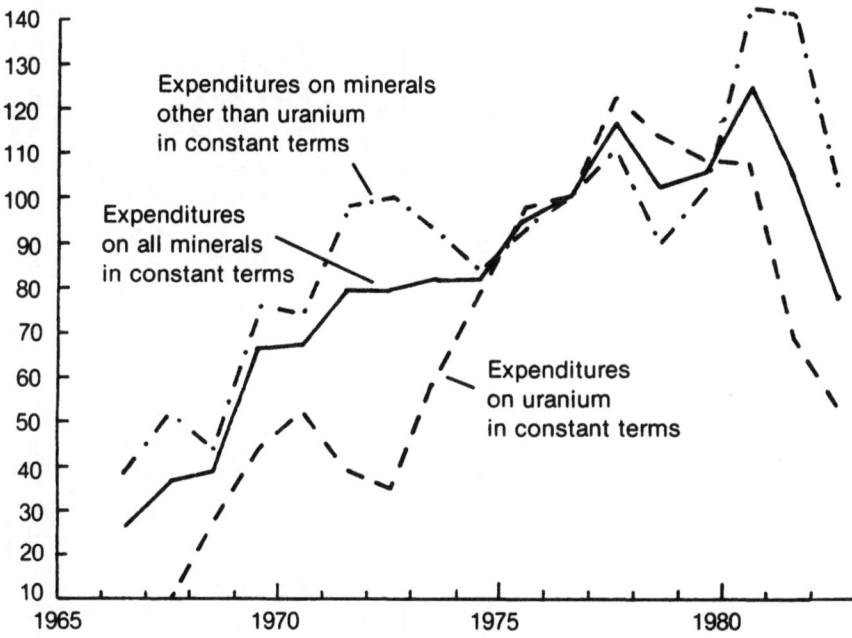

Indices of mineral exploration expenditures for uranium and other minerals by European Community companies, 1966–82 (1976 = 100). *Source:* Part 2-A-10 in this appendix.

Part 2-A-12

Mineral Exploration Expenditures of European Community Companies for Uranium (U₃O₈) and Other Minerals, by Geographic Region, 1966–82

(millions of current U.S. dollars)

Countries	Category	1966	1967	1968	1969	1970	1971	1972	1973	1974	1975	1976	1976	1977	1978	1979	1980	1981	1982
Developed																			
Africa	U_3O_8	0	0	1.28	2.50	4.96	1.51	0.20	0	0	0	0	0	0	1.14	2.95	2.10	4.47	0.07
	Other	1.32	1.40	1.01	1.02	1.32	1.66	2.76	4.23	5.36	6.51	7.30	7.30	8.20	6.51	7.47	13.86	13.63	9.81
Australia	U_3O_8	0.03	0.52	0.58	0.43	0.37	0.91	1.35	1.48	1.54	2.21	2.66	2.66	4.70	17.56	14.74	29.54	28.96	29.44
	Other	2.37	3.15	7.46	11.48	14.81	20.90	20.43	26.70	23.27	24.33	28.27	28.27	30.17	35.55	46.80	92.43	95.20	55.97
North America	U_3O_8	0.53	1.57	1.11	1.85	1.67	1.37	1.22	2.08	14.28	12.67	16.70	26.60	35.59	58.32	56.93	65.75	39.79	24.12
	Other	2.60	6.03	4.47	6.33	14.15	7.51	9.98	10.28	17.97	19.53	31.19	31.19	29.39	28.05	35.03	64.77	83.83	58.74
European Community	U_3O_8	[a]	[a]	[a]	[a]	[a]	[a]	[a]	[a]	[a]	[a]	[a]	[a]	[a]	[a]	[a]	107.89	78.42	63.09
	Other	[a]	[a]	[a]	[a]	[a]	[a]	[a]	[a]	[a]	[a]	[a]	[a]	[a]	[a]	[a]	78.57	73.43	70.74
Other European	U_3O_8	0[a]	0[a]	3.14[a]	3.33[a]	2.61[a]	2.80[a]	4.88[a]	8.04[a]	11.92[a]	18.36[a]	22.20[a]	64.35[a]	74.89[a]	70.26[a]	92.92[a]	0	1.23	0.44
	Other	7.57	9.89[a]	9.22[a]	10.44[a]	12.64[a]	28.09[a]	27.88[a]	27.42[a]	25.16[a]	35.75[a]	38.44[a]	52.15[a]	59.52[a]	52.75[a]	66.08[a]	13.79	12.13	12.63
Not allocated	U_3O_8	0	0	0.82	2.35	4.94	3.07	3.84	12.81	12.47	20.05	18.41	18.41	26.78	0	0	0	0.40	0
	Other	2.05	3.27	4.90	5.98	7.27	11.71	20.95	15.24	16.20	16.49	15.90	13.63	17.34	3.93	4.60	0.40	0.28	0.70
Subtotal	U_3O_8	0.56	2.10	6.92	10.47	14.53	9.65	11.49	24.40	40.22	53.30	59.84	112.02	141.97	147.28	167.55	205.29	153.25	117.17
	Other	15.91	23.73	27.06	35.24	50.19	69.88	82.00	83.88	87.97	102.96	121.10	132.54	144.62	126.79	159.98	263.82	278.50	208.61
Overall subtotal		16.47	25.83	33.98	45.71	64.72	79.53	93.49	108.28	128.19	156.26	180.94	244.56	286.59	274.07	327.53	469.11	431.75	325.78

Less developed

Region	Basis																		
Africa	U_3O_8	0	0	0.20	1.74	1.13	0.69	0.35	0.79	0.91	1.61	0.70	36.78	40.77	54.84	50.21	41.52	18.90	15.60
	Other	2.04	0.55	1.02	2.29	1.51	1.26	1.38	2.65	0.35	0.11	0.20	0.20	0.50	4.54	6.54	8.71	6.78	8.65
Asia	U_3O_8	0	0	0	0	0	0	0	0	0.10	0.10	0.50	5.47	3.67	0.90	0.20	0	0	0
	Other	0.38	1.66	3.33	6.08	4.37	2.68	1.31	1.76	2.44	3.61	3.62	3.62	5.73	8.10	12.61	8.13	7.80	4.81
Latin America	U_3O_8	0	0	0	0	0	0	0	0	0.10	0.60	0.90	2.94	3.26	4.54	6.26	4.53	3.83	5.99
	Other	0	0	0	0.38	0.37	0.64	0.76	0.93	3.01	7.79	6.35	6.35	13.62	12.15	16.19	29.74	44.42	28.33
Oceania	U_3O_8	2.17	4.40	4.86	13.00	0	0	0	0	0	0	0	0	0	0	0.50	0.33	0	0
	Other	0	0	0	0	0	0	0	0	0	0	0.37	0.37	1.32	0.72	0	0.25	1.25	2.00
Not allocated	U_3O_8	0	0	4.78	4.34	4.68	5.57	3.00	3.41	4.08	5.51	3.35	3.35	18.90	0	0	0	0.75	0.92
	Other	8.60	9.00	4.28	4.19	5.06	9.27	4.40	5.17	7.17	8.65	6.27	6.27	8.19	0	0	0	0	0
Subtotal	U_3O_8	0	0	4.98	6.08	5.81	6.26	3.35	4.21	5.19	7.83	5.45	48.54	66.60	60.28	56.67	46.31	23.48	22.53
	Other	13.19	15.61	13.48	25.94	11.31	13.85	7.86	10.51	12.97	20.16	16.81	16.81	29.35	27.76	35.84	47.12	60.25	43.79
Overall subtotal	All minerals	13.19	15.61	18.46	32.02	17.12	20.11	11.21	14.72	18.16	27.99	22.26	65.35	95.95	88.04	92.51	93.43	83.73	66.32
TOTAL, ALL AREAS	U_3O_8	0.56	2.10	11.90	16.55	20.34	15.91	14.84	28.61	45.41	61.12	65.29	160.56	208.57	207.58	224.22	251.60	176.73	139.70
	Others	29.10	39.34	40.54	61.18	61.50	83.73	89.86	94.39	100.94	123.12	137.91	149.35	173.97	154.55	195.82	310.94	338.75	252.40
OVERALL TOTAL	All minerals	29.66	41.44	52.44	77.73	81.84	99.64	104.70	123.00	146.35	184.24	203.20	309.91	382.54	362.13	420.04	562.54	515.48	392.10

Notes: The coverage widened as of 1976; figures are given for that year on both bases.

The companies included up to 1976 were BP Minerals, Charter Consolidated, Consolidated Gold Fields, Rio Tinto–Zinc, Selection Trust, Metallgesellschaft, Preussag, Uranerzbergbau, Urangesellschaft, Union Minière, Billiton, BRGM, Imetal, SNEA, Pechiney, Pertusola, SAMI.

Since 1976 the coverage has included Gewerkschaft Brunhilde, Saarberg–Interplan Uran GMBH, Grenex, Imperial Chemical Industries, Greek Mining Federation, and those members of the French Mining Federation not previously included.

The survey figures are expressed in constant prices for the last year they cover. They have been converted back to current dollars in the table with the price index used to take them to constant prices (the U.S. producer price index for all commodities).

The geographical coverage has varied slightly from one year to the next. There is a large "unallocated" item for both developed and less developed countries.

Source: Based on an annual survey of European companies. Until 1979 the survey was carried out by the Group of European Mining Companies; subsequent data have been collected by the Comité de Liaison des Industries de Metaux Non Ferreux, Brussels, Belgium. The statistics are not published, but are available upon request.

[a]Statistics for the European Community and other European countries have only been subdivided since 1980.

Part 2-A-13

Information on Mineral Exploration Expenditures of Other European Countries

Following are partial data on exploration expenditures for Finland, Spain, Sweden, and the United Kingdom.

Finland

Mapping and basic geophysics are national responsibilities in Finland.

Drilling averaged 40,000 to 50,000 meters per year in the 1960s and 1970s and has been nearer 70,000 to 90,000 meters per year more recently. The state accounts for 20 percent and domestic companies for the balance of drilling.

Expenditures on exploration from 1960 to 1982 were as follows:

Years	Millions of current Finnish marks	Millions of current U.S.$
1960–69 (total)	130	—
1970–79 (total)	520	—
1980	90	23
1981	100	26.8
1982	110	25.5

In recent years the state has accounted for 40 percent of mineral exploration expenditures and companies for 60 percent, of which 30 percent is for nonferrous and 30 percent for other minerals.

Source: Personal communication from the Outokumpu Oy Company, 1983.

Spain

In the 1979–81 period, total state-funded expenditures on mineral exploration (excluding hydrocarbons) were 8,346 million pesetas (U.S.$108 million). Of this total, expenditures for coal amounted to 3,228 million pesetas (U.S.$42 million). An additional 700 million pesetas (U.S.$9 million) was spent on coal by private companies. Spending on nonfuel minerals totaled 3,080 million pesetas (U.S.$40 million); 2,380 million pesetas (U.S.$31 million) was spent by the government, and some 700 million pesetas (U.S.$9 million) by companies. (Details of spending on uranium are included in part 2-A-30 of this appendix.)

Source: Ministerio de Industrial y Energia, *La Minera en España Hoy* (Madrid, October 1982).

Part 2-A-13 (continued)

Sweden

Data for all Swedish companies are not published. One major mining company spent some 3 million to 4 million Swedish kronor (Sw. Kr.) per year in the early 1960s, Sw. Kr. 9 million to 11 million per year in the early 1970s, Sw. Kr. 21 million to 26 million per year in the 1979–80 period, and Sw. Kr. 32 million to 34 million in 1981–82. Of these totals, drilling accounted for 28 percent in the early 1960s, 39 percent in the early 1970s, 28 percent in the 1979–80 period, and 40 percent in 1981–82.

United Kingdom: Coal

Expenditures of the National Coal Board (NCB) on exploration at new and existing mines from financial years 1976/77 to 1982/83 were as follows:

Year[a]	Millions of current £	Approximate equivalent in millions of U.S. dollars
1976/77	10.6	19
1977/78	11.0	19
1978/79	15.9	30
1979/80	17.0	36
1980/81	19.1	44
1981/82	7.8	16
1982/83	11.0	19

Source: Annual Report and Accounts of the National Coal Board, United Kingdom (Issues for relevant NCB financial years).
[a]Financial year, ending in late March.

Figures for earlier years were not published with the *Accounts*. Most of the spending is capitalized.

Part 2-A-14

Mineral Exploration Expenditures in the United States, 1977

There is no regular official annual survey of exploration expenditures in the United States. The spending of selected companies, as recorded in annual reports, is given in part 2-A-20 of this appendix, and part 2-A-15 presents annual statistics on drilling that cover both exploration and development.

A quinquennial census of the mineral industries is conducted by the U.S. Bureau of the Census. It includes questions about exploration and development expenditures, subdivided between capitalized and expensed expenditure. The table presented here on U.S. mineral exploration and development contains the totals from the 1977 census, the latest for which (as of this writing) data have been published.

United States: Mineral Exploration and Development, 1977
(millions of current U.S. dollars)

Standard Industrial Code (SIC) number	Ore or mineral	Capitalized		Expensed	
		Exploration and development	Mineral land and rights	Exploration and development	Mineral land and rights
1011	Iron ore	0.6	1.8	52.6	17.6
1021	Copper ore	51.0	6.3	26.2	1.5
1031	Lead and zinc ores	19.9	0.6	6.6	5.6
1041	Gold ores	3.7	0.1	0.9	1.1
1044	Silver ores	3.6	0.2	5.0	0.2
1051	Bauxite and aluminum ores	0.0	0.5	10.0	0.1
1061	Ferroalloy ores	53.6	6.6	118.6	0.2
1092/99	Mercury and miscellaneous metals	0.3	1.3	1.1	0.1
1094	Uranium-vanadium	123.0	32.0	133.5	36.0
	Total, metallic ores	255.7	49.4	354.5	62.4
145	Clay and related minerals	2.6	3.7	8.7	5.8
1411	Dimension stone	0.6	0.2	1.8	0.2
142	Crushed, broken stone	6.5	7.0	29.7	33.9
144	Sand and gravel	4.8	7.0	123.3	35.6
147	Chemical, fertilizer minerals	15.2	48.0	77.6[a]	17.7
149	Miscellaneous minerals	3.0	0.9	7.8	2.3
	Total, nonmetallic minerals	32.7	66.8	248.9	95.5

(Continued)

Part 2-A-14 (continued)

Standard Industrial Code (SIC) number	Ore or mineral	Capitalized		Expensed	
		Exploration and development	Mineral land and rights	Exploration and development	Mineral land and rights
1111	Anthracite	0.6	0.5	12.0	1.2
1211	Bituminous coal and lignite	231.0	153.8	247.7	201.6
	TOTAL, MINERAL INDUSTRIES	519.8[b]	270.5	863.1	360.7

Note: See section entitled Notes to and Discussion of Table in this part of the appendix.
Source: U.S. Bureau of the Census, *Census of Mineral Industries 1977* (Washington, D.C., 1978).
[a]Figures for 1473 fluorspar and 1479 chemical and fertilizer mining not elsewhere classified withheld to prevent disclosure.
[b]Total for 1967: $128.7 million; for 1972, $229.5 million.

Notes to and Discussion of Table

In the table "United States: Mineral Exploration and Development, 1977," *capitalized exploration and development* does not include expenditure on buildings and other structures, nor on machinery and equipment. It does, however, include capitalized expenditures on exploration and development of ore bodies at producing mines—i.e., much successful exploration would appear under this heading.

Expensed exploration and development in the table includes most unsuccessful exploration and the amount of successful exploration and development expenditure that companies are prepared to write off against annual profits.

Separate capitalized and expensed estimates are given for expenditures on land and mineral rights: most of the capitalized amount is for the acquisition of land for new mines, and much of the expensed total is for the temporary acquisition of mineral rights for exploration.

Comparison of the statistics for uranium in this table with the data on uranium expenditures for 1977 shown in part 2-A-34 of this appendix compounds the confusion as to how much of this table's totals represent genuine exploration expenses. Following are 1977 exploration expenditures on uranium in the United States as shown in the two tables:

Part 2-A-14 (continued)

From part 2-A-14[a] (millions of current U.S.$)				From part 2-A-34 (millions of current U.S.$)	
Mineral land and rights:	Capitalized	32.0	} 68.0	Land acquisition	28.2
	Expensed	36.0			
Exploration and development:	Capitalized	123.0	} 256.5	Exploration drilling	99.4
	Expensed	133.5		Development drilling	55.6
				Other expenditures	74.8
TOTAL		324.5		*TOTAL*	258.1[b]

[a]Includes vanadium.
[b]This compares with the Organisation for Economic Cooperation and Development's estimate of $293.5 million as given in part 2-A-30 of this appendix.

In open pit mines with substantial overburden, there are large preproduction expenses associated with removal of the overburden. Much of the initial stripping and subsequent overburden removal is charged against profits, and, therefore, accounts for a large share of both the capitalized and the expensed development spending shown in the table, particularly in coal mines and sand and gravel works. A fair amount of development work in underground mines also is included in the table. In conclusion, the figures in the table can provide no more than a rough guide to expenditure on exploration as such and can merely offer a check on other estimates.

Part 2-A-15

Exploration and Development Drilling in the United States, by Mineral Commodity, 1966–80

(thousands of feet, all methods)

Mineral commodity	1966	1967	1968	1969	1970	1971	1972	1973	1974	1975	1976	1977	1978	1979	1980
Copper	1,400	1,238	1,328	2,871	1,733	1,347	3,100	1,157	1,262	1,194	864	595	511	325	542
Gold	534	1,046	328	227	185	4,472	5,230	4,284	4,184	3,384	3,152	4,355	453	478	835
Iron ore	1,537	633	407	375	298	254	313	301	216	182	157	114	78	95	41
Lead	701	1,273	585	1,119	1,302	1,362	479	336	377	459	416	540	522	439	619
Molybdenum[a]	208	199	189	177	182	N.A.	N.A.	N.A.	N.A.	N.A.	153	147	184	589	184
Silver	132	274	197	293	430	312	63	127	80	174	149	117	247	101	124
Tungsten	41	47	61	54	59	59	39	49	67	108	24	107	67	60	66
Zinc	737	1,651	598	1,256	1,418	1,684	985	470	494	1,076	593	595	307	437	169
Other metals[a]	239	331	825	572	1,637	1,029	1,011	956	225	872	236	2,156	2,138	1,374	792
Barites[b]	41	6	16	51	84	136	69	92	43	33	33	28	N.A.	N.A.	27
Boron[b]	N.A.	N.A.	N.A.	N.A.	N.A.	N.A.	N.A.	N.A.	N.A.	N.A.	41	N.A.	N.A.	18	39
Fluorspar	55	81	77	52	87	87	138	137	147	127	314	184	160	0	0
Phosphate	293	263	286	399	259	195	111	159	312	124	355	87	177	174	275
Other minerals[b]	754	247	176	392	362	402	198	179	576	773	724	695	171	240	535
Subtotal	6,613	7,289	5,073	7,838	8,036	11,339	11,636	8,247	7,983	8,506	7,211	9,720	5,016	4,330	4,248
Uranium	5,417	10,102	18,665	23,873	20,730	16,271	13,300	12,540	13,343	17,600	8,909	15,500	22,540	17,920	14,650
TOTAL	12,089	17,391	23,738	31,711	28,766	27,610	24,936	20,787	21,326	26,106	16,120	25,220	27,556	22,250	18,898

Note: N.A. = not available.

Sources: U.S. Department of the Interior, Bureau of Mines, *Minerals Yearbook, Volume I* (Washington, D.C., successive issues).

a"Other metals" includes molybdenum where no separate figures are given.

b"Other minerals" includes boron and barites where no separate figures are given.

Part 2-A-15 (continued)

Exploration Drilling (Only) in the United States, by Mineral Commodity, 1973–80
(thousands of feet, all methods)

Mineral commodity	1973	1974	1975	1976	1977	1978	1979	1980
Copper	938	1,090	1,050	673	442	337	300	405
Gold	4,230	4,120	3,320	3,090	4,290	410	415	712
Iron ore	239	159	125	91	85	58	72	39
Lead	274	296	391	354	480	478	396	567
Molybdenum[a]	N.A.	N.A.	N.A.	153	147	184	589	184
Silver	97	49	132	126	66	187	55	76
Tungsten	41	45	27	1	77	54	34	47
Zinc	376	364	954	489	476	243	363	112
Other metals[a]	894	155	789	177	737	288	1,020	708
Barytes[b]	90	92	43	33	28	153	N.A.	27
Boron[b]	N.A.	N.A.	N.A.	41	N.A.	1	18	39
Fluorspar	121	137	123	308	180	N.A.	N.A.	0
Phosphate	151	305	117	348	80	172	169	260
Other minerals[b]	155	159	292	176	77	166	233	491
Subtotal	7,606	6,971	7,363	6,060	7,165	2,731	3,664	3,667
Uranium	12,400	13,000	17,300	8,480	14,100	20,800	15,000	12,600
TOTAL	20,006	19,971	24,663	14,540	21,265	23,531	18,664	16,267

Note: N.A. = not available.
Sources: U.S. Department of the Interior, Bureau of Mines, *Mineral Yearbook, Volume I* (Washington, D.C., successive issues).
[a]"Other metals" includes molybdenum where no separate figures are given.
[b]"Other minerals" includes boron and barites where no separate figures are given.

Part 2-A-16

Mineral Exploration Expenditures in Japan, 1966–80
(millions of current yen, unless otherwise noted)

Program level[a]	1966	1967	1968	1969	1970	1971	1972	1973	1974	1975	1976	1977	1978	1979	1980
A. Regional geological reconnaissance survey	132.7	250.7	316.3	361.9	395.8	414.3	488.2	582.5	661.8	773.3	806.1	831.2	853.7	863.9	864.1
No. of projects	11.0	11.0	12.0	14.0	13.0	13.0	12.0	13.0	14.0	13.0	13.0	14.0	15.0	14.0	14.0
B. Detailed geological survey total expense	199.7	236.0	288.0	346.7	484.9	682.6	935.9	1,097.4	1,193.1	1,382.5	1,459.0	1,469.1	1,474.6	1,475.5	1,505.7
Of which:															
Subsidy	119.9	141.6	172.8	208.0	290.9	409.5	561.6	658.4	795.4	921.6	972.7	979.4	983.1	983.7	1,003.8
C. Loans for domestic exploration	2,400.0	2,500.0	2,800.0	2,800.0	2,900.0	3,000.0	3,135.0	2,871.0	3,300.0	3,700.0	3,188.0	2,735.0	1,828.0	1,586.0	N.A.
Of which:															
Copper	1,754	1,742	1,711	1,657	1,862	2,094	2,166	2,154	2,162	2,239	1,883	1,751	1,281	1,115.0	N.A.
Lead and zinc	629	758	1,067	1,050	938	875	927	716	1,138	1,461	1,305	984	547	471	N.A.
Gold and manganese	18	0	22	93	100	31	42	0	0	0	0	0	0	0	N.A.

(Continued)

Part 2-A-16 (continued)

Program level[a]	1966	1967	1968	1969	1970	1971	1972	1973	1974	1975	1976	1977	1978	1979	1980
IMPLIED TOTAL DOMESTIC EXPLORATION[b]	4,365	4,545	5,090	5,090	5,270	5,455	5,700	5,220	6,000	6,725	5,795	4,970	3,325	2,885	N.A.
TOTAL EXPLORATION: A + B + C	4,698	5,032	5,694	5,799	6,151	6,552	7,124	6,900	7,855	8,881	8,060	7,270	5,654	5,224	N.A.
TOTAL equivalent in millions of current U.S.$	13.0	14.0	15.8	16.1	17.1	18.8	23.5	25.4	26.9	29.9	27.2	27.1	26.9	23.8	N.A.
TOTAL in millions of constant 1982 U.S.$	39.0	42.0	46.0	45.3	46.4	49.4	59.0	56.4	50.3	51.2	44.5	41.8	38.5	30.2	N.A.

Note: Years are fiscal, starting on April 1 of years shown. N.A. = not available.

Source: Metal Mining Agency of Japan, *1981 Report* (Tokyo, 1981).

[a] Japan's national exploration program is on three levels:

A. Basic regional geological survey work, carried out by contractors, is funded entirely by the government through the Metal Mining Agency of Japan (MMAJ). Some eleven to fifteen areas have been surveyed annually.

B. Regional surveys are followed up by detailed geological exploration of promising areas. A total of forty-eight promising districts was selected in the 1960s. The government contributes two-thirds of the cost, prefectural governments two-fifteenths, and the remaining one-fifth is paid by the mining companies that own the concessions. There were about eight prospects a year in the 1970s.

C. Subsidized loans are given to larger companies (with a paid-up capital exceeding 100 million yen or more than 1,000 employees) and outright subsidies to smaller firms (which may be subsidiaries of larger companies) for detailed prospecting of individual projects in thirteen metallic minerals including copper, lead, zinc, manganese, gold, and silver. Funds are only provided to Japanese nationals or mining companies. Loans have been up to 70 to 80 percent of budgeted cost (and usually much less), depending on the projects concerned in recent years. In the 1963–79 period the MMAJ contributed 54.7 percent of total exploration expenditure.

[b] Assuming the Metal Mining Agency of Japan bears 55 percent of cost.

Part 2-A-17

Japanese Expenditures on Overseas Exploration for Nonferrous Minerals, 1960–81

(millions of current yen, unless otherwise noted)

Country	1960	1961	1962	1963	1964	1965	1966	1967	1968	1969	1970	1971	1972	1973	1974	1975	1976	1977	1978	1979	1980	1981
Australia	0	109	61	49	0	13	17	210	18	143	231	222	196	411	79	39	17	111	214	636	1,276	189
Bolivia	26	404	171	0	0	0	0	0	0	0	109	73	69	72	132	0	0	0	0	0	0	0
Canada	142	0	0	0	97	137	210	330	703	1,019	455	344	344	223	353	382	683	717	480	783	311	552
Chile	21	141	63	228	129	151	199	233	191	41	0	0	0	0	195	544	331	439	205	20	0	0
Malaysia	0	0	0	0	0	0	1	43	576	423	22	0	0	0	0	0	0	0	0	0	0	0
Peru	137	125	30	51	55	185	61	61	236	73	81	595	11	79	765	492	392	281	130	174	93	177
Philippines	58	59	0	0	0	0	30	2	79	173	658	372	99	330	474	58	80	77	51	0	26	745
Zaire	0	0	0	0	0	0	0	0	800	2,398	1,333	1,262	1,691	1,073	906	11	0	0	0	0	0	1
Others	40	48	72	69	71	66	136	307	238	868	1,510	1,700	1,405	1,345	2,352	3,373	6,048	8,426	5,692	5,069	4,746	2,988
TOTAL	424	886	397	397	352	552	593	1,485	2,841	5,138	5,051	4,679	3,815	3,533	5,256	4,899	7,551	10,051	6,772	6,682	6,452	4,652
TOTAL (in millions of current U.S. dollars)						1.5	1.6	4.1	7.9	14.3	14.0	13.4	12.6	13.0	18.0	16.5	25.5	37.4	32.2	30.5	28.5	21.5
TOTAL (in millions of constant 1982 U.S. dollars)						4.7	5.0	12.4	23.0	40.0	38.0	35.2	31.6	28.7	33.7	28.3	41.6	57.7	46.0	38.7	31.7	21.5

Notes: Data include expenditures on copper, lead, zinc, nickel, uranium, bauxite, chrome, and manganese nodules.
There is a mixture of official subsidies and loans to encourage overseas exploration for selected minerals. The former are available for general survey work up to half the cost of initial drilling and two-thirds of other work. Detailed follow-up exploration is assisted through low-interest-bearing loans financed by annual government subventions. Up to half the cost of approved projects in base metals is covered, or 70 percent in special cases, with loans repayable over ten years. The terms for uranium exploration have been more generous.

Sources: Metal Mining Agency of Japan, *1981 Report* (Tokyo, 1981); and data provided by Ministry of International Trade and Industry (MITI).

Part 2-A-18

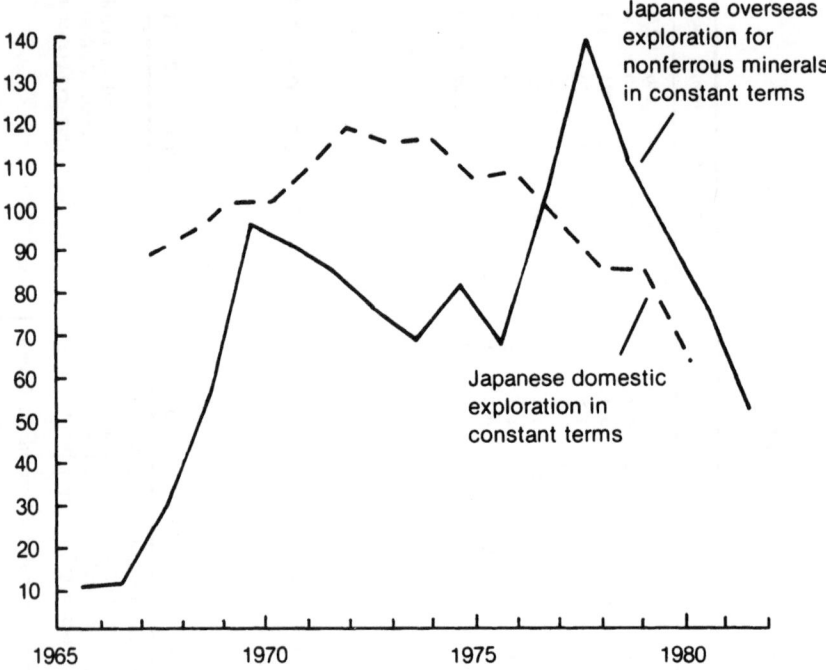

Indices of Japanese domestic and overseas mineral exploration expenditures, 1966–81 (1976 = 100). *Sources:* Parts 2-A-16 and 2-A-17 of this appendix.

Part 2-A-19

Number of Japanese Overseas Mineral Exploration Projects, by Mineral and Country, March 1982

Mineral	Country	No. of Projects	Mineral	Country	No. of Projects
Copper	Australia	2	**Uranium**	Australia	2
	Canada	4		Canada	10
	Costa Rica	1		Gabon	1
	Panama	1		Guinea	1
	Papua New Guinea	1		Mali	1
	Peru	3		Mauritania	1
	Philippines	1		Niger	3
	Zaire	1		United States	4
TOTAL		14		Zambia	1
			TOTAL		24
Lead/zinc	Canada	1	**Nickel**	Pacific Ocean	1
	Guatemala	1			
	Peru	1			
	United States	3	**Other**		
TOTAL		6	**minerals**	Mexico	1

Note: There is a mixture of official subsidies and loans to encourage overseas exploration for selected minerals. The former are available for general survey work up to half the cost of initial drilling and two-thirds of other work. Detailed follow-up exploration is assisted through low-interest-bearing loans financed by annual government subventions. Up to half the cost of approved projects in base metals is covered, or 70 percent in special cases, with loans repayable over ten years. The terms for uranium exploration have been more generous.

Source: Metal Mining Agency of Japan, *1981 Report* (Tokyo, 1981).

Part 2-A-20

Mineral Exploration Expenditures by Selected North American Companies, 1970–82

(millions of current U.S. dollars, unless otherwise noted)

Companies	1970	1971	1972	1973	1974	1975	1976	1977	1978	1979	1980	1981	1982
Canadian													
Cominco	8.6	10.9	13.8	22.6	27.8	26.7	30.5	29.4	29.7	27.8	37.7	50.9	34.0
Dome Mines	N.A.	N.A.	N.A.	N.A.	N.A.	N.A.	N.A.	N.A.	1.9	2.6	5.1	6.1	5.6
Falconbridge	8.9	10.2	6.9	8.9	11.7	7.7	8.4	9.1	6.7	12.2	24.2	31.7	N.A.
Hudson Bay Mines	N.A.	3.0	2.8	2.9	4.3	3.2	2.9	2.6	3.0	3.7	8.3	3.2	4.9
Inco	N.A.	20.4	13.8	15.2	18.6	26.1	34.6	21.1	12.4	11.9	23.0	27.3	15.6
McIntyre Mines	N.A.	N.A.	N.A.	0.9	1.2	2.0	1.7	3.0	3.8	0.3	1.2	0.4	0.4
Noranda	(5.0)	4.8	8.9	11.4	18.3	20.3	21.9	20.6	12.2	17.0	25.7	36.3	32.2
Placer	5.3	4.3	6.3	6.0	11.0	9.9	6.5	9.3	10.6	15.1	20.7	23.5	21.9
Rio Algom	2.3	2.0	2.5	3.1	5.2	6.1	6.9	6.0	5.3	8.0	9.2	10.6	8.0
Sherritt Gordon	N.A.	0.8	0.7	0.8	1.4	1.1	0.8	0.8	0.5	1.1	3.3	4.7	2.1
Teck Corporation	N.A.	N.A.	1.4	1.0	1.4	1.7	1.3	1.8	3.1	3.9	9.0	9.4	5.7
Western Mining	0.1	0.2	0.2	0.4	1.3	1.5	3.0	2.1	1.4	3.7	3.4	4.3	3.5
TOTAL	. .	58.0[a]	58.0[a]	73.2	102.2	106.3	118.5	105.8	90.6	107.3	170.8	208.4	150.0[a]
TOTAL in millions of current Canadian $	. .	58.6	57.4	73.2	100.0	108.1	116.8	112.5	103.3	125.7	199.7	250.0	185.0
TOTAL in millions of constant 1982 U.S.$	. .	130.7	124.9	147.8	189.3	179.3	187.9	157.3	125.4	136.6	199.3	221.9	150.0

U.S.

Amax	14.2	15.0	12.2	33.4	36.8	33.8	29.4	33.6	39.6	82.1	117.9	73.8	48.2
Asarco	4.7	7.5	6.1	6.7	6.6	7.1	7.2	12.3	8.0	10.7	14.7	16.8	18.5
Atlantic Richfield	N.A.	N.A.	N.A.	N.A.	N.A.	N.A.	N.A.	21.2	33.7	40.4	54.4	75.9	94.9
Conoco	6.7	8.2	8.9	12.2	22.7	25.8	29.9	32.2	27.3	31.3	37.9	N.A.	N.A.
Cyprus/Amoco	N.A.	N.A.	N.A.	5.8	6.9	5.6	6.5	4.8	N.A.	N.A.	N.A.	N.A.	N.A.
Hecla Mining	0.5	0.4	0.3	0.8	0.8	1.1	1.2	1.2	0.7	0.2	0.4	N.A.	N.A.
Homestake	1.2	1.5	0.9	1.4	2.8	5.4	7.2	9.8	6.5	11.9	18.1	22.2	21.6
Newmont	8.2	11.6	12.3	19.5	18.5	15.0	18.2	17.9	17.4	23.8	23.6	25.7	19.1
Phelps Dodge	6.1	8.2	8.8	8.8	8.9	9.6	6.9	7.2	5.5	6.7	9.9	10.6	11.1
St. Joe Minerals/Fluor	N.A.	N.A.	N.A.	2.3	2.3	3.7	5.9	12.9	13.0	16.0	19.8	N.A.	N.A.
TOTAL for seven U.S. companies[b]	41.6	52.4	49.5	82.8	97.1	97.8	100.0	114.2	105.0	166.7	222.5	..	..
TOTAL in millions of constant 1982 U.S.$	99.1	118.1	106.6	167.2	179.9	164.9	158.6	169.8	145.4	212.1	259.7	..	..

Notes: Data exclude expenditures for oil and gas. N.A. = not available; . . = lack of data in column precludes meaningful total.
  The table shows expenditures as recorded in annual reports. Accounting conventions vary widely in their treatment of exploration; individual accounting methods can even change over time. In many instances the cost of successful exploration is capitalized, and details may not be given separately. Hence, although these are the best data available, readers are cautioned that the figures in the table give an incomplete and possibly misleading picture, even for individual companies.

Sources: Company annual reports.
[a]Includes approximations for data that are not available.
[b]Amax, Asarco, Conoco, Hecla, Homestake, Newmont, Phelps Dodge.

Part 2-A-21

Mineral Exploration Expenditures by Major South African Companies, 1970–83
(millions of current rand, unless otherwise noted)

Company	1970	1971	1972	1973	1974	1975	1976	1977	1978	1979	1980	1981	1982	1983
Anglo American Corporation[a]	N.A.	N.A.	3.5	3.5	4.4	7.3	9.5	15.2[a]	11.9	17.2	28.5	35.6	39.5	N.A.
Anglovaal[b]	N.A.	N.A.	N.A.	0.2	0.2	0.5	1.0	1.4	1.4	2.7	5.0	5.2	6.4	7.8
De Beers Consolidated[c]	N.A.	N.A.	N.A.	N.A.	N.A.	17.6	20.3	25.9	29.4	38.8	44.5	62.5	59.0	N.A.
Gold Fields of South Africa[d]	N.A.	N.A.	N.A.	N.A.	1.2	2.0	3.3	4.1	2.4	2.2	4.2	7.5	14.7	14.5
General Mining Union Corporation[e]	1.6	1.1	0.8	1.5	2.2	4.5	10.1	9.7	12.6	11.0	13.9	21.6	14.8	N.A.
Union Corporation[f]	N.A.	1.6	1.4	1.8	3.2	6.1	5.9	6.7	4.9	5.3	(5.0)	0.0	0.0	N.A.
Johannesburg Consolidated Investment Company (JCI)[g]	N.A.	N.A.	N.A.	N.A.	3.3	3.8	5.9	3.5	4.0	2.8	3.6	6.1	9.5	6.8
Barlow Rand[h]	N.A.	N.A.	N.A.	N.A.	N.A.	N.A.	N.A.	N.A.	N.A.	N.A.	9.3	12.3	12.3	12.2
TOTAL[i]	..	..	..	..	29.0 (est.)	41.8	56.0	66.5	66.6	80.0	104.7	138.5	143.9	..

TOTAL in millions of current U.S.$^i					43.0	45.1	63.9	77.1	76.2	95.8	135.0	158.5	133.6	..
TOTAL in millions of constant 1982 U.S.$^i	..	..	..		79.7	76.1	101.3	114.6	105.5	121.9	157.5	168.8	133.6	..

Notes: N.A. = not available; . . = lack of data in column precludes meaningful total.

Data cover worldwide expenditures, not just expenditures in South Africa. Consequently, the totals shown exceed those given in table 5-A-1 in the appendix to chapter 5 of this volume. Data in this table exclude expenditures for oil and gas.

The table shows expenditures as recorded in annual reports. Accounting conventions vary widely in their treatment of exploration; individual accounting methods can even change over time. In many instances the cost of successful exploration is capitalized, and details may not be given separately. Hence, although these are the best data available, readers are cautioned that the figures in the table give an incomplete and possibly misleading picture, even for individual companies.

Sources: Company annual reports.

[a]Group costs of prospecting; time periods are calendar years from January 1, 1970, to December 31, 1976; 1977 = 15 months, ending March 31, 1978; 1978 to 1983 are financial years ending March 31 following the year shown.

[b]Consolidated totals for years ending on June 30 (e.g., 1973 = July 1, 1972, to June 30, 1973, inclusive).

[c]Consolidated totals for calendar years. Expenditures are for prospecting and research; a substantial portion of them was spent outside South Africa.

[d]Consolidated totals of drilling and prospecting costs for years ending June 30 (e.g., 1973 = July 1, 1972, to June 30, 1973, inclusive).

[e]Consolidated group totals for calendar years; Union Corporation expenditures are included from 1981 on.

[f]Consolidated totals for calendar years. Data include expenditures subsequently capitalized from 1977 to 1979.

[g]Consolidated group totals for years ending June 30 less amounts capitalized.

[h]Group totals for exploration, research, and development for years ending September 30. Barlow Rand data are not included in the totals of this table because pre-1980 data do not appear in the company's annual reports.

[i]Excluding Barlow Rand data.

Part 2-A-22

Mineral Exploration Expenditures in Brazil

Docegeo (1971–82), Brazilian Department of Mineral Production (1977–80), and Working Concessions (1977–80)

Various apparently contradictory estimates of Brazilian exploration expenditures have been published. Figures based on mining companies' applications for exploration claims are usually grossly exaggerated for two reasons:

1. Each time a mining company applies for an exploration claim, it must annex an exploration budget estimate. In order to gain access to large areas, a mining company typically sets up perhaps five shell companies, each of which applies for the maximum ten claims of 10,000 hectares each. Upon initial investigation, only a few of the fifty claims may show potential; the barren areas are dropped or allowed to expire over a period of years, but the computer of the Department of Mineral Production—Departamento Nacional da Produção Mineral (DNPM)—sums the budgets for all fifty areas. The budgets are really a legal formality, and may be exaggerated for good effect as well.

2. Because a large exploration budget is politically desirable, producing mines classify mining development work and technological studies (pilot plants, milling tests, and so forth) as if they were exploration.

The company Docegeo (the exploration arm of Companhia Vale do Rio Doce [CVRD]) published a full series of figures showing its exploration expenditures for 1971 (the year it was founded) through 1982. Those figures (which exclude heavy spending on the Brazilian iron ore deposit Carajas that would have distorted the data) are as follows:

Year	Amount (millions of current U.S.$)	Year	Amount (millions of current U.S.$)
1971	1.2	1977	23.6
1972	10.1	1978	29.0
1973	16.4	1979	20.2
1974	18.2	1980	20.3
1975	19.1	1981	22.8
1976	24.3	1982	23.4

Brazil's annual mineral report *Anuario Mineral Brasileiro* includes estimates for exploration and prospecting in working concessions, and also details of DNPM's spending, as shown in the two tables in this part.

Part 2-A-22 (continued)

Mineral Exploration Expenditures of the Brazilian Department of Mineral Production, 1977–80

(millions of current cruzeiros, unless otherwise noted)

Expenditure	1977	1978	1979	1980
Basic geological mapping	242.0	518.2	338.8	355.1
Mineral prospecting and appraisal,	260.4	345.6	211.6	642.5
Of which:				
Coal	76.7	33.1	25.6	479.1
Underground water	4.3	15.4	21.0	29.9
Metallic minerals	105.0	165.6	98.5	69.2
Nonmetallic minerals	14.5	21.3	8.4	1.3
Geophysical	36.1	61.6	26.7	51.1
Geochemical	23.8	48.6	30.2	11.9
TOTAL EXPLORATION SPENDING	502.4	863.8	550.4	997.6
TOTAL in millions of current U.S.$	35.5	47.8	20.4	18.9
TOTAL in millions of constant 1982 U.S.$	52.8	66.2	26.0	22.1

Source: Departamento Nacional da Producão Mineral, Ministry of Mines and Energy, *Anuario Mineral Brasileiro* (Brasilia, 1980, 1981 editions).

Part 2-A-22 (continued)

Mineral Exploration Expenditures in Working Concessions in Brazil, 1977–80
(millions of current cruzeiros, unless otherwise noted)

Expenditure	1977	1978	1979	1980
TOTAL	332.0	432.7	771.5	1,211.1
Of which:				
Bauxite			68.0	42.1
Limestone			138.6	100.2
Coal			10.7	53.9
Copper			10.4	85.3
Chrome			0.0	10.9
Diamonds			8.2	16.9
Tin			74.1	140.0
Iron			72.6	229.0
Potash			76.0	156.2
Fluorspar			1.0	27.1
Magnesite			4.5	19.5
Manganese			11.8	22.0
Niobium			10.6	29.1
Nickel			23.9	21.0
Gold			54.3	43.2
Tungsten			58.3	68.8
Zinc			39.4	25.7
Dolomite			27.3	45.7
Phosphate			20.2	16.6
Lead			10.1	20.6
Other minerals			51.5	37.3
TOTAL in millions of current U.S.$	23.5	23.9	28.6	23.0
TOTAL in millions of constant 1982 U.S.$	34.9	33.1	36.4	26.8

Note: The figures in this table are not comprehensive, and informed opinion (personal communications from RTZ Brazil, 1983) assesses genuine exploration expenditure for metallic and nonmetallic minerals, including coal, in 1983 as follows: Docegeo—U.S.$13 million; Companhia de Pesquisa de Recursos Minerais (CPRM, a government–private sector partnership)—U.S.$20 million; international mining companies—U.S.$45 million; state mine development companies—U.S.$14 million; Carajas—U.S.$15 million; others—U.S.$25 million; total—U.S.$132 million (figures in current dollars). In addition, technical and development spending, including pilot plant work, varies in the $45 million to $85 million range. Total spending was down by about 30 percent in 1983 from that in 1982, and was much lower compared with the boom years of the late 1970s.

Source: Departamento Nacional da Producão Mineral, Ministry of Mines and Energy, *Anuario Mineral Brasilerio* (Brasilia, 1980 and 1981 editions).

Part 2-A-23

Mineral Exploration Expenditures in New Caledonia for Nickel (by Company), 1969–81, and for Chrome and Other Products, 1979–81

(millions of C.F.P. francs, unless otherwise noted)

Expenditures for:	1969	1970	1971	1972	1973	1974	1975	1976	1977	1978	1979	1980	1981
Nickel, by:													
Groupe Ballande				24.0	25.0	23.0	18.2	22.6	31.8	11.8	9.7	21.5	24.2
Groupe Bataille				13.5	13.0	0.0	0.0	0.0	0.0	0.0	0.0	0.0	0.0
Cofimpac-Inco				104.0	35	0.0	0.0	0.0	0.0	5.5	0.0	0.0	0.0
Patino-Cofremmi				84.5	90	41.8	0.0	0.0	30.5	62.4	37.8	88.5	79.1
Groupe Montagnat				13.4	15.1	9.0	12.7	14.1	16.7	10.6	6.4	17.8	6.1
Panamax Cie				110.0	2.2	4.1	2.8	1.3	0.0	0.0	0.0	0.0	0.0
Groupe Pentecost				84.0	40.9	44.4	47.1	43.5	51.1	30.7	38.0	55.8	68.0
M. R. De Rouvray				10.0	36.7	20.9	23.1	4.8	6.2	1.5	1.5	5.0	4.0
De Rouvray-Granges				240.0	0.0	0.0	0.0	0.0	0.0	0.0	0.0	0.0	0.0
SLN				213.0	148.0	165.0	203.8	198.2	201.0	169.8	183.4	214.0	228.0
Noumea Entreprises				35.1	11.0	15.0	8.5	12.7	14.0	1.3	4.9	0.0	0.0
Amax				0.0	0.0	21.4	324.0[a]	33.5	36.2	0.0	0.0	0.0	0.0
BRGM				28.0	35.0	18.0	14.5	27.0	1.4	12.7	8.5	5.8	0.0
T. Lafleur J.				25.0	18.2	5.5	9.0	7.3	7.7	0.0	6.0	12.2	12.0
Other explorers				55.5	24.9	50.3	15.0	17.2	10.5	0.7	3.6	0.2	11.0
Total	400.0	780.0	1,572.0	1,040.0	495.0	418.4	678.7[b]	382.2	407.1	307.0	299.8	420.8	432.4
Total in millions of current U.S.$	4.3	7.8	15.7	11.4	6.2	4.8	8.8	4.4	4.6	3.8	3.9	5.5	4.4
Total in millions of constant 1982 U.S.$	10.8	18.6	35.4	24.5	12.5	8.9	14.8	7.0	6.8	5.3	5.0	6.4	4.7

(Continued)

Part 2-A-23 (continued)

Expenditures for:	1969	1970	1971	1972	1973	1974	1975	1976	1977	1978	1979	1980	1981
Chrome											89.6	29.2	1.1
Other products											0.6	2.0	0.9
Mineral inventory											9.0	63.6	91.0
OVERALL TOTAL											399.0	515.6	525.4
OVERALL TOTAL *in millions of current* U.S.$											5.2	6.7	5.3
OVERALL TOTAL *in millions of constant 1982 U.S.$*											6.6	7.9	5.7

Source: Rapport Annuel des Services des Mines et d'Energie 1981 (New Caledonia).
[a]Trial mining.
[b]Includes 324 for Amax trial mining.

Part 2-A-24

Zimbabwe: Exclusive Prospecting Orders Granted, by Type of Mineral Commodity, 1950–78

Year	Precious metals	Precious stones	Base metals	Fuels	Industrial minerals	Total
1950	1	0	6	2	0	9
1951	0	0	2	1	0	3
1952	0	0	1	1	0	2
1953	0	1	5	1	0	7
1954	1	0	1	3	0	5
1955	0	0	6	0	1	7
1956	1	0	6	3	0	10
1957	4	1	12	5	0	22
1958	1	0	8	0	0	9
1959	3	0	6	0	1	10
1960	3	4	4	0	0	11
1961	1	0	2	0	0	3
1962	7	8	9	1	1	26
1963	7	4	5	0	0	16
1964	12	0	8	0	3	23
1965	6	1	2	0	3	12
1966	3	4	17	0	1	25
1967	1	0	5	0	0	6
1968	5	2	33	2	2	44
1969	29	8	54	0	0	63
1970	26	1	42	0	4	48
1971	20	1	34	2	3	42
1972	9	1	19	1	1	25
1973	9	2	12	1	2	22
1974	33	2	36	3	2	43
1975	3	0	3	0	0	22
1976	N.A.	N.A.	N.A.	N.A.	N.A.	20[a]
1977	N.A.	N.A.	N.A.	N.A.	N.A.	6[a]
1978	N.A.	N.A.	N.A.	N.A.	N.A.	0[b]

Note: Some orders were granted for more than one category. N.A. = not available.
Source: Rhodesia Geological Survey, *Bulletin No. 82* (Salisbury, 1978).
[a]Full breakdown not available.
[b]No exclusive prospecting orders were granted during the first half of 1978; source does not provide information on the last half of the year.

Part 2-A-25

Zimbabwe: Exploration Expenditures Under Exclusive Prospecting Orders, by Type of Mineral Commodity, 1947–75
(thousands of current Zimbabwe dollars)

Mineral commodity	Expenditures under:		
	Orders 1 to 250 (1947–69)	Orders 251 to 400 (1968–72)	Orders 401 to 500[a] (1971–75)
Precious metals	**2,902**	**93**	**1,228**
Alluvial gold	136	0	1
Gold	1,768	33	1,181
Gold plus copper	83	0	0
Gold plus antimony	212	36	0
Silver	2	0	5
Platinoids	701	24	41
Base metals	**17,620**	**5,772**	**3,537**
Iron	217	0	5
Chromium	962	76	33
Molybdenum	54	29	29
Tantalum	37	13	0
Tungsten	385	65	295
Nickel plus copper	6,599	4,470	1,577
Copper	8,351[b]	980	783
Lead	409	27	336
Zinc	387	36	479
Tin	194	2	0
Mercury	25	0	0
Fluorite	0	28	0
Vanadium	0	46	0
Fuels	**1,785**	**121**	**126**
Uranium	113	28	0
Coal	1,672	93	126
Industrial minerals	**218**	**23**	**63**
Kyanite	93	0	5
Lithium	0	0	22
Limestone	32	0	0
Graphite	3	0	0
Asbestos	9	0	0
Fireclay	79	11	23
Olivine	2	0	0
Diaspore	0	7	0
Magnesite	0	5	0
Bauxite	0	0	13

(Continued)

Part 2-A-25 (continued)

	Expenditures under:		
Mineral commodity	Orders 1 to 250 (1947–69)	Orders 251 to 400 (1968–72)	Orders 401 to 500[a] (1971–75)
Precious stones	**342**	**527**	**229**
Diamonds	116	492	229
Emeralds	206	35	0
Topaz	20	0	0
TOTAL	22,866	6,536	5,183

Sources: Rhodesia Geological Survey, *Bulletin Nos.* 72, 74, and 82 (Salisbury, 1974, 1975, and 1978, respectively).
[a]Excludes expenditures under orders 446, 452, 462, 486, 492, 493, 497, and 499. Either these orders had been extended, or the war prevented completion of work.
[b]Excludes 9 million Rhodesian dollars spent by the Messina Group on development of the Mangula Copper Mine under order 42.

Part 2-A-26

Zimbabwe: Expenditures Under Exclusive Prospecting Orders (EPOs), by Company, 1947–75

Company or group	No. of EPOs	Percent of total spending
Anglo American Group	90	12.8
Anglovaal Group	6	1.4
Blanket Mine (Pvt.) Ltd.	31	1.6
General Mining Union Corporation	5	1.4
Gold Fields of South Africa Group	18	1.9
Johannesburg Consolidated Investment (JCI) Group	45	17.4
Kimberlitic Searches, Ltd.	12	1.5
Lonrho Group	24	2.7
Messina Group	49	18.5
Rand Mines Group	3	1.8
Rhodesian Chrome Mines, Ltd.	36	7.3
Rio Tinto Group	41	9.1
RST Group	23	9.2
SA Manganese, Ltd.	13	2.2
Union Corporation Group	3	1.8
Others	101	9.4
TOTAL	500	100.0

Source: Rhodesia Geological Survey, *Bulletin No. 82* (Salisbury, 1978).

Part 2-A-27

Exploration for Tin by Producing Members of International Tin Agreement, Except Australia

The producing members of the International Tin Council regularly submit details of their production costs to the council. One category of costs is "on property exploration and development." It is expressed per unit of output and in most countries is subdivided according to mining methods. The data, derived from sample surveys, are not regularly published. The Tin Council, however, issued some figures in its 1979 report entitled *Tin Production and Investment* (London, International Tin Council, 1979). The same report includes a limited amount of information about individual countries' expenditures:

Bolivia. In Bolivia, exploration for tin is carried out by the Geological Service of Bolivia (Geobol); by Comibol (Corporacion Minera de Bolivia), the national mining company; and by private-sector companies with the encouragement of the National Mining Exploration Fund (NMEF). Geobol does basic geologic surveying and mapping. It has prospected in association with the World Bank, the United Nations Development Program, and the United Kingdom's Institute of Geological Sciences.

NMEF was established in 1977 to finance prospecting and exploration programs throughout the country. Its advances are only repayable from successful exploration in the form of a royalty. NMEF's initial $12 million capital was provided by the Bolivian government, and subsequent contributions have come from France, West Germany, Japan, and the World Bank.

Indonesia. Most exploration for tin in Indonesia is carried out by the national tin corporation, PT Tambang Timah, and in 1979 three foreign mining companies were also prospecting on specific leases. The national program, particularly offshore, intensified starting in the late 1960s.

Malaysia. In Malaysia, preliminary exploration—detailed geologic mapping and mineral resources evaluation of those areas with better mineral prospects—is done by the Department of Geological Survey. Its work stops short of actual drilling, which is carried out by the Department of Mines. The latter also explores in areas within State Land and Malay Reservations through the agency of the Mines Research Institute. Most exploration for tin in Malaysia is in the hands of private companies, but there is a growing trend toward joint state–private company ventures. Prospecting activity in the early 1970s is shown below in the table "Prospecting for Tin in Malaysia, 1970–76."

(Part 2-A-27 continues on next page.)

Part 2-A-27 (continued)

Prospecting for Tin in Malaysia, 1970–76
(thousands of acres)

Year	Total acreage under prospecting rights	Total acreage prospected	Total acreage selected for mining
1970	221.1	Negl.	13.6
1971	251.2	Negl.	13.0
1972	220.1	Negl.	12.1
1973	465.8	103.0	11.9
1974	761.6	58.2	11.6
1975	981.0	293.3	6.3
1976	916.1	181.5	9.8

Note: Negl. = negligible.
Source: International Tin Council, *Tin Production and Investment* (London, 1979).

Nigeria. The Nigerian tin industry was historically fragmented into roughly 100 companies, which carried out their own prospecting. The Nigerian Mining Corporation (NMC) established in 1972 was charged with exploration, among other things. By 1978 a majority of the industry had been brought under NMC's control. Exploration in this country has lagged behind the exploitation of known reserves.

Thailand. The Department of Mineral Resources has responsibility for geologic surveys and exploration in Thailand and for the granting of prospecting licenses. Private organizations have made significant finds of tin in this country.

Existing mines. The shares of "on property exploration and development" expenditures in operating costs at existing tin mines are shown below in the first two columns of the table "Tin Mines: Percentage of Operating Costs . . ." In the last two columns of the table, the percentages are multiplied by the costs per unit and then by the production of tin in concentrates by each mining method to give inferred total costs (in current U.S. dollars). The 1978 total of $25.7 million is equivalent to almost $37 million in constant 1982 U.S. dollars. Part of this expenditure is on the costs of advance development, which are written off in the year they are incurred. However, the data do not include the costs of exploration outside existing mines.

Part 2-A-27 (continued)

Tin Mines: Percentage of Operating Costs for Exploration and Development "on Property," by Country, 1971 and 1978

Country and type of operation	Exploration and development as percentage of operating costs		Implied total costs in millions of current U.S.$[a]	
	1971	1978	1971	1978
Bolivia				
Lode underground	0.0	4.2	Negl.	12.3
Dredge	Negl.	2.9	Negl.	0.3
Indonesia				
Dredge offshore	0.9	3.4	0.1	1.5
Dredge onshore	0.9	2.9	0.1	0.8
Gravel pump	0.9	4.1	0.2	5.5
Malaysia				
Dredge	1.4	0.4	0.6	0.3
Gravel pump	1.3	1.6	1.3	3.8
Open cast	0.0	0.0	0.0	0.0
Underground	5.6	0.3	0.3	Negl.
Thailand				
Dredge offshore	4.0	0.0	0.2	0.0
Dredge onshore	1.1	0.1	Negl.	Negl.
Gravel pump	3.0	2.2	0.8	1.2

Note: Negl. = negligible.
Source: International Tin Council, *Tin Production and Investment* (London, 1979).
[a]Implied total costs are the percentages in the first two columns multiplied by costs per unit and then by the production of tin in concentrates by each mining method.

Expenditures on exploration and development, even in existing mines, fluctuate widely from year to year. Some estimates for Malaysia, calculated in the same way as those in the table above, are as follows:

Year	Millions of current U.S.$	Millions of constant 1982 U.S.$
1971	2.2	5.8
1972	2.5	6.3
1973	3.0	6.7
1974	5.1	9.5
1975	3.4	5.8
1976	1.7	2.8
1977	N.A.	N.A.
1978	4.1	5.9
1979	9.9	12.6
1980	N.A.	N.A.
1981	26.2	26.7
1982	7.7	7.7

Source: Personal communications, Rio Tinto-Zinc Corporation, 1983.

Part 2-A-28

Mineral Exploration Expenditures in Botswana, Mexico, Papua New Guinea, and Saudi Arabia

Botswana

Following is a list of estimated mineral exploration expenditures in Botswana from 1970 to 1982 (based on data in chapter 4 of this volume). The totals exclude geologic surveys financed from bilateral government aid amounting to approximately $1 million to $2 million per year. Some of the spending shown here is also included in data for the European companies in parts 2-A-10 and 2-A-12 of this appendix.

Year	Millions of current U.S.$	Millions of constant 1982 U.S.$
1970	5.5	13.5
1971	5.6	13.3
1972	3.0	7.0
1973	3.5	7.8
1974	7.6	14.9
1975	7.0	12.2
1976	4.5	7.3
1977	7.2	11.3
1978	6.2	9.0
1979	7.3	9.7
1980	9.5	11.2
1981	12.5	12.3
1982	16.4	16.4

Mexico

Under the Mining Law of 1975, the federal government of Mexico, through its Mineral Resources Council, has exclusive exploration rights for fertilizer minerals (phosphate, potash, sulfur), iron ore, and coal. A paragovernmental agency, Uranex, is responsible for all uranium exploration. In addition, incentives are offered to private companies to explore for base metals. In the 1961–77 period, 35 important new deposits of a wide variety of minerals (including manganese, silica, tungsten, and molybdenum) were discovered, but most of these await development. In 1974 the government created the Mexican Non-Metallic Minerals Fund with an initial capital of 350 million pesos (U.S.$28 million) to help promote the exploration and development of Mexico's nonmetallic minerals. In 1980, incentives were given to small and medium-sized mining companies to explore, and a large percentage of the national reserves were returned to the private sector.

Part 2-A-28 (continued)

The major private mining companies in Mexico's Chamber of Mines agreed to devote 2.4 billion pesos to exploration in the 1977–82 period under the Lopez Portillo administration, and the federal government undertook to provide a similar sum. In the three years and one month from December 1, 1976, to December 31, 1979, the companies spent 873 million pesos (U.S.$38.4 million) on exploration, and they expended a further 0.9 billion to 1 billion pesos (U.S.$30.6 million) in 1980. Allowing for expenditure by the state and by smaller companies, total exploration spending in Mexico probably averaged some $50 million per annum in the early 1980s.

Papua New Guinea

Following are estimated mineral exploration expenditures in Papua New Guinea from 1964 to 1982. (The data are from chapter 4 in this volume.) Part of the expenditure shown here is included in the totals for European companies in parts 2-A-10 and 2-A-12 of this appendix.

Year	Millions of current U.S.$	Millions of constant 1982 U.S.$	Year	Millions of current U.S.$	Millions of constant 1982 U.S.$
1964	1.6	5.0	1974	4.9	9.0
1965	3.2	9.6	1975	1.5	2.5
1966	5.6	16.0	1976	1.4	2.2
1967	5.6	15.5	1977	1.2	1.8
1968	3.3	8.8	1978	2.9	4.0
1969	6.8	17.0	1979	2.1	2.7
1970	10.1	24.0	1980	3.1	3.6
1971	7.8	17.5	1981	7.5	8.0
1972	6.7	14.4	1982	9.0	9.0
1973	4.8	9.6			

Saudi Arabia

The Saudi Arabian Deputy Minister of Mineral Resources has financed a substantial minerals exploration program in that country since the early 1960s. The program has included expenditure on basic infrastructure, buildings and associated works, and on technical education, as well as on exploration as it is usually defined. There are four main exploration units under contract: (1) the Saudi Arabian Directorate General of Mineral Resources, (2) the French Bureau de recherches géologiques et minières (BRGM), (3) the U.S. Geological Survey, and (4) Riofinex of the United Kingdom. In addition, several contractors, including British Steel, BP/Seltrust, Preussag, and Minatome have worked on specific targets. In all cases the programs are completely financed by Saudi Arabia. It is very

Part 2-A-28 (continued)

difficult to obtain detailed analyses of total exploration expenditures, but basic geologic survey work accounted for almost three-fifths of the total budget in Saudi Arabia's first five-year plan and only one-third in the latest (third) plan; subsequent exploration accounted for the balance of the total. Annual spending on exploration broadly defined but excluding oil and gas, building construction, and training and education, probably rose in constant 1982 U.S. dollars from some $30 million in the mid-1960s to about $50 million in the early 1970s, $60 million in the mid-1970s, and some $125 million in the early 1980s. Because of the physical nature of the country, the lack of infrastructure, shortage of skilled labor, and need to import all stores and equipment, the costs of all aspects of exploration, but particularly of drilling, are roughly four to five times as great in Saudi Arabia as in Australia, so these figures are not directly comparable with those of other countries.

Part 2-A-29

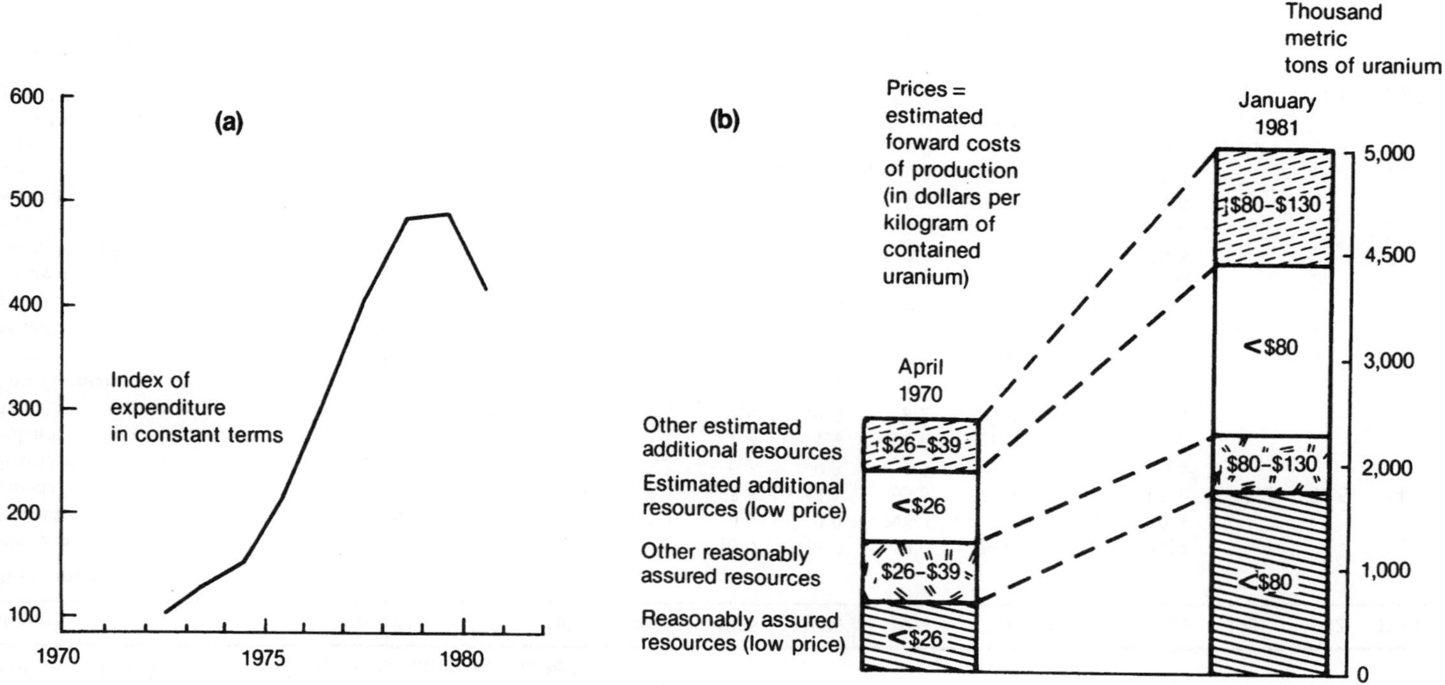

(a) **Index of Western world expenditures on uranium exploration, 1972–80 (1972 = 100).** Expenditures deflated to constant 1982 terms with U.S. producer price index for all commodities. *Sources:* Organisation for Economic Cooperation and Development/International Atomic Energy Agency, Uranium "Red Books" (various issues). (b) **Estimated total uranium resources for a group of selected countries (listed in part 2-A–30 of this appendix), 1970–81.** *Source:* Part 2-A–32 of this appendix.

Part 2-A-30

Uranium Exploration Expenditures for Selected Countries, 1972–83

(millions of current U.S. dollars, unless otherwise noted)

Countries	Pre-1972	1972	1973	1974	1975	1976	1977	1978	1979	1980	1981	1982	1983
Developed													
Australia[a]	31.0	12.8	16.3	14.6	10.0	14.0	18.0	29.0	32.0	49.4	..	..	..
Belgium	0.0	0.0	0.0	0.0	0.0	0.0	Negl.	0.1	0.4	0.5	..	..	..
Canada[b]	4.9	6.0	6.0	8.0	23.5	43.1	65.5	75.9	110.9	107.1	85.2	57.5	34.5
Denmark	0.4	0.0	0.2	0.2	0.3	0.3	1.1	0.1	0.5	0.4	..	..	..
Finland	0.7	0.5	0.6	0.9	1.4	1.4	1.0	1.1	1.1	1.1	..	..	..
France	87.0	4.8	5.4	8.2	14.6	19.6	18.2	26.7	61.7	89.5	..	..	..
West Germany	N.A.	1.2	1.7	2.0	2.7	3.1	6.6	8.1	10.0	11.0	..	..	..
Greece	0.0	0.1	0.1	0.1	0.2	0.4	0.8	0.8	1.2	1.4	..	..	..
Ireland[c]	0.0	0.0	0.0	0.2	0.4	0.4	0.4	0.6	2.0	2.5	..	..	..
Japan	10.6	0.6	0.5	0.5	0.4	0.5	0.4	0.6	0.7	0.7	..	..	..
Portugal	5.6	0.1	0.2	0.2	0.2	0.3	0.3	0.3	0.5	0.6	..	..	..
South Africa[d]	0.8	N.A.	0.4	0.8	3.4	7.3	12.3	24.6	31.3	28.9	..	..	..
Spain	13.3	1.8	1.9	2.6	6.5	7.1	6.8	8.0	12.0	11.0	..	..	..
Sweden	1.0	1.2	1.5	2.1	2.9	5.0	5.4	5.9	5.0	5.5	..	..	..
Switzerland	0.7	Negl.	0.2	0.1	0.1	0.1	0.3	0.4	0.4	0.4	..	..	..
Turkey	3.5	0.4	0.5	0.6	0.8	0.9	2.0	1.2	0.7	0.9	..	..	..
United Kingdom	0.6	0.1	0.1	Negl.	Negl.	Negl.	Negl.	0.4	0.5	0.6	..	..	..
United States	238.9	32.4	49.5	80.9	130.2	195.9	293.5	371.5	385.6	332.6	180.3	97.8	56.7
Total, developed countries	399.0	62.0	85.1	122.0	197.8	299.4	432.6	555.3	656.5	644.1	..	..	..

Less developed

Argentina	N.A.	N.A.	N.A.	N.A.	N.A.	N.A.	4.8	7.2	5.6	7.4
Bolivia	0.2	0.1	0.1	1.2	2.0	3.3	2.2	1.0	1.6	N.A.
Botswana	0.0	0.0	0.0	0.6	N.A.	0.1	0.1	0.1	0.1	N.A.
Brazil	10.0	1.5	8.5	10.0	12.0	24.4	25.6	29.3	12.5	8.6
Central African Republic	15.0	0.0	0.0	0.0	1.7	1.5	1.2	1.2	1.2	N.A.
Chile	N.A.	N.A.	N.A.	N.A.	0.2	0.6	0.8	0.8	0.8	0.8
Colombia	0.0	0.0	0.0	0.5	0.5	1.5	1.5	1.8	2.5	3.0
Ecuador	0.0	0.0	0.0	0.0	Negl.	Negl.	Negl.	Negl.	Negl.	Negl.
Gabon[c]	13.7	4.0	4.0	4.0	4.0	4.0	4.8	8.5	9.6	8.6
Ghana	N.A.	0.0	0.0	0.0	0.0	0.0	N.A.	N.A.	N.A.	N.A.
India	39.1	2.6	2.4	3.2	3.4	3.4	4.4	4.9	7.8	N.A.
Jordan	0.0	N.A.	Negl.	Negl.	N.A.	N.A.	N.A.	N.A.	N.A.	N.A.
Korea	0.0	0.1	Negl.	Negl.	0.0	0.2	0.2	0.3	N.A.	N.A.
Madagascar	N.A.	0.0	0.0	0.0	Negl.	Negl.	0.3	0.8	N.A.	N.A.
Mexico	N.A.	0.5	1.6	1.7	2.2	2.8	1.4	1.2	1.7	5.0
Namibia	N.A.	N.A.	N.A.	N.A.	1.9	3.9	3.8	4.8	2.1	3.7
Niger[f]	N.A.	N.A.	N.A.	N.A.	N.A.	N.A.	N.A.	N.A.	N.A.	N.A.
Nigeria[g]	0.8	0.5	0.5	0.5	0.9	0.6	1.0	1.2	1.0	N.A.
Peru	Negl.	Negl.	Negl.	Negl.	Negl.	0.1	0.2	0.1	0.2	N.A.
Philippines	0.3	Negl.	Negl.	Negl.	Negl.	Negl.	N.A.	N.A.	N.A.	N.A.
Somalia[h]	2.2	1.5	2.0	2.0	2.3	0.0	0.0	0.0	N.A.	N.A.
Sudan	N.A.	N.A.	N.A.	0.0	0.0	0.2	N.A.	N.A.	N.A.	N.A.
Thailand	N.A.	N.A.	N.A.	N.A.	N.A.	N.A.	0.1	0.3	0.3	N.A.
Total, less developed countries[i]	81.3	10.8	19.1	23.7	31.1	46.6	52.4	63.5	47.0	(37.1)

(Continued)

Part 2-A-30 (continued)

Countries	Pre-1972	1972	1973	1974	1975	1976	1977	1978	1979	1980	1981	1982	1983
OVERALL TOTAL[i]	480.0	72.8	104.2	145.7	228.9	346.0	485.0	618.8	703.5	681.2	481.2	286.6	189.5
OVERALL TOTAL in millions of constant 1982 U.S.$		156.7	210.4	269.9	386.0	548.6	721.0	856.6	895.3	795.0	512.4	286.6	182.6

Notes: N.A. = not available; Negl. (i.e., "negligible") = less than $0.1 million; . . = available data not consistent with data for earlier years.

Sources: Organisation for Economic Cooperation and Development, Nuclear Energy Agency (NEA), and International Atomic Energy Agency (IAEA), *Uranium: Resources, Production and Demand* (December 1983 and earlier issues).

[a] Data for 1980 interpolated from Australian statistics in this appendix.
[b] Total for 1971–74 allocated across years to 1974. Pre-1972 total therefore equals 1971.
[c] Pre-1978 total allocated across years 1974–78.
[d] Pre-1974 figures cover government expenditure only.
[e] Pre-1977 total of $33.7 million allocated over entire period to 1976.
[f] Pre-1975 total was $34 million.
[g] Pre-1978 total of $2.3 million allocated.
[h] Pre-1973 total of $3.7 million allocated and 1973–75 total of $6.3 million allocated.
[i] Pre-1973 total of $3.7 million allocated and 1973–75 total of $3.7 million allocated, and accordingly the overall totals for those years, are based on incomplete data.
The totals for "Less developed countries," especially for pre-1972 and 1980, are based on incomplete data.

Part 2-A-31

Uranium Exploration: Overseas Expenditures by Selected Countries, 1972–83
(millions of current U.S. dollars)

Country	Pre-1972	1972	1973	1974	1975	1976	1977	1978	1979	1980	1981	1982	1983
Australia	0.2	0.0	0.0	0.0	0.0	0.0	0.0	0.0	0.0	0.0	0.0	0.0	0.0
Belgium	0.0	0.0	0.0	0.0	0.0	0.0	0.0	1.1	1.2	0.9	0.5	0.5	0.3
France	58.2	8.6	14.9	19.0	22.7	31.9	32.0	36.9	52.3	68.2	66.2	40.4	26.5
Germany	——	36.0[a]	6.0	25.0	14.0	21.0	21.0	28.0	30.0	30.0	26.0	26.0	26.0
Japan	8.3	1.2	1.2	1.4	2.9	8.4	24.0	24.4	24.5	29.3	30.2	24.2	24.2
Korea	N.A.	N.A.	N.A.	N.A.	N.A.	N.A.	N.A.	1.0	2.0	4.2	4.0	3.7	3.3
Spain	0.0	0.0	0.0	0.0	4.1	1.3	1.4	2.5	4.0	2.0	2.0	3.0	0.2
Switzerland	0.0	0.0	0.0	1.5	1.5	0.8	2.6	2.2	2.9	3.2	2.8	1.7	N.A.
United States	15.0	2.0	3.0	5.0	5.0	19.0	31.2	35.9	43.2	39.0	35.3	14.9	6.1

Notes: The expenditures shown in this table are included in the host countries' total expenditures in part 2-A-30 of this appendix.
N.A. = not available.

Sources: Organisation for Economic Cooperation and Development, Nuclear Energy Agency (NEA), and International Atomic Energy Agency (IAEA), *Uranium: Resources, Production and Demand* (December 1983 and earlier issues).
[a]Pre-1973.

Part 2-A-32

Estimated Uranium Resources for Selected Countries, 1970 and 1981
(thousand metric tons of uranium)

Countries	Reasonably assured resources						Estimated additional resources						Overall totals	
	April 1970			January 1981			April 1970			January 1981			April 1970	January 1981
	<$26 per kg U^a	$26–$39 per kg U^a	Total at <$39 per kg U^a	<$80 per kg U^a	$80–$130 per kg U^a	Total at <$130 per kg U^a	<$26 per kg U^a	$26–$39 per kg U^a	Total at <$39 per kg U^a	<$80 per kg U^a	$80–$130 per kg U^a	Total at <$130 per kg U^a	(at $39 per kg U)^a	(at $130 per kg U)^a
Developed														
Australia	16.7	7.1	23.8	294	23	317	5.1	5.1	10.2	264	21	285	34	602
Austria	N.A.	N.A.	N.A.	0.0	0.3	0.3	N.A.	N.A.	N.A.	0.7	1.0	1.7	0.0	2
Canada	178.0	100.0	278.0	230.0	28.0	258.0	177.0	130.0	307.0	358.0	402.0	760.0	585.0	1,018
Denmark	N.A.	3.8	3.8	0.0	27.0	27.0	N.A.	N.A.	N.A.	0.0	16	16	3.8	43
Finland	N.A.	N.A.	N.A.	0.0	3.4	3.4	N.A.	N.A.	N.A.	0.0	0.0	0.0	0.0	3.4
France	35.0	6.9	41.9	59.3	15.6	74.9	19	11.9	30.9	28.4	18.1	46.5	72.8	121.4
Greece	N.A.	N.A.	N.A.	1.4	4.0	5.4	N.A.	N.A.	N.A.	2.0	5.3	7.3	N.A.	12.7
Italy	1.2	N.A.	1.2	0.0	2.4	2.4	N.A.	N.A.	N.A.	0.0	2.0	2.0	1.2	4.4
Japan	2.1	3.5	5.6	7.7	0.0	7.7	N.A.	N.A.	N.A.	0.0	0.0	0.0	5.6	7.7
Portugal	7.4	N.A.	7.4	6.7	1.5	8.2	6.0	11.5	17.5	2.5	0.0	2.5	24.9	10.7
South Africa	154.0	50.0	204	247	109	356	11.5	26.9	38.4	84.0	91.0	175.0	242.4	531.0
Spain	8.5	7.7	16.2	12.5	3.9	16.4	N.A.	N.A.	N.A.	8.5	0.0	8.5	16.2	24.9
Sweden	N.A.	269.0	269.0	0.0	38.0	38.0	N.A.	38.0	38.0	0.0	44.0	44.0	307.0	82.0
Turkey	N.A.	N.A.	N.A.	2.5	2.1	4.6	N.A.	N.A.	N.A.	0.0	0.0	0.0	0.0	4.6
United Kingdom	N.A.	N.A.	N.A.	0.0	0.0	0.0	N.A.	N.A.	N.A.	0.0	7.4	7.4	0.0	7.4
United States	192.0	108.0	300.0	362.0	243.0	605.0	390.0	230.0	620.0	681.0	416.0	1,097.0	920.0	1,702.0
West Germany	N.A.	N.A.	N.A.	1.0	4.0	5.0	N.A.	N.A.	N.A.	1.5	7.0	8.5	0.0	13.5
Others	2.8	1.2	4.0	0.0	0.0	0.0	8.5	0.0	8.5	0.0	0.0	0.0	12.5	0.0
Rounded total, developed countries	600.0	553.0	1,153.0	1,224.0	505.0	1,729.0	617.0	453.0	1,070.0	1,430.0	1,030.0	2,460.0	2,223.0	4,189.0

Less developed

Algeria	N.A.	N.A.	N.A.	26.0	0.0	26.0	N.A.	0.0	N.A.	0.0	0.0	0.0	0.0	26.0
Angola	0.0	0.0	0.0	0.0	0.0	0.0	0.0	11.5	11.5	0.0	0.0	0.0	11.5	0.0
Argentina	7.7	8.5	16.2	25.0	5.3	30.3	17.0	25.0	42.0	3.8	9.6	13.4	58.2	43.7
Brazil	0.8	N.A.	0.8	119.1	0.0	119.1	0.8	N.A.	0.8	81.2	0.0	81.2	1.6	200.3
Central African Republic	8.0	N.A.	8.0	18.0	0.0	18.0	8.0	N.A.	8.0	0.0	0.0	0.0	16.0	18.0
Chile	N.A.	N.A.	N.A.	0.0	Negl.	Negl.	N.A.	N.A.	N.A.	0.0	6.7	6.7	N.A.	6.7
Egypt	N.A.	N.A.	N.A.	0.0	0.0	0.0	5.0	5.0	N.A.	0.0	5.0	5.0	N.A.	5.0
Gabon	10.4	N.A.	10.4	19.4	2.2	21.6	N.A.	0.8	10.0	0.0	9.9	9.9	20.4	31.5
India	N.A.	2.3	2.3	32.0	0.0	32.0	N.A.	N.A.	0.8	0.9	24.2	25.1	3.1	57.1
Korea	N.A.	N.A.	N.A.	Negl.	11.0	11.0	N.A.	N.A.	N.A.	N.A.	N.A.	N.A.	N.A.	11.0
Mexico	N.A.	0.9	1.9	2.9	0.0	2.9	N.A.	N.A.	N.A.	3.5	2.6	6.1	1.9	9.0
Namibia	N.A.	N.A.	N.A.	119.0	16.0	135.0	N.A.	N.A.	N.A.	30.0	23.0	53.0	N.A.	188.0
Niger	20.0	10.0	30.0	160.0	0.0	160.0	29.0	10.0	39.0	53.0	53.0	53.0	69.0	213.0
Somalia	N.A.	N.A.	N.A.	0.0	6.6	6.6	N.A.	N.A.	N.A.	0.0	3.4	3.4	N.A.	10.0
Zaire	N.A.	N.A.	N.A.	1.8	0.0	1.8	N.A.	N.A.	N.A.	1.7	0.0	1.7	N.A.	3.5
Rounded total, less developed countries	48.0	22.0	70.0	623.0	41.0	564.0	60.0	52.0	112.0	175.0	85.0	260.0	182.0	824.0
ROUNDED TOTAL (BOTH)	648.0	575.0	1,223.0	1,747.0	546.0	2,293.0	677.0	505.0	1,182.0	1,605.0	1,115.0	2,720.0	2,405.0	5,013.0

Notes: See section, "General Notes to Table," p. 98 in this appendix. For detailed notes see original sources.

N.A. = not available; Negl. = negligible.

Sources: Organisation for Economic Cooperation and Development, Nuclear Energy Agency (NEA), and International Atomic Energy Agency (IAEA), *Uranium: Resources, Production and Demand* (February 1982 and earlier issues).

[a] Current dollars.

(Part 2-A-32 continues on next page.)

Part 2-A-32 (continued)

General Notes to Table

The data on uranium resources of individual countries presented in the table were collated by an Organisation for Economic Cooperation and Development (OECD) working party from national submissions. The table shows two major categories of resources, reflecting different levels of confidence in the quantities reported:

• *Reasonably assured resources* refers to uranium that occurs in known mineral deposits of such size, grade, and configuration that it could be recovered within the given production-cost ranges with present technology. Estimates of tonnage and grade are based on specific sample data.

• *Estimated additional resources* refers to additional uranium that is expected to occur, mostly on the basis of direct geologic evidence, in extensions of existing deposits, little-explored deposits, and undiscovered deposits along well-defined geologic trends. Estimates of tonnage and grade may be inferred from the best-known parts of particular or similar deposits. These resources have a potential for later conversion to the more certain category with further exploration. Some countries do not include undiscovered deposits in this category.

Most countries record the recoverable uranium rather than the in situ totals.

The two years 1970 and 1981 given for each major category are subdivided into two categories based on the estimated costs of exploitation. When allowance is made for inflation, the two subdivisions for each year given in the table are broadly comparable. Although there are some differences from country to country in the bases for computing costs, most bases are similar, and in general take in all forward costs of production including a rate of return on invested capital; past exploration costs are not included. The cost categories do not necessarily reflect the prices at which uranium will be available to the user.

The lower-cost subdivision of reasonably assured resources (e.g., less than $80 per kilogram in 1981) is considered as reserves.

In conclusion, the "Overall totals" (far right-hand columns) shown for each year—particularly in their inclusion of the less certain category of estimated additional resources—are merely illustrative. They are shown solely to give some indication of the broad success of exploration over the decade.

Part 2-A-33

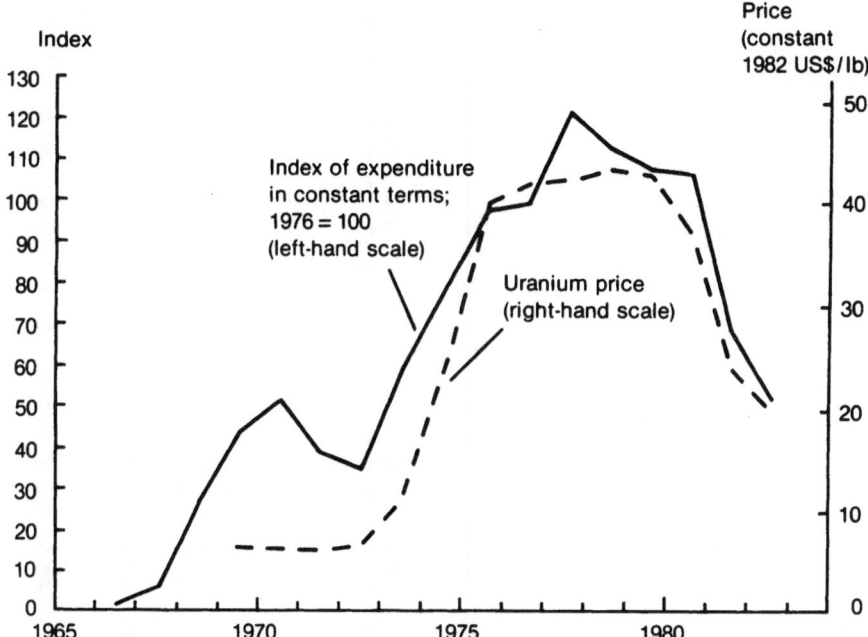

Index of uranium exploration spending worldwide by European Community companies (1976 = 100), and uranium prices, 1966–82. *Sources:* Exploration: European companies' unpublished responses to a questionnaire from the Comité de Liaison des Industries de Metaux Non Ferreux, Brussels, Belgium; uranium prices: *Nuexco Monthly Report on Uranium* (Atlanta).

Part 2-A-34

Uranium Exploration in the United States, 1966–82

	Total, 1966 to 1970	1971	1972	1973	1974	1975	1976	1977	1978	1979	1980	1981	1982
Exploration expenditures (millions of current U.S.$)													
Land acquisition	53.0	9.8	4.7	7.7	12.6	16.7	13.9	28.2	30.7	44.5	35.1	11.4	11.3
Surface drilling													
Exploration[a]	72.6	17.0	15.4	19.5	35.0	51.9	70.7	99.4	113.3	119.6	94.8	56.4	20.9
Development[b]	14.2	3.9	2.7	5.8	9.8	21.9	38.3	55.6	56.4	43.4	30.9	11.5	6.9
Other[c]	57.8	10.4	9.6	16.5	21.7	31.5	47.8	74.8	113.9	108.4	106.2	65.5	34.5
TOTAL in millions of current U.S.$	197.6	41.1	32.4	49.5	79.1	122.0	170.7	258.1	314.3	315.9	267.0	144.8	73.6
TOTAL in millions of constant 1982 U.S.$		93.3	69.8	99.9	146.5	205.7	270.7	383.7	435.1	402.0	311.6	154.2	73.6

Surface drilling (million feet)

Exploration	54.3	11.80	11.95	11.76	14.72	15.69	20.36	27.96	28.95	28.07	19.60	10.87	4.23
Development	16.3	3.08	3.08	5.25	6.84	9.73	14.44	17.62	19.15	13.01	8.59	3.35	1.13

Surface drilling costs ($/foot)

Exploration (current U.S.$)	1.34	1.44	1.29	1.66	2.37	3.31	3.47	3.56	3.91	4.26	4.84	5.19	4.96
Exploration (constant 1982 U.S.$)		3.24	2.78	3.35	4.39	5.58	5.50	5.29	5.41	5.42	5.65	5.53	4.96
Development (current U.S.$)	0.87	1.28	0.88	1.10	1.43	2.25	2.65	3.16	2.95	3.34	3.60	3.43	6.13
Development (constant 1982 U.S.$)		2.88	1.89	2.22	2.65	3.79	4.20	4.70	4.08	4.25	4.20	3.65	6.13

Note: Different estimates of U.S. expenditures on uranium exploration are given in parts 2-A-14 and 2-A-30 of this appendix. The section on Notes to and Discussion of Table in part 2-A-14 of this appendix highlights the differences in 1977.

Sources: U.S. Department of Energy, *Statistical Data of the Uranium Industry,* GJO-100(83) and earlier issues from an annual survey.

[a]Includes drilling done in search of new ore deposits or extensions to known deposits, and drilling at the location of a discovery up to the time that the company decides that sufficient ore reserves are present to justify commercial exploration.

[b]Includes all drilling of a uranium deposit to determine more precisely size, grade, and configuration after the determination is made that the deposit can be commercially exploited.

[c]Includes all other costs (exclusive of land acquisition and drilling programs) directly associated with a company's uranium exploration effort. Included are items such as geologic and geophysical investigations and costs incurred by field personnel in the course of exploration effort, expenditures for geologic research, and overhead and administrative charges directly associated with supervising and supporting field and exploration activities. The overhead and administrative charges reported do not include internal corporate charges, such as directors' salaries that are not directly associated with a company's exploration effort.

Part 2-A-35

Copper: Operating Mines, Development Projects, and Exploration Ventures in the Western World, 1980
(reserves in million metric tons of contained copper)

Country	Operating mines		Development projects		Exploration ventures	
	Number	Reserves	Number	Reserves	Number	Reserves
Africa						
Algeria	1	0.01	1	0.02	1	0.05
Angola	—	—	—	—	1	0.12
Botswana	1	0.47	—	—	—	—
Burundi	—	—	—	—	1	0.20
Congo PRO	1	0.01	—	—	—	—
Ethiopia	—	—	—	—	1	0.06
Morocco	4	0.24	1	0.05	1	0.03
Mozambique	1	—	—	—	—	—
Namibia	5	0.49	—	—	5	1.40
South Africa	9	3.79	—	—	2	0.66
Sudan	—	—	—	—	1	0.38
Upper Volta	—	—	—	—	1	0.32
Zaire	8	16.05	—	—	1	2.91
Zambia	10	26.18	—	—	3	1.22
Zimbabwe	9	0.30	—	—	1	0.19
Asia						
Burma	—	—	1	0.73	—	—
India	4	1.78	1	0.79	11	0.53
Indonesia	1	1.15	—	—	1	1.32
Iran	1	0.06	—	—	—	—
Japan	16	1.31	—	—	—	—
Jordan	—	—	—	—	1	0.82
Malaysia	1	0.37	—	—	—	—
Oman	—	—	1	0.24	—	—
Pakistan	—	—	—	—	3	1.36
Philippines	17	10.85	4	1.97	12	2.98
Saudi Arabia	—	—	—	—	2	0.48
Thailand	—	—	—	—	1	0.80
Australia						
Australia	9	5.01	2	0.07	24	7.43
Fiji	—	—	—	—	1	2.21
Papua New Guinea	1	2.80	1	3.29	4	5.61

(Continued)

Part 2-A-35 (continued)

Country	Operating mines Number	Operating mines Reserves	Development projects Number	Development projects Reserves	Exploration ventures Number	Exploration ventures Reserves
Europe						
Finland	8	0.60	—	—	—	—
France	1	—	—	—	3	0.08
Germany	1	0.01	—	—	—	—
Greece	—	—	—	—	1	0.09
Ireland	2	0.02	—	—	2	0.21
Italy	1	0.01	1	0.02	—	—
Norway	8	0.93	—	—	—	—
Portugal	2	1.06	1	0.70	1	0.41
Spain	4	1.73	1	0.25	1	0.10
Sweden	7	1.72	1	0.28	—	—
Turkey	5	0.93	1	0.81	3	0.94
United Kingdom	1	0.00	—	—	1	0.11
Yugoslavia	3	3.14	1	2.80	1	0.46
Latin America						
Argentina	1	0.00	—	—	4	10.63
Bolivia	1	0.02	—	—	—	—
Brazil	—	—	1	1.08	4	9.57
Chile	13	83.07	2	0.17	15	40.40
Colombia	—	—	1	0.57	2	6.84
Dominican Republic	—	—	—	—	1	0.03
Ecuador	1	0.00	—	—	1	0.50
Guatemala	1	0.03	—	—	1	0.01
Mexico	15	10.84	2	0.01	9	7.50
Panama	—	—	—	—	2	11.83
Peru	18	6.47	3	7.62	12	19.16
Venezuela	—	—	—	—	2	0.07
North America						
Canada	63	14.38	12	4.30	64	18.10
United States	56	43.81	3	0.85	36	31.94
TOTAL	311	239.64	46	26.62	245	190.06

Notes: The table excludes mines that were closed down temporarily in 1980. In some cases the reserves of a number of smaller mines have been grouped together. Dash = not applicable.

This table is based on data in annual reports and press announcements; because it shows the reserves reported by companies, there is probably little consistency among estimates—what one company treats as a reserve might be regarded as an uneconomic resource elsewhere. Indeed, many of the deposits included under the heading "Exploration ventures" are presently uneconomic. The table is merely intended to show the large number of copper deposits that have been discovered and await development. Their combined reserves are similar to those of operating mines. Further exploration may be needed to delineate the ore bodies and to take many projects to the feasibility stage, but in a number of cases full feasibility studies already exist.

Source: Unpublished estimates provided by Economics Department, Rio Tinto-Zinc Corporation PLC, London.

3

Base and Precious Metals Exploration by Major Corporations

RODERICK G. EGGERT

Exploration is the search for new mineral wealth. It acts in concert with technologic change and other factors to augment mineral supply. Geologic mapping, geochemical sampling, geophysical surveys, diamond drilling, and other exploration activities provide information about unknown resources. This information helps identify mineral deposits, thereby generating new mineral reserves. The new reserves provide ore for smelting and refining operations, which in turn provide unfinished metal to fabricators, who fashion products used ultimately by metal consumers. Exploration influences the availability and prices of minerals, the geographic location of mining and processing, and the international flow of minerals among producing and consuming countries.

This chapter analyzes base and precious metals exploration by one of the principal participants in the world of exploration—major corporations. The objective is to identify and assess the factors responsible for changes in the level and distribution of corporate exploration over the last fifteen to twenty years. The first section of the chapter discusses exploration as an economic activity, providing a conceptual framework for the analysis. The next part looks at changes in exploration expenditures over time for a number of North American and European companies. This is followed by an examination of the distribution of exploration funds among minerals, paying particular attention to porphyry copper, massive sulfide, molybdenum, and gold and silver deposits. The next section analyzes the distribution of funds among countries, and in so doing examines the validity of the widely held belief that recent changes in this distribution have been determined largely by government policies and political risks. The final section of the chapter reviews the findings and discusses their implications.

Exploration As an Economic Activity

Corporate exploration is an economic activity. As much as some geologists might like to conduct exploration purely for the advancement of knowledge or the sake of enjoyment, those who finance exploration view it as an investment. Companies will invest in this activity when expected net returns, discounted appropriately for time and risk, exceed returns from alternative investments. These include other ways of augmenting mineral supply, such as purchasing a known deposit or using new extraction and processing techniques to improve productivity at existing operations. One would expect, therefore, that variations in exploration activity reflect changes in the components of net returns—expected revenues, costs, and risks associated with exploration.

What factors determine these components of net returns? Exploration revenues are the net returns from the subsequent development and mining of the deposits discovered during exploration. These revenues obviously are influenced by expected metal prices, output levels, and future demand, as well as development and mining costs (which, in turn, are affected by a variety of considerations including mineral taxation and cost of capital). Exploration costs are influenced by similar considerations that affect exploration productivity and prices for exploration goods and services. These considerations include government policies that directly affect exploration, such as land regulations.

As for risk, three types are distinguished here—political, economic, and geologic. Political risk reflects the variability in returns to exploration due to government actions. Economic risk reflects the variability in returns due to changing commodity prices and costs of extraction and processing. Geologic risk reflects the variability in returns due to the physical nature of ore bodies, which consists of two parts: variability associated with the probability of any exploration target becoming an ore deposit, and variability in returns during production because of the physical nature of the ore.

Reality, however, rarely mirrors this neat scheme. Potential returns from exploration are often so difficult to estimate—because of the difficulty in anticipating the timing and risks involved in discovering a deposit and then bringing it into production—that changes in the level and distribution of exploration are in some instances determined more by habit and simple rules of thumb than by careful calculations of potential returns from exploration.

Levels of Exploration Expenditure

Figure 3-1 traces metal exploration expenditures over some fifteen years for six U.S. and four Canadian companies, and two groups of U.S. and European firms. The data, it should be noted, are not completely comparable across firms. The U.S. mining and the U.S. oil companies, which preferred to remain anonymous, provided actual expenditures by the corporate exploration group within each firm. The other figures are from published sources. Some represent exploration expenses claimed for tax purposes on income statements, whereas others represent the figures cited as exploration expenditures in annual reports. The aggregate numbers for U.S. and European companies were tabulated by other authors according to their own definitions of exploration.[1]

Despite these inconsistencies, care has been taken to ensure that the data are as comparable as possible over time, to permit analysis of changes in exploration levels.

Taken as a group, the curves display several common features. First, the general trend of real expenditures (1982 U.S. dollars) is upward. This is more pronounced for the group of U.S. and European firms and most of the individual U.S. companies than for the Canadian firms. Second, expenditures are cyclical. This is particularly true for the Canadian companies and for several U.S. firms, less so for European firms. Three recent boom and bust periods can be identified, although not all firms exhibit the pattern.

The first boom occurred in the early 1970s, most noticeably for the Canadian firms. Subsequent expenditures by the Canadians dropped sharply, while expenditures by the European companies leveled off after a period of steadily increasing expenditures. The second boom occurred in the middle 1970s—for three of four Canadian firms the peak year was 1976, whereas for five of the six U.S. firms and the European group the peak was in 1977. The years 1978 and, to a lesser extent, 1979 were trough years. The third boom occurred in 1980 and 1981, followed by dramatic cutbacks in 1982.

Despite these broad similarities, there are pronounced differences among firms. For example, expenditures by some firms change only gradually from year to year, whereas expenditures by others swing sharply over the short term. Expenditures by several firms follow a generally downward trend, even though the overall trend has been upward. The factors that account for these similarities and differences among firms include mineral price trends, exploration success, and corporate goals.

---

[1]Throughout this chapter, the U.S. GNP implicit price deflator was used to convert expenditures and prices into 1982 U.S. dollars, except where noted otherwise.

108

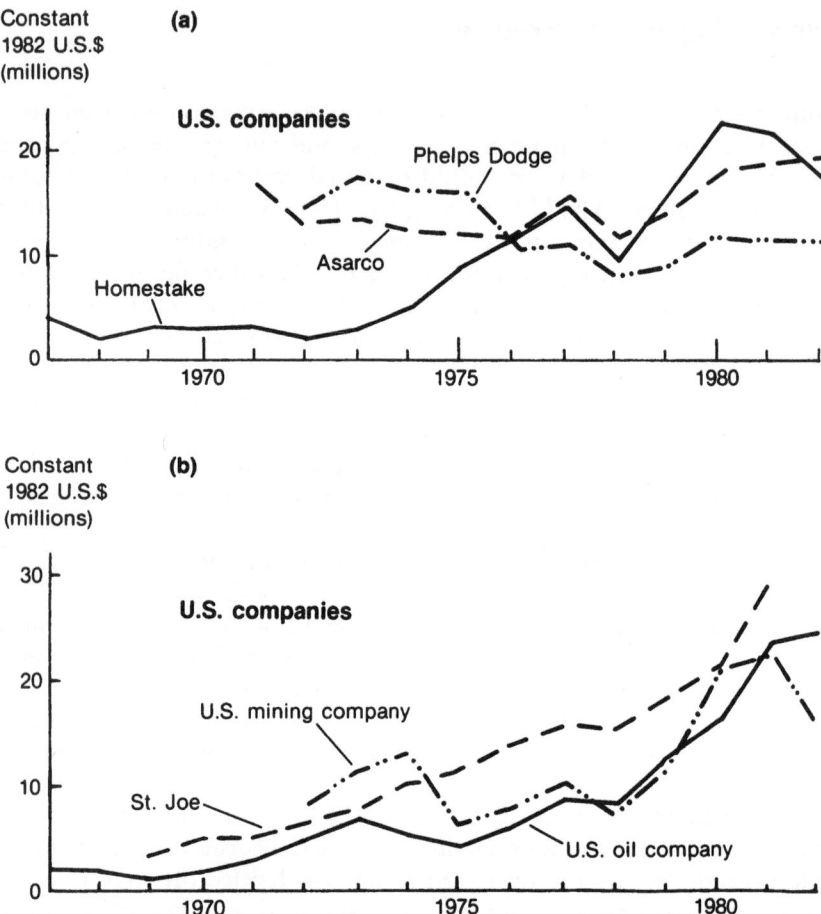

Figure 3-1 (a–d). **Mineral exploration expenditures for selected North American and European companies between 1965–71 and 1982.** Data for Phelps Dodge, the U.S. oil company, and the group of European companies do not include uranium expenditures. The other data undoubtedly include sizable uranium expenditures, but it is impossible to exclude these from the totals. Expenditures for oil and gas exploration are excluded in all cases, with the possible exception of Asarco, whose data may include a small oil and gas component after 1980. The data from published sources include any expenditures for iron ore and bauxite, but such expenditures appear to have been relatively small or nonexistent for most firms. *Sources:* Chapter 2 in this volume; G. A. Barber, "Foreign Exploration Pros and Cons," *Mining Congress Journal* vol. 67, no. 2 (1981) pp. 20–23; Phelps Dodge, *Annual Report* (New York, 1970–82); Asarco, *Annual Report* (New York, 1970–82); Homestake, *Annual Report* (San Francisco, 1970–81); Inco, *Annual Report* (Toronto, Ontario, 1974–82); Cominco, *Annual Report* (Vancouver, British Columbia, 1968–82); Noranda, *Annual Report* (Toronto, Ontario, 1967–82); Falconbridge, *Annual Report* (Toronto, Ontario, 1969–82); confidential written surveys for the U.S. mining company and U.S. oil company.

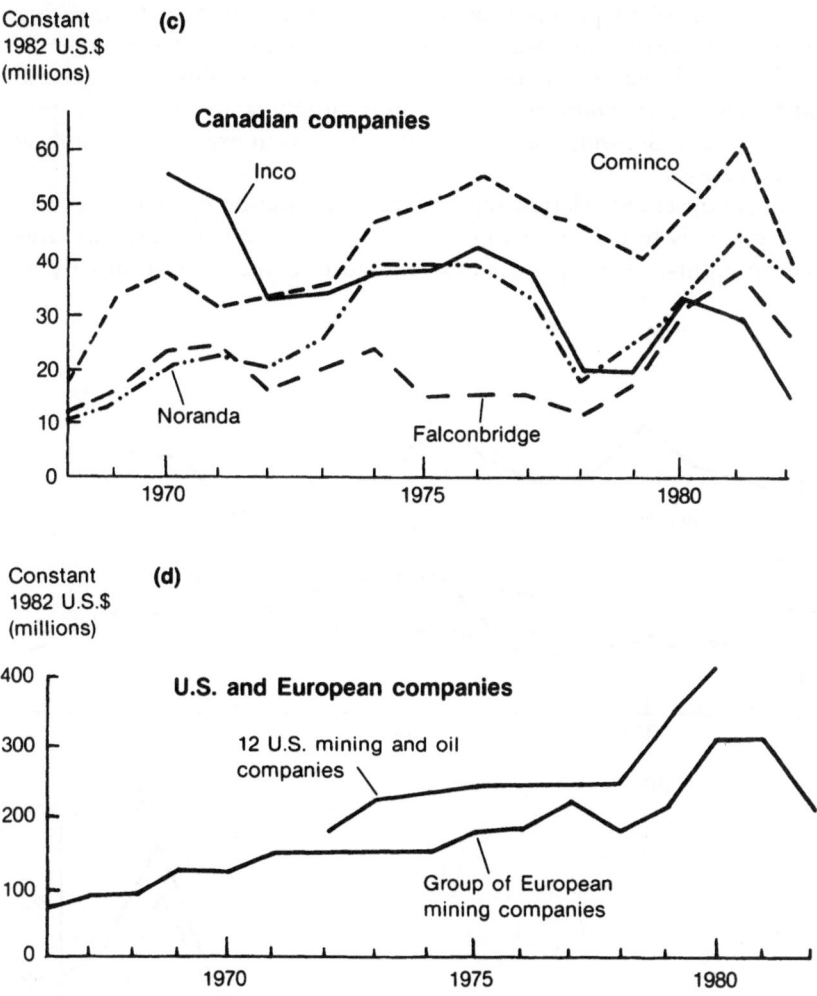

Figure 3-1 (continued)

Mineral Price Trends

Recent mineral prices can explain much of the similarity among company expenditure trends. These prices alter expected net returns from exploration in two very different ways. First, they influence expected future prices and thus revenues from mining and, in turn, exploration. Second, they affect the availability and cost of the internal funds used for financing exploration. In the first case, prices influence the demand for exploration funds; in the second, the supply of such funds.

Figure 3-2 displays price trends for several major metals. Comparing real mineral prices (figure 3-2) with real corporate exploration expenditures (figure 3-1) lends support to two observations. First, the general upward trend in expenditures corresponds closely with the rise in gold and silver prices. Second, many of the fluctuations in expenditures follow price fluctuations.

When the observed relationship between exploration expenditures and mineral prices is tested more rigorously by regressing real expenditures against an eight-mineral price index, lagged one year, the results (table

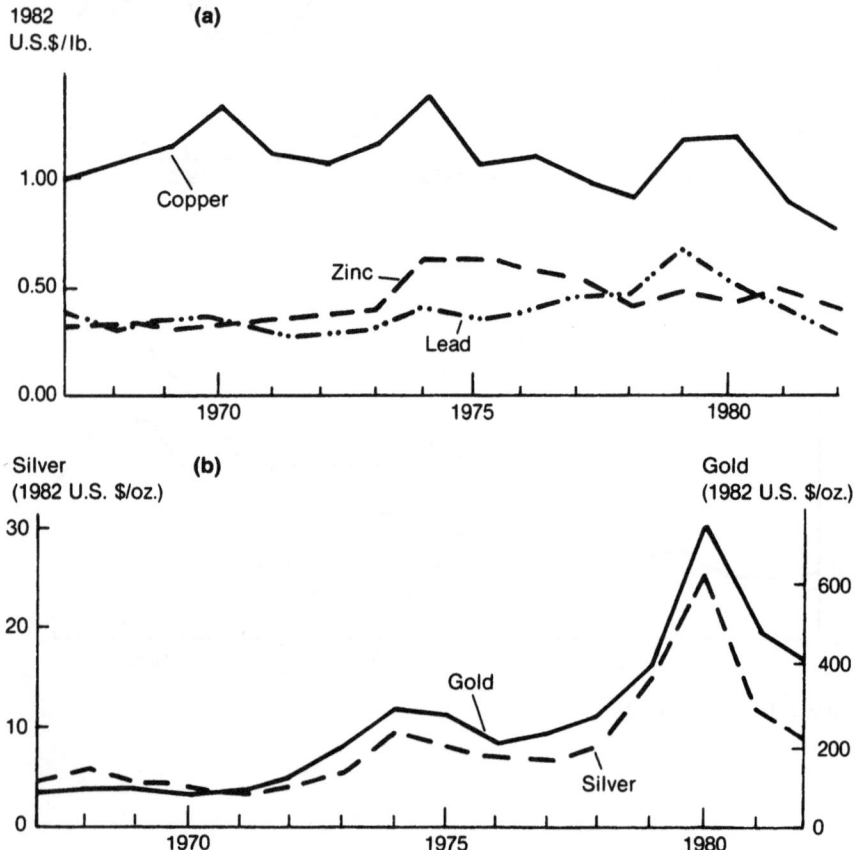

Figure 3-2 (a, b).   Annual average prices for selected mineral commodities, 1967–82. Prices quoted are as follows: copper—U.S. domestic refinery price; lead—U.S. primary delivered price; zinc—U.S. high-grade delivered price; gold and silver—annual average selling price. *Sources:* "Average Annual Metal Prices, 1920–1982," *Engineering and Mining Journal* vol. 184, no. 3 (1983) p. 51; *Metal Bulletin Handbook*, vol. 1 (Surrey, England, Metal Bulletin Books, 1983); *Metal Statistics, 1971–1981*, 69th ed. (Frankfurt am Main, Metallgesellschaft, 1982); *Roskill's Metals Databook*, 3d ed. (London, Roskill Information Services, 1982).

3-1) provide statistical support for the theory that exploration expenditures are influenced by mineral prices. The coefficient for lagged mineral prices is positive in all cases and significant at the 95 percent confidence level (using a one-tail t-statistic to test for significance) in nine out of twelve instances. The mineral price index, described further in appendix 3-A, reflects percentage changes in real prices for the five minerals displayed in figure 3-2, as well as molybdenum, nickel, and tin. These eight metals were selected because they represent the most important base and precious metals in terms of mine-production value.

As suggested earlier, recent mineral prices may be an important determinant of exploration expenditures for two very different reasons. First,

Table 3-1. Regression Results: Exploration Expenditures Related to a Mineral Price Index Lagged One Year

Firm(s) and years of expenditures	Constant	Coefficient for lagged price index	n	R^2
Asarco	5.95*	.035**	11	.57
(1971–82)	(2.18)	(.012)		
Homestake	−7.14*	.099**	14	.85
(1968–82)	(2.69)	(.016)		
Phelps Dodge	3.46	.025	10	.59
(1972–82)	(5.79)	(.021)		
St. Joe	14.13	.051*	12	.90
(1969–81)	(8.83)	(.024)		
U.S. mining company	−1.21	.067*	10	.49
(1972–82)	(5.70)	(.031)		
U.S. oil company	27.77	.021	14	.88
(1968–82)	(15.25)	(.023)		
Cominco	17.43**	.098**	14	.59
(1968–82)	(5.24)	(.031)		
Falconbridge	4.56	.073*	14	.47
(1968–82)	(6.14)	(.035)		
Inco	−18.89	.127*	12	.49
(1970–82)	(17.78)	(.061)		
Noranda	7.67	.096**	14	.68
(1968–82)	(6.11)	(.035)		
12 U.S. firms	424.56	.527	8	.75
(1972–80)	(230.02)	(.521)		
European firms	43.32	.954**	14	.79
(1968–82)	(30.80)	(.183)		

Notes: n = number of observations. R^2 = coefficient of determination = the percent of the variation in exploration expenditures that can be explained by variations in the price index (lagged one year). Standard errors are displayed in parentheses. The method of Cochrane-Orcutt was used to correct for first-order autocorrelation. Checks were done with the Hildreth-Lu method to ensure that local solutions for rho had not been found.

*Significant at the 95 percent confidence level.

**Significant at the 99 percent confidence level (a one-tail t-statistic was used to test for significance because the coefficient is expected to be positive).

they influence expected revenues from exploration by shaping expectations of future prices and revenues from mining, which in turn strongly affect expected revenues and net returns from exploration. Although a number of factors influence estimates of potential returns from future exploration and mining, including expected costs, demand, and other factors already discussed, it is only argued here that recent prices may be a very important determinant of such estimates. Many of the available price forecasting techniques depend on recent price trends; other techniques may assume a correlation or other relationship between recent and future price trends, and they generally assume that the underlying behavioral relationships governing price determination are maintained over time. Changes in exploration expenditures are likely to follow price changes with a short lag, because expenditures in any particular year are the result of a budgeting process that begins at least six months prior to the beginning of that year. Changes in budget allocations and expenditures for exploration in any year, therefore, are likely to reflect changing price expectations and to a lesser extent prices from the previous year.

The second reason for presuming exploration expenditures are related to recent mineral prices is that mineral prices strongly influence exploration costs—not costs in the usual sense of salaries or diamond drilling, but rather the cost of exploration funds. Mineral prices are one important determinant of mining revenue and the level of internal funds available for financing exploration.[2] The argument assumes that the cost of external funds is greater than the cost of internal funds, and figure 3-3 shows how this might work.

SS_1 is a supply curve for investment funds, and the cost of using the limited amount of internal funds is approximately constant. When a firm calls on external sources of debt and equity for additional financing, however, the cost of capital increases as the firm's debt-to-equity ratio increases or as additional stocks dilute the management's control over the firm (Branson, 1979, pp. 219–220). DD is a demand curve for exploration funds. During periods of low prices and correspondingly low levels of internal funds, the equilibrium exploration expenditure corresponds to q_1. During periods of higher prices and correspondingly higher levels of internal funds, the supply schedule shifts to the right, as depicted by SS_2; in this case, expenditures increase to q_2, even though the demand curve for exploration funds has remained stationary. (Note that in the first argument concerning the relationship between mineral prices and exploration expenditures, mineral price changes cause a shift in the demand curve for exploration funds by altering expected exploration revenues.) Furthermore, rivalry among competing exploration groups may make them

---

[2]The relationship between exploration expenditures and the availability of internal funds was suggested by Ventura (1982, pp. 12–13), based on graphical similarities between exploration expenditures and corporate cash flow, lagged one year.

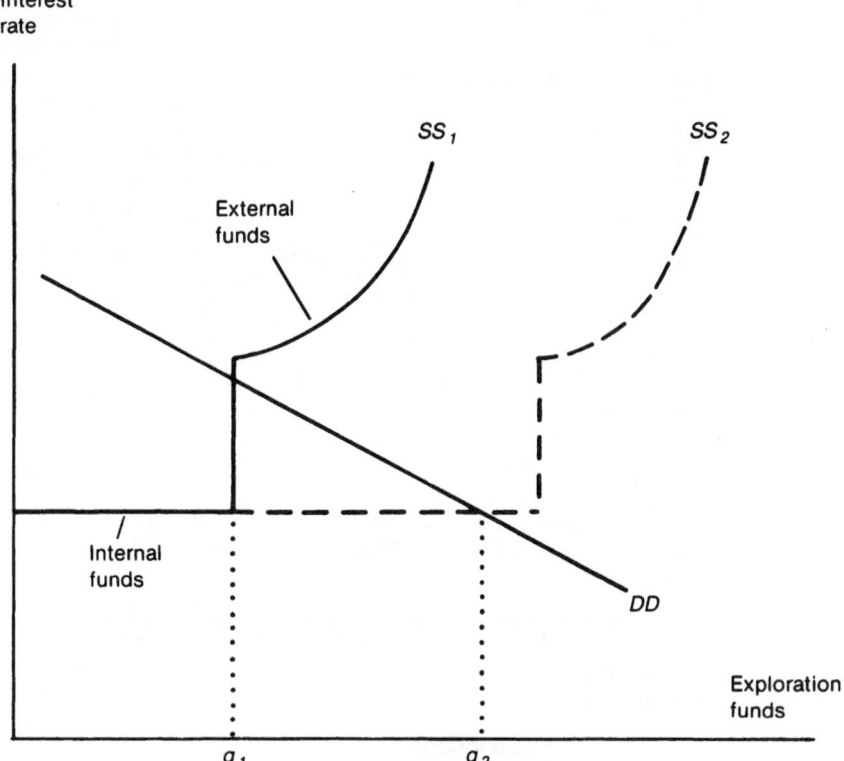

Figure 3-3.   Possible supply and demand schedules for exploration funds.

reluctant to call on external sources of finance, if it means divulging information about the exploration program that they would prefer to keep confidential.

For five of seven companies tested in this study, exploration expenditures correspond closely with net income, which is used as a measure of the availability of internal funds. The graphs in figure 3-4 imply a lagged relationship between expenditures and net income (expenditures lagging income), and this qualitative inference is reinforced by the regression results (displayed in the figure) in the form of positive and significant coefficients for the independent variable, net income, lagged one year.

The preceding theoretical argument and empirical results support the conventional wisdom that large North American firms use internal funds for exploration. According to Paul Bailly, former president of Occidental Minerals (personal communication), large U.S. companies do not use *any* external funds for exploration; only some small companies finance exploration with money raised in the stock markets.

Therefore, mineral prices influence exploration expenditures in two very different ways—by influencing potential revenues from mining ventures and

114

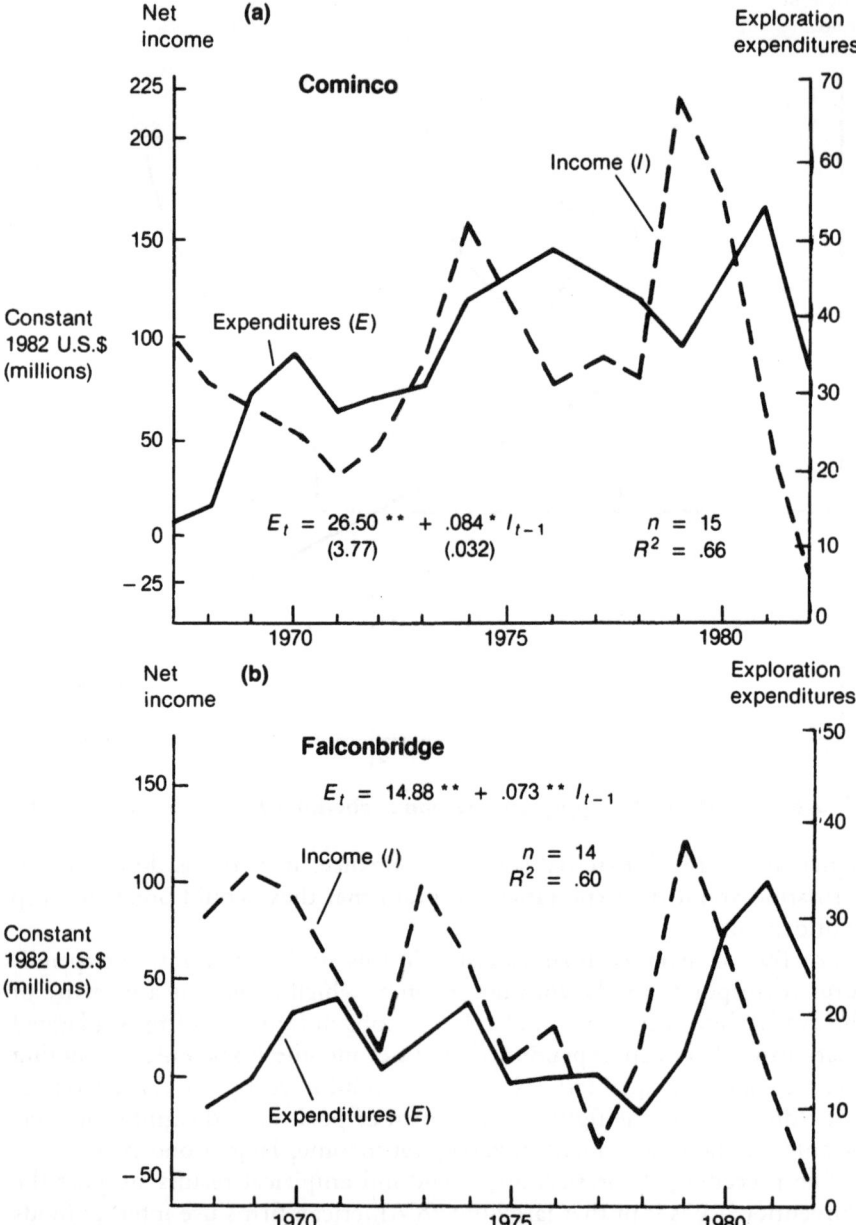

Figure 3-4 (a–e). Exploration expenditures and net income for five mining companies between 1967–71 and 1982. For a description of n, R^2, and other symbols, see notes to table 3-1 in this chapter. *Sources:* Cominco, *Annual Report* (Vancouver, British Columbia, 1968–82); Falconbridge, *Annual Report* (Toronto, Ontario, 1969–82); Inco, *Annual Report* (Toronto, Ontario, 1974–82); Noranda, *Annual Report* (Toronto, Ontario, 1967–82); Phelps Dodge, *Annual Report* (New York, 1970–82).

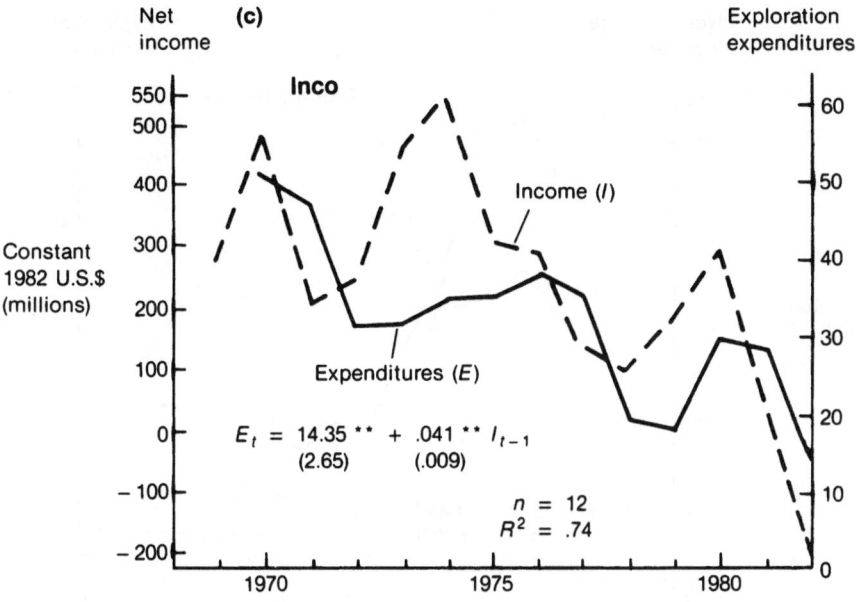

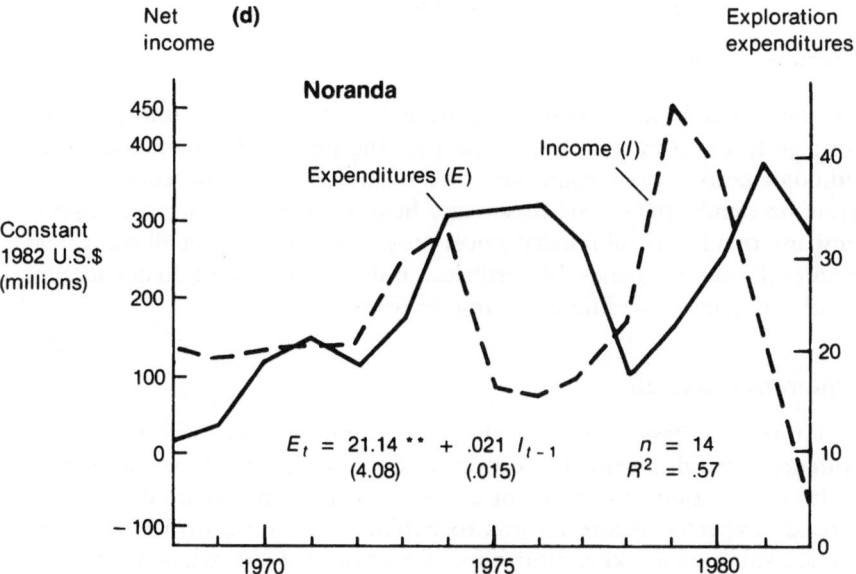

Figure 3-4 (continued)

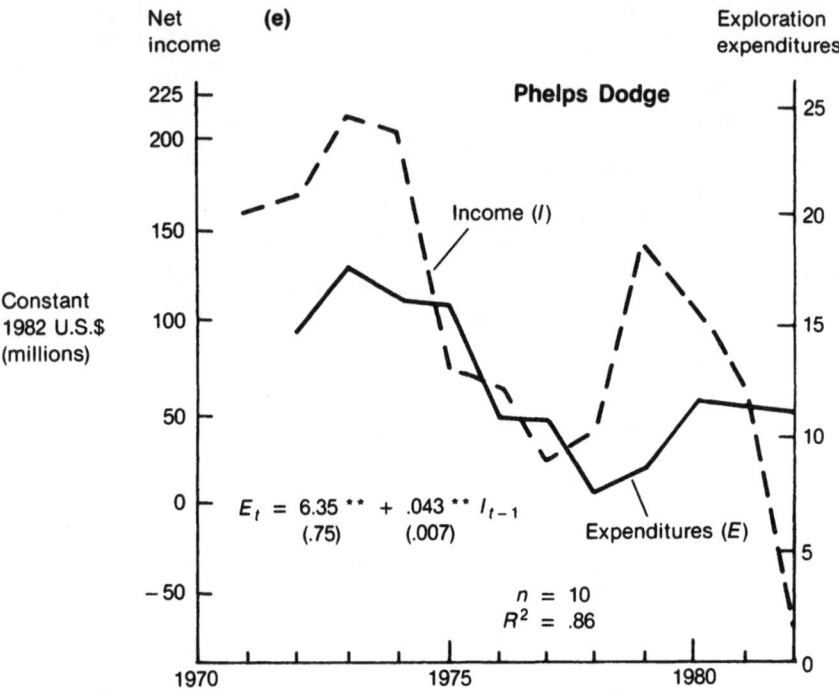

Figure 3-4 (continued)

thus the demand for exploration funds, and by altering costs (through their effect on levels of internal funds) and thus the supply of exploration funds. Although both reasons contribute to the similarities among company expenditure trends, the second reason also helps explain the differences among company trends: not all mineral prices move together and not all companies produce the same minerals. Nevertheless, mineral prices alone do not account for all similarities and differences among firms.

Exploration Success

Exploration success is perhaps the most important nonprice factor that influences overall levels of expenditure by a company. A period of successful exploration, when one or several mineral deposits are discovered, increases expected future returns to exploration, and a firm is likely to increase subsequent expenditures for two reasons. First, when a company discovers promising mineralization during reconnaissance or initial drilling, it has a strong incentive to undertake a detailed drilling and evaluation program. This detailed program is usually more expensive than reconnaissance exploration, and the result is an increase in total expenditures for the company. Second, after a company discovers or confirms promising mineralization, it may very well raise its estimates of potential returns from exploration in general and in response raise budget allocations and overall expenditures for exploration. Successful exploration enhances the bargaining position of the exploration manager when it comes to competing within the corporation for investment funds.

In both cases, a company lowers its estimates of geologic risk, which reflects the probability of any prospect becoming a mine. At the same time, a firm may also raise its estimates of potential revenues from exploration and mining. The result is an increase in potential returns from exploration, and the company increases subsequent expenditures. During 1981 and 1982, for example, Asarco spent approximately $7 million (in current terms) on underground exploration on a promising gold prospect discovered in 1980 near Timmins, Ontario, Canada; these expenditures account for much of the increase in Asarco expenditures during these years. At St. Joe, detailed evaluation at two discoveries, El Indio (Chile) and Pierrepont (New York), may be partly responsible for the increase in St. Joe expenditures from 1979 to 1981. Furthermore, a steady stream of St. Joe discoveries since the late 1960s, including the Woodlawn massive sulfide deposit in Australia, diamonds and tin in Brazil, gold in California, and zinc in Tennessee, may account for the steady rise in expenditures during the period.

Exploration failure, on the other hand, may cause a company to lower estimates of expected returns and thus expenditures for exploration. At Phelps Dodge, for example, no significant discoveries were listed in annual reports between 1973 and 1978, and during this period real exploration expenditures declined. In 1978 Phelps Dodge announced the discovery of porphyry molybdenum mineralization near Pine Grove, Utah, and expenditures increased during 1979 and 1980, coinciding with detailed evaluation of this target.

Corporate Goals

Corporate goals also influence overall levels of exploration expenditures. If, for example, a company decides to diversify and that the best way to reach that goal is through discovery of new mineral deposits, rather than through acquisition of known deposits, then it may reasonably decide to increase exploration expenditures. At St. Joe, for instance, annual report statements suggest that efforts to diversify production away from lead have been partly responsible for increased exploration expenditures during the 1970s. At the U.S. oil company, the general rise in expenditures from 1967 to the mid-1970s also reflects, at least in part, a corporate desire for diversification. This company is one of a number of U.S. oil companies that became active in nonfuel mineral exploration in the 1960s and 1970s. Realizing that diversification into minerals production through exploration would take a number of years, it was prepared to increase funding gradually for a few years, even if no immediate discoveries were made.

Exploration Targets

In conjunction with deciding how much to spend on exploration, a company must determine for which mineral it will explore. This section examines how funds have been allocated over time for porphyry copper, massive sulfide, porphyry molybdenum, and gold and silver deposits and identifies several factors that have been responsible for these changes. It then describes a typical cycle in the exploration for any particular deposit-type. The section closes with a discussion of diversification, a phenomenon that has characterized many exploration programs in recent years.

Data from two unnamed companies, which provided historical breakdowns of their exploration expenditures according to commodity or deposit-type, form the quantitative basis of this section. For gold and silver, data are available from five companies. The quantitative information is evaluated in light of additional evidence available from published sources about other companies. In this analysis, the role of changing relative prices is taken into account. The relative price for a mineral reflects how its price has changed in comparison with other major mineral prices. It is calculated by dividing the price of a mineral commodity by the mineral price index introduced above. (For further details, see appendix 3-A.)

Porphyry Copper Deposits

Porphyry copper deposits are estimated to contain 40 to 45 percent of world copper reserves (Gluschke, Shaw, and Varon, 1979, p. 45) and have been important targets for many exploration groups over the last twenty-five years. As shown in figure 3-5, the share of overall exploration expenditures spent on porphyry copper has declined markedly for two firms during the period from 1972 to 1982. Porphyry copper's share for the industry as a whole probably followed a similar declining trend, based on data found in the statistical appendix for chapter 2 (particularly figure 2-A-15) and on anecdotal remarks in trade journals and elsewhere (for example, Tremblay and Descarreaux, 1978, p. 94; and Hodge and Oldham, 1979, p. 97).

The decline in the relative price of copper has roughly paralleled porphyry copper's decline in share of expenditures. One might reasonably infer that as copper's relative price has fallen, estimates of future prices and potential revenues have declined. This, in turn, has made copper less attractive as an exploration target.

Nevertheless, porphyry copper's share of expenditures has not followed changes in copper's relative price exactly. In particular, the share increased from 1978 to 1981, while the relative price of copper fell. This probably is due in part to rising relative prices for three metals commonly found with copper in porphyry deposits—gold and silver, whose relative prices rose dramatically from 1977 to 1980, and molybdenum, whose relative

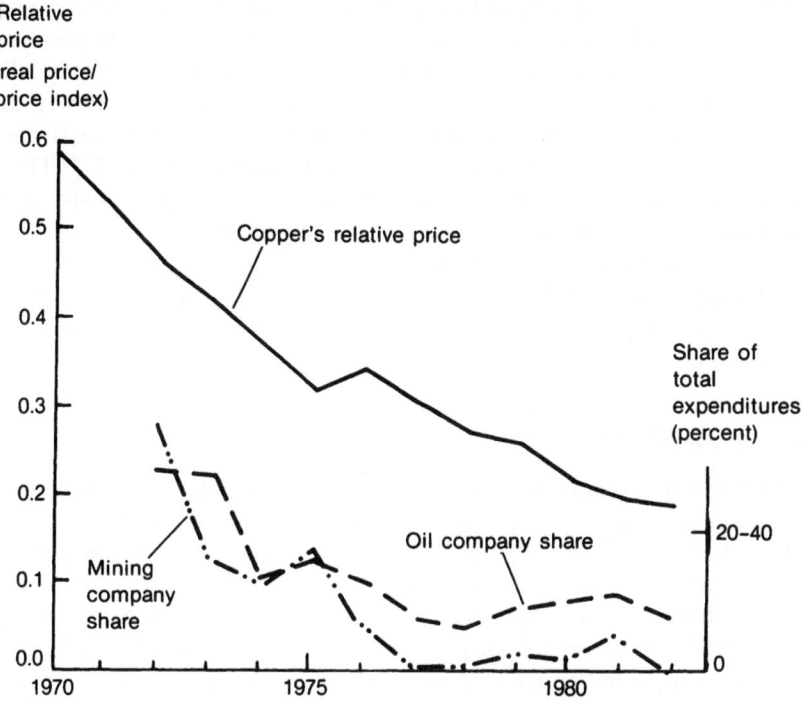

Figure 3-5.   Porphyry copper's share of total exploration expenditures for two firms, and copper's relative price, 1970–82. To protect the identities of the two firms, only a range of percentages is provided. See appendix 3-A regarding the price index.

price increased sharply from 1977 to 1979. As prices for the associated minerals rose, expected future prices presumably did too. This, in turn, increased potential revenues from deposits containing the associated minerals, resulting in an increase in the relative importance of porphyry copper exploration. As an example, in 1979 the oil company initiated a substantial porphyry copper exploration program in South America, at a time when relative molybdenum, gold, and silver prices were increasing. Since then, relative prices for all three associated minerals have fallen substantially, and the relative importance of porphyry copper deposits has declined for both exploration groups.

In addition to the falling relative prices of copper, three other factors may have been responsible for porphyry copper's declining share of expenditures. First, porphyry copper deposits were important exploration targets from the mid-1950s to the mid-1970s. Many discoveries were made, greatly expanding copper reserves, and a number of these deposits await production. Accordingly, the possibility has declined for discovering deposits with lower costs than the existing stock of deposits, barring the emergence of new ore-deposit concepts or advances in producton technologies. Second, capital and energy costs for large, low-grade porphyry copper deposits may have risen more rapidly than similar costs for smaller and higher-grade metal deposits; real costs have increased as newly discovered porphyries have declined in grade, been farther below the surface, and located in more remote areas (see United Nations, 1981, p. 17). Third, multimetallic deposits, such as massive sulfide-type deposits, appear to have gained in importance in many exploration programs because production of several metals can occur simultaneously. Companies have been attracted by the potential for significantly diversified production from one deposit, reducing fluctuations in earnings caused by price instability of any one mineral commodity.

Massive Sulfide Deposits

Generally higher-grade and smaller than porphyry copper deposits, massive sulfide deposits typically contain significant concentrations of copper, lead, and zinc (as well as silver or gold) and have received considerable attention over the last twenty years. As shown in figure 3-6, massive sulfide's share of total expenditures for both firms declined moderately through 1974 and rose abruptly in 1975 and 1976. This increase in share for both companies appears to conform to an industry-wide surge in massive sulfide exploration. As noted by Mullins, Lawrence, and Deschamps (1977, p. 105), "While porphyry copper lost its glamour, massive sulfides became a favorite of explorationists."

After 1976, however, massive sulfide shares followed significantly different paths for each firm. For the oil company, massive sulfide exploration grew to be a very important part of its program. For the mining company, massive sulfide's share declined so that by 1980 it represented a very minor component of its overall program; by 1982, however, the share had grown to its highest level for the eleven-year period. It is unlikely that relative mineral prices have been the major driving force behind these changes. During this period, copper's relative price fell considerably, whereas relative prices for lead and zinc fluctuated around a generally declining trend.

For the oil company, massive sulfide exploration developed into an important part of the overall exploration program during a period from 1975 to 1977, when relative lead and zinc prices were generally steady or increasing. Massive sulfide's share then declined somewhat, roughly paralleling changes in the relative prices. Nevertheless, the decline was not greater because of the company's commitment to copper, lead, and zinc exploration. This commitment was based on the following criteria: expected demand must be large and growing (or at least level), current production capacity must be insufficient to meet these demands over the next five to twenty years, and deposit-types must be large enough so that

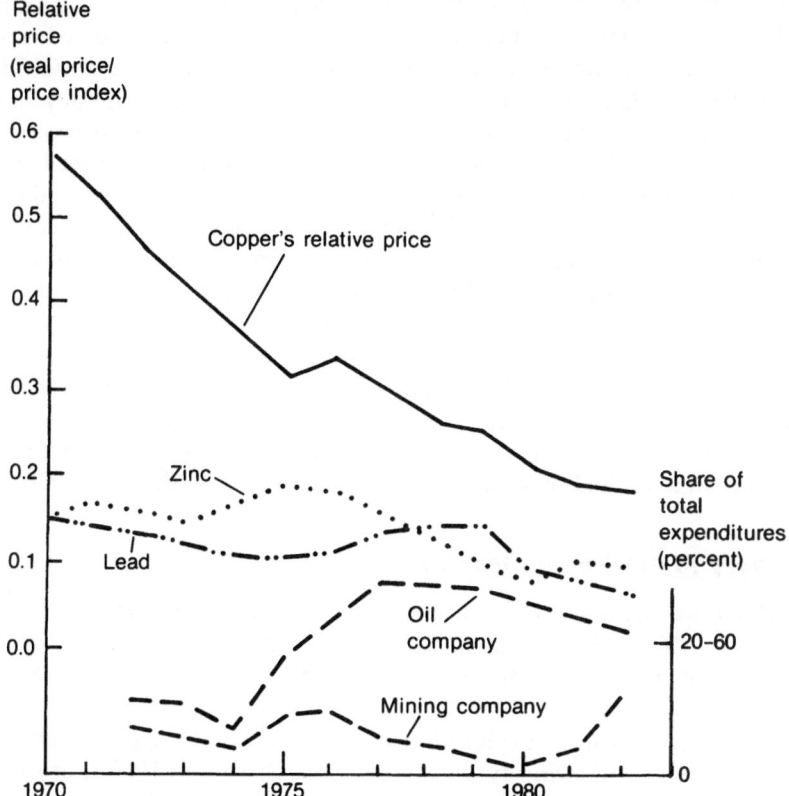

Figure 3-6. Massive sulfide's share of total exploration expenditures for two firms, and relative prices for copper, lead, and zinc, 1970–82. To protect the identities of the two firms, only a range of percentages is provided. See appendix 3-A regarding the price index.

production will have a noticeable impact on the oil parent's income statement. The company contends that profits can be made over the next five to twenty years from minerals that satisfy these criteria, as long as deposits are found with low costs relative to current producers. Thus copper, lead, and zinc have been and remain an important part of the oil company's metal exploration efforts despite falling relative prices, because estimates of potential returns from production of these minerals have not diminished significantly.

For the exploration industry as a whole, if massive sulfide's share of overall expenditures did indeed increase during a period when relative copper, lead, and zinc prices fell generally, this phenomenon could be explained by several factors. First, relative prices for gold and silver, metals often present in significant amounts in these deposits, rose during the same period. Second, the typical massive sulfide deposit provides for significant production of several minerals, as noted earlier and will be discussed more fully later. Third, and perhaps most important, because massive sulfides are smaller and higher grade than porphyry copper deposits, they have lower capital and energy costs per unit of metal. After 1973, future mineral prices and demand became increasingly difficult to forecast. Faced with these uncertainties, mining firms have become reluctant to burden themselves with the large capital repayment costs typical of large porphyry deposits, and they have turned increasingly to massive sulfide and other deposits that promise a quicker return. Finally, several significant massive sulfide deposits, discovered in the mid-1970s, created an aura of excitement about the possibilities for more successful exploration. The discoveries near Izok Lake in Canada's Northwest Territories and near Crandon, Wisconsin, are but two examples. This flurry of massive sulfide discoveries undoubtedly increased perceptions of potential returns from similar exploration and resulted in an increase in expenditures in the mid-1970s.

Porphyry Molybdenum Deposits

Molybdenum became an important target for many exploration groups during the 1970s. As shown in figure 3-7, molybdenum's share of the mining company's expenditures grew steadily from 1972 to 1978, remained fairly level until 1980, and then fell precipitously. For the oil company, molybdenum exploration commenced in 1979, grew substantially during the following two years, and fell sharply in 1982. For major corporations as a whole, molybdenum's share may very well have followed a similar path—steadily increasing in the late 1970s, in 1980, and in some instances 1981, followed by a drastic drop thereafter. During the same period, the relative price of molybdenum declined somewhat from 1972 to 1974, increased moderately during the next three years, rose sharply during 1978 and 1979, and thereafter fell significantly.

Molybdenum exploration in the late 1970s certainly was encouraged by its rising relative prices, reflecting rising real prices. The spot price went from $2.90/lb. in 1975 to $9.22/lb. in 1979 and $22.00/lb. in 1980; in constant 1980 U.S. dollars, these translate into prices of $4.12, $10.08, and $22.00—an overall increase of more than 500 percent. The importance of molybdenum in many uses relating to energy production also encouraged rosy perceptions of its potential revenues in the late 1970s, a time when energy-related exploration was booming.

The precipitous decline in molybdenum's share of exploration expenditures in the early 1980s was caused by a saturated market. Between 1960 and 1980, mines at Climax, Colorado, and Questa, New Mexico, satisfied U.S. demand and allowed for significant exports. Three new mines have opened at Henderson, Colorado; Tonopah, Nevada; and Thompson Creek, Idaho; numerous additional deposits, all on indefinite hold, include those

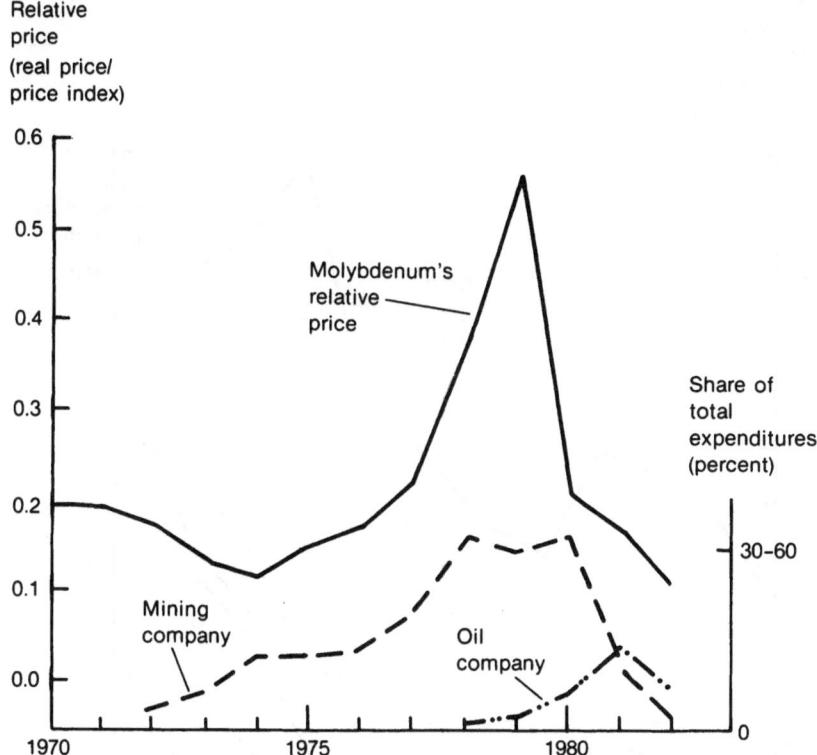

Figure 3-7.   Molybdenum's share of total exploration expenditures for two firms, and molybdenum's relative price, 1970–82. To protect the identities of the two firms, only a range of percentages is provided. See appendix 3-A regarding the price index.

at Quartz Hill, Alaska; Mount Hope, Nevada; Mount Emmons, Colorado; Mount Tolman, Washington; and Pine Grove, Utah. This specter of over-supply is reinforced by the potential for significantly more by-product production of molybdenum from many porphyry copper mines.

Gold and Silver Deposits

Gold and silver deposits became important exploration targets for many companies during the 1970s and early 1980s. As shown in figure 3-8, gold and silver's share of overall exploration expenditures rose dramatically in the early 1980s for the five companies that provided data for this study. Schreiber and Emerson (1984) substantiate the trend: in 1983 exploration expenditures for gold in the United States by thirty large companies accounted for 55 percent of total expenditures for metals and industrial minerals, compared with 33 percent in 1980 and 14 percent in 1975.

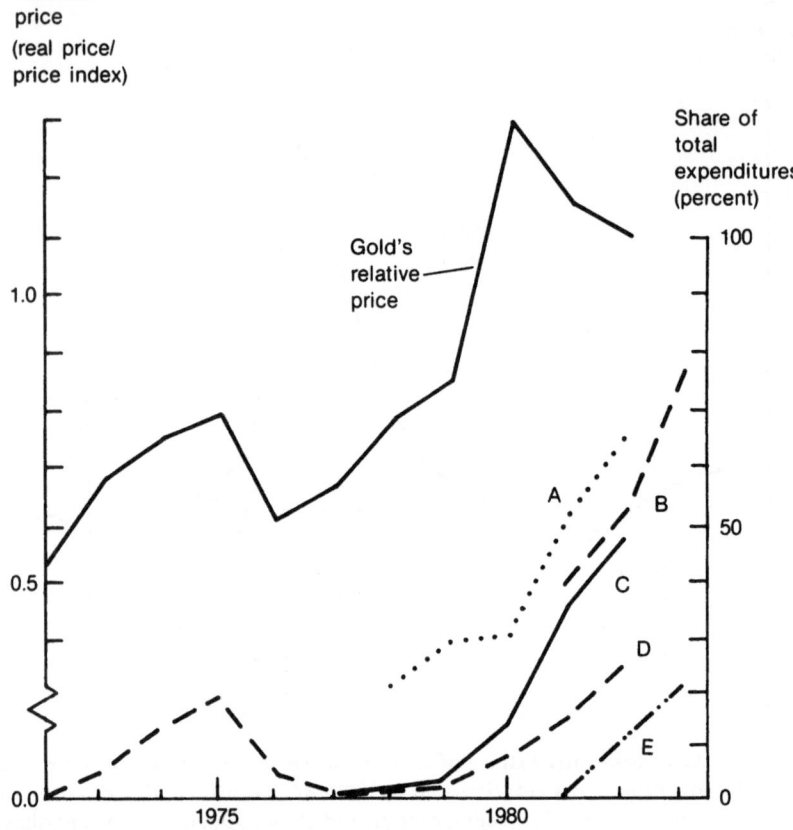

Figure 3-8. Gold and silver's share of total exploration expenditures for five firms, and gold's relative price, 1973–83. Firms A, B, and C are U.S. mining companies, and firms D and E are U.S. oil companies. See appendix 3-A regarding the price index. Share data for 1983 are budget allocations, not actual expenditures. *Sources:* Company exploration departments.

Gold and silver prices relative to other major metals have risen dramatically since the early 1970s. Gold's relative price, displayed in figure 3-8, experienced short-term highs in 1975 and 1980. Silver's relative price followed a similar but not identical path. These relative prices no doubt prompted the increased share of exploration expenditures devoted to gold and silver exploration, by creating enormous expectations about potential revenues from future production of these metals. Two other factors, however, acted in concert with the high prices to stimulate the gold and silver exploration boom. First, as noted by Brown (1983), improvements in heap leaching and carbon-in-pulp recovery of gold have permitted profitable production from the low-grade ores that typify many of the more recent discoveries. Second, the desire to diversify seems to have contributed to the growing importance of precious metals exploration. St. Joe and Inco, for example, initiated sizable gold and silver exploration efforts as part of diversification programs.

Finally, data from company D in figure 3-8 show how the interplay of prices and exploration success can influence a particular mineral's share of overall exploration expenditures. The initial interest in gold and silver exploration was due primarily to rising prices in the early 1970s. The subsequent share increases in 1974 and 1975 were prompted by costs associated with detailed drilling and evaluation of a promising gold prospect, which had been detected during the initial exploration phase. In 1976, gold and silver's share fell sharply, at the same time that the detailed evaluation was completed. In response to the drop in gold prices, the deposit was put on hold awaiting higher prices. Little, if any, new gold and silver exploration was initiated until 1980 when the company responded to the relative price increases of the previous three years.

The Episodic Nature of Exploration and Discovery

It appears that exploration and then discovery of particular minerals follow episodic patterns over time. The following discussion generalizes from the preceding analysis and draws on chapter 10 in this volume by Rose and Eggert, which presents more detailed information on the discovery of specific minerals.

The idealized episode begins with incentives for stepped-up exploration for particular minerals. Initial incentives take a number of forms, including (1) increased mineral prices and demand, which raise estimates of potential revenues from future production, (2) new or revised geologic models,[3] and (3) improved exploration and production technologies. The last two reduce exploration and mining costs, and all three increase the chances for profit.

The cycle continues with a flurry of exploration activity, leading to discoveries, which in turn provide incentives for more exploration over the short term. A bandwagon effect occurs: a successful innovator in exploration is followed by a legion of imitators. Over the longer term, discoveries of a particular mineral or deposit-type discourage further exploration by expanding known sources of supply. This leads to the final stage in the cycle: declining exploration activity and discovery. Falling prices and demand also may contribute to the final stage. With regard to the minerals already discussed in this chapter, the following episodes can be identified. Porphyry copper exploration experienced a long cycle that lasted from the mid-1950s to the mid-1970s. Steadily growing copper

[3] A geologic model is an idealized set of geologic characteristics common to most or all deposits of a particular type. It is used as a guide for reconnaissance exploration and subsequent follow-up work.

demand and refinement of a geologic model of ore occurrence contributed to the interest in porphyry copper deposits. Discoveries, the leveling of copper demand, and increases in capital costs for porphyry deposits relative to other deposit-types contributed to a decrease in interest.

Porphyry molybdenum exploration, on the other hand, experienced a much shorter episode in the middle and late 1970s. Price increases, as well as optimistic projections of future demand and a new geologic model, contributed to the boom. Falling prices and a spate of discoveries were primarily responsible for the cycle's end.[4]

Today, in the mid-1980s, gold and silver are in the midst of an exploration boom. Price increases appear to have been the boom's major determinant, although improvements in processing technology have been a contributing factor, as discussed earlier.

Massive sulfides, in contrast to the other deposit-types considered in this study, experienced less-pronounced exploration episodes. Nevertheless, a small boom occurred in the early 1980s due mainly to the opportunity for diversified production, lower capital and energy costs relative to porphyry copper deposits, new discoveries, and a geologic model of massive sulfide occurrence. In addition, as discussed in chapter 9, much massive sulfide exploration occurred during the 1950s in Canada following the development of airborne and ground electromagnetic techniques, which permitted assessment of large areas that lack surface rock exposure.

Diversification

An important element of many recent corporate exploration programs has been diversification. Any analysis of how exploration funds have been allocated among commodities, therefore, would be incomplete without some discussion of this phenomenon. Diversification reduces fluctuations in earnings due to price instability. In addition, companies have been driven to diversify by price increases for several commodities, such as gold and oil and gas, that have raised potential financial returns from an investment in them.

Exploration, however, is only one of three alternative means of achieving diversified production. The other options are to acquire a known deposit, which then will be developed and brought into production, or to purchase an operating company. Most companies have used a combination of these three methods.

The diversification of exploration programs has occurred on at least four levels. First, a number of companies that historically explored for a limited number of minerals, usually in the vicinity of existing mines for minerals

---

[4]Actually, a few exploration groups became interested in molybdenum as far back as the early 1960s—Kennecott and Amax worked on geologic models then and began large-scale exploration that continued through the 1970s.

to extend the lives of existing operations, have begun to explore for other minerals in various geographic and geologic conditions. For example, prior to the 1970s St. Joe was primarily a producer of lead and zinc with operations in the United States, and most of the company's exploration efforts were aimed at expanding the lead-zinc reserves of existing operations. During the 1970s, St. Joe significantly increased exploration expenditures and began to explore for a much wider range of metals. It became a gold producer in Chile in 1979 (a St. Joe discovery) and California in 1980 (a St. Joe acquisition), and an iron ore producer in 1979. At the same time, St. Joe also made diversification investments in energy resources—they purchased CanDel Oil Limited (1973), Massey Coal (1974), Tennessee Consolidated Coal (1976), and Coquina Oil Corporation (1977). Inco is another example of a mining company that historically had a limited exploration focus, nickel, and now has a much broader focus, including gold and oil and gas.

Second, as noted previously, multimetallic exploration targets, such as volcanogenic massive sulfides, have become increasingly important as companies have tried to reduce fluctuations in earnings caused by price instability. Multimetallic deposits allow for diversification into more than one mineral through only one mine.

Third, energy minerals, including uranium, coal, and oil and gas, became increasingly important exploration targets for mining companies in the 1970s. Figure 3-9 shows how exploration for energy minerals evolved over time for Amax and Phelps Dodge. Other mining companies that invested heavily in oil and gas exploration in recent years include Asarco, Homestake, Newmont, Noranda, and, as mentioned previously, Inco and St. Joe.

Finally, in perhaps the most well-known example of exploration diversification, metals became important exploration targets for many major oil companies, although by the mid-1980s some firms had pulled out of metals. Companies such as Exxon and Getty entered metals exploration by forming new exploration groups within the parent company. Others, such as Arco, Pennzoil, and Sohio, relied primarily on acquisitions of existing companies to enter metals exploration. Still others, such as Amoco, drew heavily on both methods. In many instances, oil companies initially were attracted to metals exploration by the energy metal uranium and subsequently expanded their programs to include other metals. Figure 3-10 displays this pattern for one oil company.

Location of Exploration

In conjunction with how much to spend and what to look for, companies must decide where to explore. This section examines how the geographic allocation of corporate exploration expenditures has changed over time

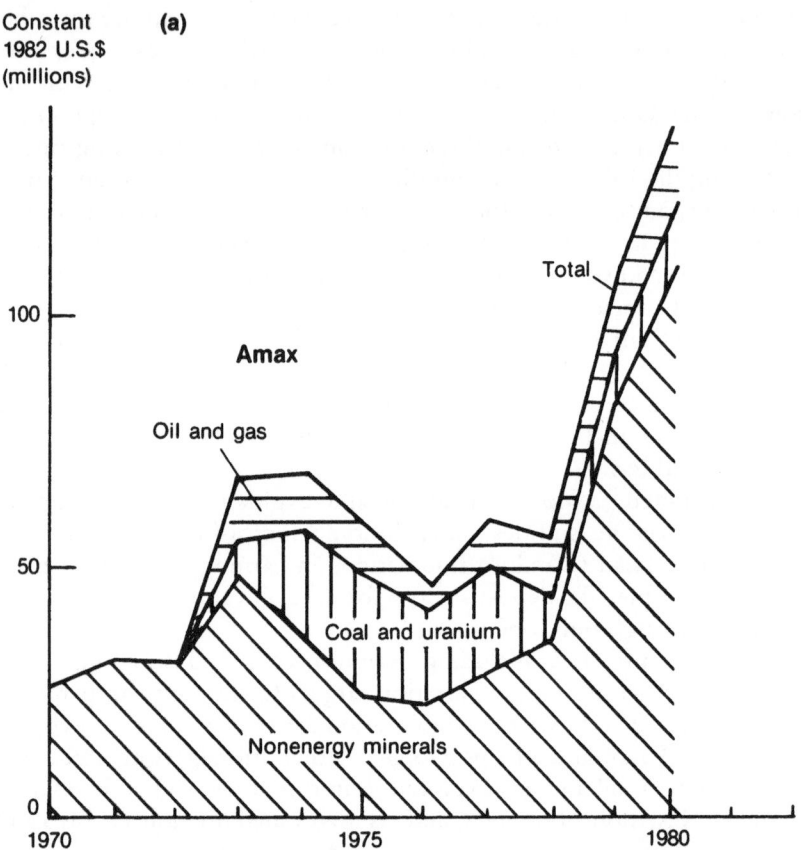

Figure 3-9.   Exploration expenditures for energy and nonenergy minerals by (a) Amax and (b) Phelps Dodge, 1970–82. *Sources:* Amax, *Annual Report* (Greenwich, Conn., 1970–81); Phelps Dodge, *Annual Report* (New York, N.Y., 1970–82).

for Australia, the United States, and developing countries, after reviewing the results of previous research for Canada.

Although many factors presumably influence the location of exploration, including geologic potential and availability of infrastructure, it frequently is argued that government policies and political risks are of overriding importance to corporate explorers. For example, changes in Australian and Canadian policies in the 1970s concerning taxation, ownership, and other issues have been blamed for reductions in exploration there. For the United States, Barber (1981, p. 23) suggests that the

greatest stimulus for expanding foreign exploration by U.S. companies . . . [is] our federal government. The precipitous withdrawal of our public lands prior to adequate evaluation of their mineral potential has provoked reduction or termination of domestic exploration programs.

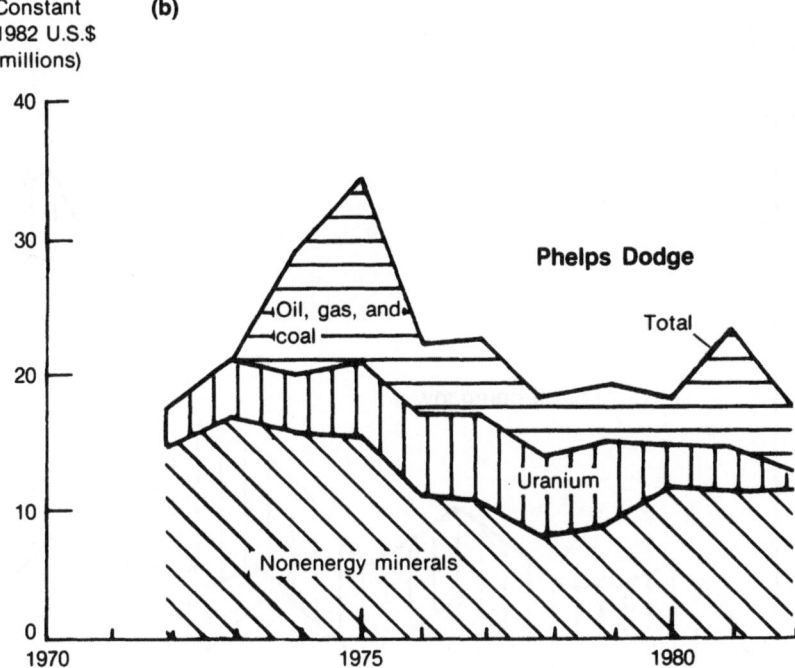

Figure 3-9 (continued)

In developing countries, increases in political risk often are cited as the primary cause of what has been perceived to be a dramatic decline in recent corporate exploration there. For example, the Brandt Commission report (1980, p. 155) states:

The mining companies place much of the blame [for the shift away from exploration in developing countries] on the instability of concession agreements in the Third World, and the erosion of what they regard as their contractual rights by nationalization.

The rationale is that government policy changes raise or lower the costs and thus the investment attractiveness of operating in a country by providing incentives or disincentives for exploration and mining. Incentives take a number of forms, including tax holidays for new mines, exclusive exploration rights for large blocks of land, and concessionary power rates for mineral-processing facilities. Disincentives include host-country ownership requirements, restrictive land and tax policies, and environmental regulations. Political risk alters the investment attractiveness of a country by raising or lowering the risk premium that is required by firms to invest in an area.

This section, therefore, examines the extent to which changing government policies and political risks have been the driving forces behind

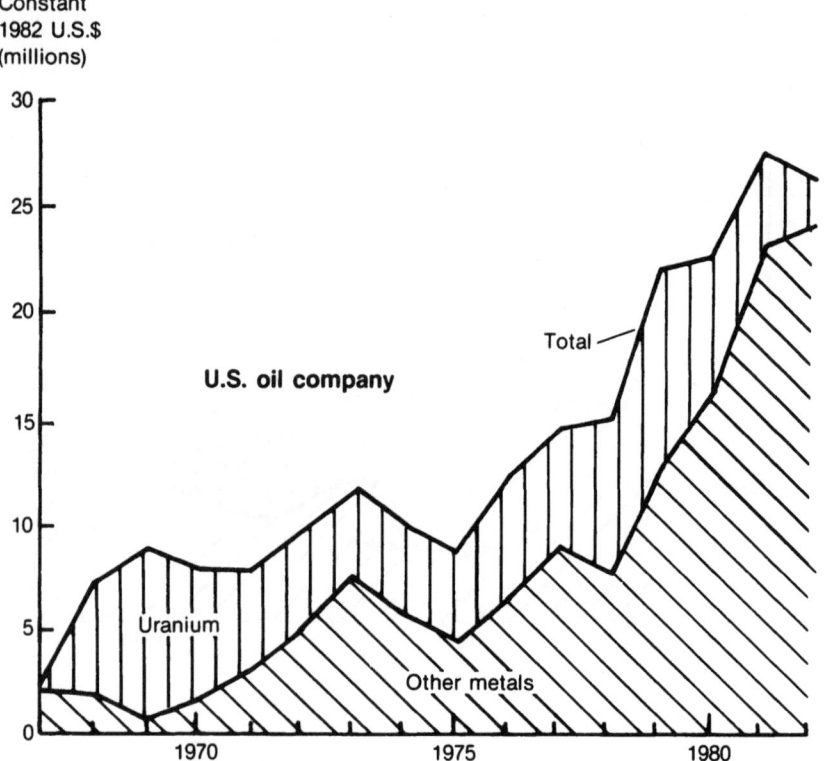

Figure 3-10. Metal exploration expenditures for uranium and other metals by a U.S. oil company, 1967–82. *Source:* Oil company exploration department.

recent changes in the geographic allocation of corporate exploration expenditures.

Canada

DeYoung (1977, 1978) studied the effects of Canadian tax law changes in the early 1970s on mineral exploration there, and his results support the theory that policies and risks are of considerable importance to the location of exploration. By comparing mineral-exploration trends in various Canadian provinces with those in neighboring regions, he concluded that the major effect of tax changes was to shift the location of exploration from one political region to another.

The most striking example is British Columbia, where exploration expenditures, drilling footage, and claim staking declined in 1974 and 1975, following the introduction of particularly onerous provincial tax changes, including two layers of royalties; at the same time, exploration increased in the neighboring U.S. states of Alaska, Washington, and Montana, and in Canada's Yukon and Northwest Territories, which, as territories under federal control, were not significantly influenced by federal-provincial tax battles (DeYoung, 1978, p. 31). At the international level, DeYoung noted that Canadian mining companies are estimated to have spent 60 percent of their exploration funds abroad in 1975, compared with 20 percent in 1971 (1977, p. 102). Nevertheless, DeYoung admitted that it is clearly impossible to separate the effect of changing tax laws from other influences upon exploration budget allocations, such as mineral prices and geologic concepts.

Australia

In December 1972 the Labour Party was elected to power in Australia, and, as shown in figure 3-11, private exploration expenditures declined in real terms over the next three to four years. Defeated in 1975, the Labourites relinquished power to the Liberal-National party coalition and

Figure 3-11.   Mineral exploration expenditures in Australia by private organizations, 1965–84. Data were originally in Australian dollars, which were then converted to 1982 U.S. dollars using appropriate exchange rates and price deflators. The data include expenses and capitalized expenditures; represent private expenditures outside of production leases and exclude all government expenditures; include exploration for copper, lead, zinc, silver, nickel, cobalt, gold, iron ore, mineral sands, tin, tungsten, scheelite, wolfram, uranium, other metals, coal, construction materials, diamonds, and other nonmetallic materials. *Source:* Australian Bureau of Statistics, as reported in the appendix to chapter 2 in this volume.

thereafter exploration expenditures rose steadily. Commenting in 1974 on the Labour government, Derry, Michener, and Booth (1974, p. 48) maintain "that the degree of uncertainty which now hangs over every aspect of mining and exploration in Australia has virtually halted significant exploration and mining development." Three years later, Mullins, Lawrence, and Deschamps (1977, p. 104) wrote that "Australia is rapidly gaining in stature following the return to power of the Liberal-National Party coalition."

Exploration expenditure data from four sources are used here in conjunction with additional information to examine these and other changes in Australia. If expenditure changes represent reactions to changing policies and risks, then one would expect the share of total world expenditures devoted to exploration in a country to increase during periods of favorable policies and low risks and to decline during periods of adverse policies and high risks. If, on the other hand, expenditure changes are unrelated to policies and risks, then one would expect the country's share of total expenditures to change in response to other influences.

As shown in figure 3-12, the share of total exploration expenditures devoted to exploration in Australia has varied widely over time for three firms and a group of European mining companies. In general, Australia's share increased substantially from the mid-1960s to 1973, declined from 1973 to 1977 (with the exception of the oil company), increased to record levels by 1980, and declined moderately thereafter.

The increase in share during the late 1960s and early 1970s, displayed in the European data, was prompted by more optimistic perceptions of Australia's geologic potential. As reported by Roberts (1977) and King (1973), significant discoveries of bauxite, manganese, lead, zinc, and particularly iron ore in the 1950s and early 1960s led to renewed interest in Australian exploration. Iron ore supplied the initial impetus. King (1973, p. 15) lists twelve important iron ore discoveries (out of twenty-nine total mineral discoveries in Australia) between 1956 and 1965 that were made in part because of the "recognition of the great iron ore province in northwestern Australia" (Sullivan, 1974, p. 20) and in part because of the proximity of the emerging Japanese steel industry, lying far from existing ore sources. During the same period, the Weipa bauxite deposit, the Macarthur and Woodcutters lead-zinc deposits, and the Groote Eylandt manganese deposit were discovered.

Discoveries made in 1966, most notably the Kambalda nickel region in Western Australia and several phosphate deposits in Queensland, sparked the exploration surge that lasted until the early 1970s. King lists fifty-eight mineral discoveries made in Australia between 1966 and 1973, including fourteen nickel sulfide deposits near Kambalda, in the area that, as noted by Sullivan (1974, p. 20), "was previously considered a gold province, and any mention that nickel might be present was generally regarded with great skepticism."

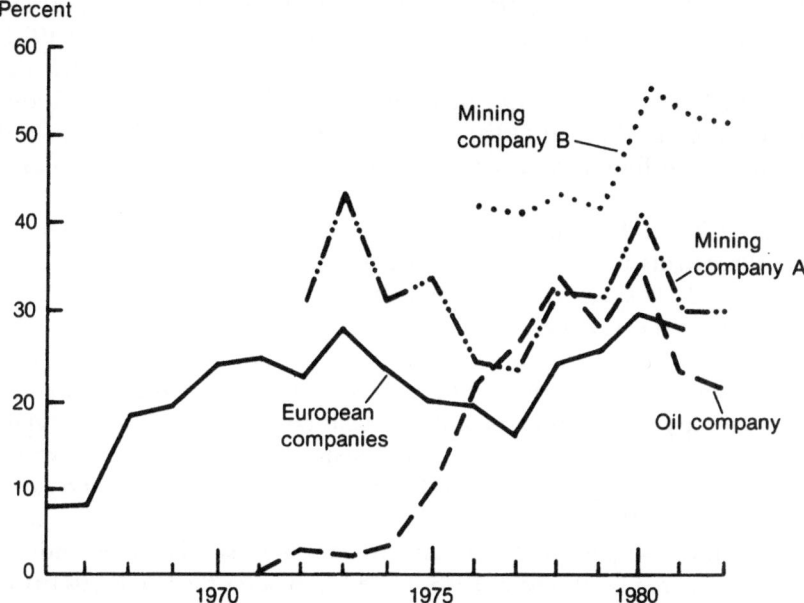

Percent

Figure 3-12.   Share of total exploration expenditures by European mining companies and three other firms spent in Australia, 1966–82. The data for the European companies and mining company B exclude expenditures for oil, gas, and uranium but include coal. The other data include only expenditures for base and precious metals exploration. *Sources:* For the European companies: chapter 2 of this volume; for the three other companies, data were obtained directly from the companies by the author.

Returning to the effects of the Labour party on metals exploration in Australia, it is readily apparent from figure 3-11 that much of the decline in exploration expenditures occurred before the inception of its rule in December 1972. Labour policies put in place, therefore, could not have been directly responsible for most of the decline in exploration expenditures that occurred during the early and mid-1970s.

Nevertheless, there may be some merit to the view that expectations, that is, uncertainty over future policy, as well as Labour policy changes themselves, discouraged exploration in Australia relative to other countries. From 1973 to 1977, Australia's share of total world exploration declined for the group of European mining companies and mining company A, as shown in figure 3-12. Proposed restrictions on foreign-owned equity, ownership conflict between the federal and state governments over mineral rights in certain regions, establishment of the Takeover Review Board to approve any arrangements between Australian resource owners and foreigners, and the consent required from aboriginal people to work on their land in the future all contributed to a sense of uncertainty surrounding the future of foreign-controlled mining investment in Australia.

However, an equally important cause of the decline in share appears to have been the success of exploration in Australia between 1956 and 1973. By 1973, the exploration boom may have run its course. Moreover, the companies were worried about marketing what they already had discovered. As noted by Barnett (1980, p. 5), "So much iron ore, bauxite, and nickel had been found there was little point in funding further exploration."

From 1977 to 1980, Australia's share of exploration increased generally for the three firms and the group of European mining companies, as shown in figure 3-12. This resurgence began about a year after the return of the Liberal-National party coalition. Although the coalition eased some of the Labour government restrictions, most notably those on foreign-owned equity, other factors contributed to renewed interest in Australian exploration. Geology in several parts of Australia is favorable for different types of gold deposits. With prices high, a gold rush took off in Australia, where it became perhaps the most sought-after mineral in the late 1970s and early 1980s (Oldham, 1981, pp. 54, 59). As an example, gold exploration represented almost 40 percent of the oil company's Australian expenditures in 1982, compared with zero in 1978. In addition, after the late 1970s firms appear to have adjusted to new ground rules for exploration in Australia and other places around the world and are more willing to participate in joint ventures with local partners, even in minority positions.

The United States

Certain government policies, particularly land management regulations adopted in the 1970s, have made exploration and mining in the United States more difficult to undertake. First, the withdrawal of public lands in general from mineral development in the mid-1970s attracted much attention. In a much-quoted article, Bennethum and Lee (1975) estimate that the percentage of public lands withdrawn from locatable mineral development increased from 17 percent in 1968 to 53 percent in 1974. The following quote from Mullins, Lawrence, and Lalande (1976, p. 91) exemplifies the frustration of many mineral explorers: "The practice of withdrawing federal lands from exploration and mining has accelerated to the degree that by 1980, if continued, all public lands could be closed." Second, in the late 1970s concern centered around the closing of many millions of acres of Alaskan lands to any development, following a time when Alaska had been a favorite mineral frontier (see, for example, Mullins, Lawrence, and Deschamps, 1977, p. 110; and Hodge and Oldham, 1979, p. 97). Many were disturbed by what they viewed as misguided policy.

Nevertheless, it is not clear that U.S. federal and state policies on balance have been more onerous or restrictive than policies in other countries. Table 3-2 displays how the share of total world exploration expenditures

Table 3-2.   Share of Total Mineral Exploration Expenditures Spent in the United States by Nine Companies, 1971–82
(percent)

Year	Amax	Asarco	Noranda	U.S. mining companies				U.S. oil companies	
				1	2	3	4	1	2
1971	—	—	13	—	—	34	—	48	90
1972	54	>50	16	—	38	22	49	34	71
1973	45	approx. 50	17	80	13	18	49	30	53
1974	64	—	26	83	19	29	53	40	57
1975	68	>63	31	—	21	31	49	36	55
1976	54	—	34	100	35	35	55	35	66
1977	48	—	29	87	47	38	68	33	66
1978	57	>60	33	62	30	42	67	30	60
1979	58	>75	39	73	29	46	62	27	57
1980	58	approx. 75	36	58	20	48	56	22	47
1981	—	approx. 75	38	65	28	51	54	29	40
1982	—	—	33	76	34	27	49	29	37

Sources: Amax, *Annual Report* (Greenwich, Conn., 1970–81); Asarco, *Annual Report* (New York, N.Y., 1970–82); Noranda, *Annual Report* (Toronto, Ontario, 1967–82); personal communications with the U.S. mining and oil companies, which required anonymity.

spent in the United States changed between 1971 and 1982 for nine companies. Taken together, the data reveal no common and distinctive trends. For oil company 1, the U.S. share declined in the early 1970s, as the company initiated exploration efforts in Canada and Australia; the share continued to fall from 1974 to 1980, and then rose in 1981. At Amax, the large increase in the U.S. share in 1974 probably resulted from greatly expanded exploration for energy minerals; in the late 1970s, the share increased because of large outlays on detailed exploration and evaluation at Mt. Emmons, Colorado, and Mt. Tolman, Washington, molybdenum prospects. Asarco expenditures increased gradually, although this observation is based on incomplete data.

For foreign exploration groups, on the other hand, some have argued that U.S. policies on balance are relatively favorable for exploration. According to Cook (1983, p. 12): "The entry of foreign exploration companies into the U.S. will continue and possibly accelerate. . . . The attraction of the U.S. for exploration is the political stability and generous foreign investment policies."

Exploration expenditures by Noranda in the United States rose from 13 percent of total expenditures in 1971 to nearly 33 percent in 1982. Finally, Consolidated Gold Fields, based in London, spent more than 80 percent of its direct or corporate exploration (that is, excluding expenditures by subsidiaries and partially owned companies, such as Newmont) in North America, much of it in the United States (Kramer, 1983). Nevertheless, these shifts toward the United States could well be due to other factors that offset any adverse effects of policy.

The implication, based on fragmentary data and incomplete analysis, is that there is no evidence to support the view that U.S. government policies are more onerous than those of other governments. They may even be more favorable than most.

Developing Countries

There is a widespread belief that exploration in developing countries by major corporations has declined markedly in recent years because of ill-conceived policies on the part of host governments and associated uncertainties regarding the future. This has led to fears of future shortages in mineral supply from developing countries. The purpose of this section is to assess the truth of this proposition, using the limited evidence, and more generally to evaluate recent trends in the location of exploration activity in developing countries.

A group of European mining companies has collected statistics on the geographic breakdown of exploration expenditures by member firms for the period 1966–84 (for details, see chapter 2 in this volume). As shown in figure 3-13, the developing-country share of overall exploration (excluding uranium) fell from 44 percent in 1966 to 21 percent in 1970, with the steepest decline occurring in 1970 (from 41 percent to 21 percent). After 1970 the share fluctuated between 10 and 25 percent of total exploration expenditures. Can it be inferred that 1970 was a watershed year in perceptions of political risk?

Resource nationalism in mineral-rich countries had been simmering since the early 1960s and appears to have reached a head in 1970 in its effect on exploration. In Mexico a policy instituted in 1961 required that "all new mining ventures including exploration had to be 51 percent Mexican" (Walthier, 1976, p. 46); operating mines were given twenty years to comply. In 1964 President Eduardo Frei of Chile called for nationalization of foreign-owned copper mines, which ultimately occurred following the election of President Allende in 1969 (the nationalizations were not completed until 1971) (see Moran, 1973). In Africa in the late 1960s, Zambia acquired controlling interest in its two largest mining groups, Anglo-American Corporation and Roan Selection Trust, and Zaire nationalized Union Minière's holdings there.

Thus, one might expect the big drop in shares in 1970 to be reflected in expenditures in Latin America and Africa (excluding South Africa). However, the share of exploration expenditures in African developing countries fell from only 5.2 percent in 1969 to 3.2 percent in 1970, and Latin America's share remained the same (at 0.5 percent). The big decline in the developing-country share reflects primarily a decline in Oceania,

Percent

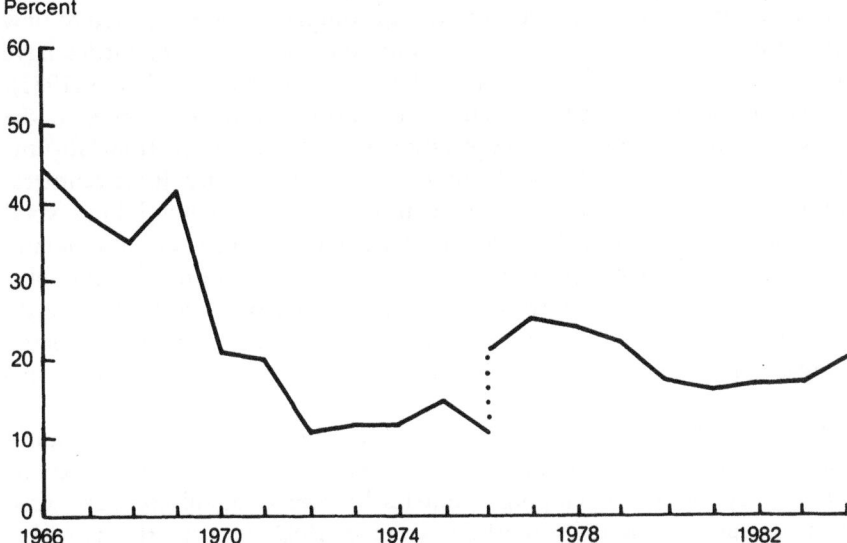

Figure 3-13. Share of exploration expenditures spent in developing countries for a group of European mining companies, 1966–84. Excluded are expenditures for oil, gas, and uranium exploration; the discontinuity in 1976 results from a change in the composition of the European companies from which data were collected. *Source:* Chapter 2 of this volume.

excluding Australia, from 17.0 percent in 1969 to zero in 1970. Radetzki (1982, p. 41) maintains that if the Bougainville Island copper project in Papua New Guinea is excluded from the statistics from 1966 to 1969, the decline in developing-country share is much less dramatic—to 21 percent from slightly more than 30 percent rather than from more than 40 percent. Thus, a large part of the decline in 1970 reflects the movement of one project from the exploration to development stage, and not necessarily increased perceptions of political risk.

Still, it is impossible to dismiss completely the role of government policies and political risks. After 1970 the developing-country share was consistently lower than it had been during much of the 1960s. Although these trends cannot be attributed solely to one factor, the resource nationalism of some developing countries in earlier years (highlighted above) must be considered a plausible contributing factor. Adverse policy changes toward mining companies and the specter of future changes can do nothing but raise the costs and risks and thus lower potential returns for a firm considering an investment in such a country, other factors remaining the same.

Just as it is difficult to demonstrate an adverse link between policies and expenditures in the 1970s, so it is difficult to document a recent reversal. But there is some indication that corporate exploration in developing

countries has increased in recent years, as companies have adjusted to new
rules for operating in many developing countries and as countries have
moderated some of their views and policies. As discussed by Davies (1983),
many developing countries seem to be more willing to adhere to the
provisions of a contract and to use national courts and international forums
to settle disputes, thereby lowering the risks to a company. Joint ventures,
quite rare before the 1970s, are now commonplace, often involving several
companies as well as the host government and offering a way of reducing
the perceived risks of operating in a country. Furthermore, recent exploration in developing countries may have been encouraged by the steady
increase in geologic data available for these countries and the associated
increased potential for high-grade discoveries, or bonanzas, in poorly explored areas.

Data from U.S. companies on their exploration activities in developing
countries are very incomplete. Table 3-3 shows that the share of expenditures devoted to developing countries by one U.S. oil company rose
from 0.5 percent in 1975 to 10.2 percent in 1982. Perhaps this is representative of an increasing share of expenditures in developing countries
by U.S. oil companies in general since the mid-1970s, but the evidence is
not solid. Porphyry copper exploration in Chile has constituted a large
part of these expenditures. This development appears representative of a
trend toward increased exploration efforts in the Andean Cordillera, particularly Chile. A recent survey of large U.S. exploration groups by Barber
(1981) indicates that 70 percent of the respondents either had active programs in Chile or were planning them, making Chile the third most
preferred foreign country for exploration, after Australia and Canada. For
the group of European mining companies, the exploration share for Latin

**Table 3-3. Share of Exploration Expenditures Spent
in Developing Countries by a U.S. Oil Company,
1973–82**
(percent)

Year	Developing country share
1973	2.3
1974	3.5
1975	0.5
1976	0.7
1977	0.7
1978	1.8
1979	8.3
1980	7.8
1981	11.6
1982	10.2

Source: Company exploration department.

America rose from 3.0 percent in 1976 to 9.4 percent in 1981 and 11.5 percent in 1984, and most of this activity appears to have occurred in Chile and Brazil.

What caused the renewed interest in Chilean exploration? First, the Pinochet government has been much more receptive to direct foreign investment than was the Allende regime (which nationalized the foreign-owned copper companies in the early 1970s). A foreign firm in Chile now can own 100 percent of the equity in new mining interests, and new mining legislation, which became effective in December 1983, grants real property rights to foreign concessionaires (see *Engineering and Mining Journal*, 1984). Second, although Chile long has been geologically attractive for exploration, several recent discoveries—including the La Escondida porphyry copper deposit and the El Indio gold-copper deposit—have stimulated exploration in areas that were previously believed to have relatively little geologic potential.

Returning to the fear that future mineral supplies from developing countries will be inadequate because of inadequate current exploration efforts there, the limited data are clearly insufficient for making such an inference. The European data represent only part of overall corporate exploration expenditures. North American and other private corporations conduct significant exploration efforts in developing countries, and some evidence suggests that mineral exploration there by U.S. oil companies increased after the mid-1970s. Furthermore, even if the developing-country share of total private exploration expenditures has declined, exploration by public organizations, such as state-owned mining companies, may have partially replaced private exploration (see chapter 2 in this volume).

Summary and Conclusions

Corporate mineral exploration occurs in response to changing financial incentives. Economic, political, and technical factors influence expected revenues, costs, and risks—and ultimately expected returns—from an investment in mineral exploration.

The level of exploration expenditures varies with the *overall* expected returns to exploration. Mineral prices strongly influence these returns by affecting expected revenues and by altering costs and the supply of funds, through their effect on the availability of internal funds for financing exploration. In addition, a company's exploration success and corporate goals, such as diversification, influence perceived risks and revenues, and therefore expected returns from exploration.

The distribution of funds with respect to both commodities and countries tends to vary with *relative* expected returns. First, relative prices for mineral commodities are an important consideration because of their effect on potential mineral revenues. But other factors, such as changing dis-covery rates, advances in exploration and mining technologies, new ex-ploration models, and corporate goals such as diversification, also influence expected returns and may be equally responsible for changes in this dis-tribution of funds. In the second case, which concerns the allocation of exploration funds among countries, although government policies and political risks are an important determinant of expected costs and risks in a country, their importance appears to have been exaggerated. Geologic criteria, particularly recent discoveries in an area, are probably the most important determinants of changes in the geographic distribution of funds. Nevertheless, to the extent that government policies and political risks discourage exploration in otherwise geologically and economically attrac-tive areas, they create a suboptimal allocation of exploration funds, unless other investors, such as state enterprises, successfully adopt the role for-merly played by private corporations.

A rational view to take of corporate mineral exploration is that of an activity that competes for funds with other forms of investment that also serve to augment mineral supply. Exploration occurs in response to the desire for lower-cost mineral production, a goal that also can be reached through improved production and processing techniques and further geo-logic research.

However, corporate exploration can help limit the effects of depletion of mineral resources, if prices of mineral commodities accurately reflect their relative scarcity. As deposits of a mineral are depleted, the mineral's relative scarcity and price tend to increase. As expected returns from the production of the mineral increase, incentives to carry out more explo-ration also rise. Assuming that increased exploration leads to additional discoveries, relative scarcity declines and the specter of depletion subsides. Thus, to the extent that exploration by major corporations represents exploration by all participants, exploration acts systematically and in con-cert with material substitution and other forms of technologic change to help offset the cost-increasing effects of depletion.

∎

This chapter would not have been possible without the generous help of many companies and individuals. Good exploration data are difficult to find, and so I am particularly grateful for the information provided by several companies that preferred anonymity to the spotlight. The presentation benefited greatly from thoughtful reviews of an earlier draft by Paul Bailly, Robert Cairns, John DeYoung, Jr., Hans Landsberg, Pierce Parker, Arthur Rose, and John Tilton, as well as the participants at the International Institute for Applied Systems Analysis workshop, where the paper was presented originally. Much of the research for this chapter was conducted at IIASA, in Laxenburg, Austria, where I was a research scholar. Any remaining errors of fact or judgment are mine.

References

Amax Incorporated. 1970–81. *Annual Report* (Greenwich, Conn.).

Asarco Incorporated. 1970–82. *Annual Report* (New York, N.Y.).

Barber, G. A. 1981. "Foreign Exploration—Pros and Cons," *Mining Congress Journal* vol. 67, no. 2, pp. 20–23.

Barnett, D. W. 1980. "National and International Management of Mineral Resources." Paper presented at the joint meeting of the Institution for Mining and Metallurgy, the Society of Mining Engineers (SME) of the American Institute of Mining, Metallurgical, and Petroleum Engineers (AIME), and the Metallurgical Society of AIME, London, May (London, Institution of Mining and Metallurgy).

Bennethum, G., and L. C. Lee. 1975. "Is Our Account Overdrawn?" *Mining Congress Journal* vol. 61, no. 9, pp. 33–48.

Brandt Commission. 1980. *North-South: A Program for Survival* (London, Pan Books).

Branson, W. H. 1979. *Macroeconomic Theory and Policy*, 2d ed. (New York, Harper and Row).

Brown, W. K. 1983. "Exploration for Gold: Costs and Results." Paper presented at the SME-AIME annual meeting, Atlanta, Georgia, March 6–10 (Littleton, Colorado, SME-AIME).

Cominco Limited. 1968–82. *Annual Report* (Vancouver, British Columbia).

Cook, D. R. 1983. "Exploration or Acquisition: The Options for Acquiring Mineral Deposits," *Mining Congress Journal* vol. 69, no. 21, pp. 10–12.

Davies, W. 1983. "Sanctity of Contract," *Engineering and Mining Journal* vol. 184, no. 1, pp. 87–89.

Derry, D. R., C. E. Michener, and J. K. B. Booth. 1974. "World Mining Exploration Trends—Politics of Prospecting," *World Mining* vol. 27, no. 3, pp. 46–50.

DeYoung, J. H., Jr. 1977. "Effects of Tax Laws on Mineral Exploration in Canada," *Resources Policy* vol. 3, no. 2, pp. 96–107.

————. 1978. "Measuring the Economic Effects of Tax Laws on Mineral Exploration," *Proceedings of the Council of Economics of AIME, 107th Annual Meeting* (New York, AIME).

Engineering and Mining Journal. 1983. "Average Annual Metal Prices 1920–1982," vol. 184, no. 3, p. 51.

————. 1984. "Chile's New Mining Code," vol. 184, no. 1, p. 78.

Falconbridge Nickel Mines Limited. 1969–82. *Annual Report* (Toronto, Ontario).

Gluschke, W., J. Shaw, and B. Varon. 1979. *Copper: The Next Fifteen Years* (Boston, D. Reidel Publishing Company).

Hodge, B. L., and O. L. Oldham. 1979. "Mineral Exploration in 1978," *World Mining* vol. 32, no. 8, pp. 89–99.

Homestake Mining. 1970–81. *Annual Report* (San Francisco, Calif.).

Inco Limited. 1974–82. *Annual Report* (Toronto, Ontario).

Kennecott Copper. 1970–80. *Annual Report* (New York, N.Y.).

King, H. F. 1973. "A Look at Mineral Exploration 1934–1973," *The Western Australia Conference* (Parkville, Victoria, Australia, Australasian Institute of Mining and Metallurgy).

Kramer, D. 1983. "Gold Fields Is Focusing Exploration in North America," *American Metal Market*, September 14, pp. 1, 20.

Metal Bulletin. 1983. "Will More Moly Mines Close?" January 11, p. 21.

Metal Bulletin Books Limited. 1983. *Metal Bulletin Handbook*, vol. 1 (Surrey, England, Metal Bulletin Books Limited).

Metallgesellschaft Aktiengesellschaft. 1982. *Metal Statistics 1971–1981*, 69th ed. (Frankfurt am Main, Metallgesellschaft).

Moran, T. H. 1973. "Transnational Strategies of Protection and Defense by Multinational Corporations: Spreading the Risk and Raising the Cost for Nationalization in Natural Resources," *International Organization* vol. 27, no. 2, pp. 273–287.

Mullins, W. J., R. D. Lawrence, and P. Lalande. 1976. "Exploration Rush to Energy Minerals: Canadian Arctic Search and Discovery," *World Mining* vol. 29, no. 7, pp. 86–93.

Mullins, W. J., R. D. Lawrence, and D. W. Deschamps. 1977. "Explorationists Seek U_3O_8, Porphyry, Silver, Sulphide Deposits," *World Mining* vol. 30, no. 7, pp. 104–111, 235.

Newmont Mining Corporation. 1968–82. *Annual Report* (New York, N.Y.).

Noranda Mines Limited. 1967–82. *Annual Report* (Toronto, Ontario).

Oldham, L. 1980. "Mineral Exploration in 1979," *World Mining* vol. 33, no. 8, pp. 46–53.

———. 1981. "Mineral Exploration in 1980," *World Mining* vol. 34, no. 9, pp. 54–65.

Phelps Dodge. 1970–82. *Annual Report* (New York, N.Y.).

Radetzki, M. 1982. "Has Political Risk Scared Mineral Investment Away from the Deposits in Developing Countries?" *World Development* vol. 10, no. 1, pp. 39–48.

Roberts, P. J. 1977. "Mineral Exploration in Australia: 1965 to 1973," *Australian Mineral Industry—Quarterly Review* vol. 29, no. 2 (issued by Australian Bureau of Mineral Resources, Geology, and Geophysics), pp. 15–25.

Roskill Information Services Limited. 1982. *Roskill's Metals Databook*, 3d ed. (London, Roskill Information Services Limited).

St. Joe Minerals Company. 1968–80. *Annual Report* (New York, N.Y.).

Schreiber, H. W., and M. E. Emerson. 1984. "North American Hardrock Gold Deposits: An Analysis of Discovery Costs and Cash Flow Potential," *Engineering and Mining Journal* vol. 185, no. 10, pp. 50–57.

Sullivan, C. J. 1974. "Mineral Investment Decisions for the Coming Decades," *CIM Bulletin* vol. 67, no. 747, pp. 19–22.

Tremblay, M., and J. Descarreaux. 1978. "Mineral Exploration in 1977—Emphasis on Small, Rich Deposits," *World Mining* vol. 31, no. 7, pp. 90–97.

United Nations Centre on Transnational Corporations. 1981. "Transnational Corporations in the Copper Industry," ST/CTC/21.

U.S. Bureau of Mines. 1950–81. *Minerals Yearbook* (Washington, D.C., Government Printing Office).

Ventura, D. 1982. "Structures de Financement de la Prospection Miniere: Le Probleme Specifique des Pays en Voie de Developpement" ("Financial Arrangements for Mineral Exploration: The Specific Problem of the Developing Countries") *Chronique de La Recherche Miniere* no. 464, pp. 5–25.

Walthier, T. N. 1976. "The Shrinking World of Exploration," part 2. *Mining Engineering* vol. 28, no. 5, pp. 46–50.

Appendix 3-A
Mineral Price Index

The mineral price index reflects percentage changes in real (inflation-adjusted) prices for the following minerals: copper, lead, zinc, gold, silver, molybdenum, nickel, and tin. Price changes are weighted according to the average value of mine production during the period from 1971 to 1981. For each mineral, this average value was calculated by multiplying the eleven-year average annual mine output by the eleven-year average annual real price. Mine-production figures represent production from market-economy countries and, for all minerals other than molybdenum and gold, were obtained from *Metal Statistics 1971–1981*, 69th ed. (Frankfurt am Main, Metallgesellschaft, 1982). Gold and molybdenum figures were estimated from U.S. Bureau of Mines, *Minerals Yearbook* (Washington, D.C., Government Printing Office, various years).

As an example, consider a hypothetical index with two minerals, A and B. Assume that the average value of mine production for A represents 75 percent and for B represents 25 percent of the total value of mine production in the two-mineral universe. If, in year t, A's real price increases by 10 percent and B's real price increases by 20 percent, then the price index will increase by 12.5 percent, as shown below:

(.75) (10 percent) + (.25) (20 percent) = 12.5 percent.

If the price index was 100 in year $t - 1$, then it would be 112.5 in year t. The following values for the real price index were calculated:

1967—100.00	1975—162.16
1968—103.31	1976—153.50
1969—105.19	1977—158.15
1970—109.47	1978—163.08
1971—102.60	1979—220.11
1972—110.54	1980—266.00
1973—135.16	1981—222.83
1974—182.80	1982—192.64

4

Mineral Exploration in Developing Countries: Botswana and Papua New Guinea Case Studies

CHARLES J. JOHNSON
ALLEN L. CLARK

There are 155 developing countries in the world, of which 78 produce minerals (other than common commodities such as sand, gravel, clay, and salt). Only 30 of these countries contribute 5 percent or more of the world's production of any mineral commodity. Developing countries account for roughly half the total land area of the world and 36 percent of its estimated reserves of forty-six nonfuel minerals (Kursten, 1983). It is generally believed that over the long term an increasing share of internationally traded mineral commodities will come from developing countries as the developed countries deplete their higher-grade mineral deposits.

Fairly dependable information is available on mineral production, reserves, and investments in major mineral projects for most mineral-producing developing countries. However, it is difficult to obtain an accurate estimate of the worldwide distribution of expenditures on mineral exploration. It is even more difficult to know which factors have influenced exploration expenditures in specific developing countries. Estimates of these expenditures can be attempted by examining reports in mining journals, annual reports of major mining companies, and reports by government and international agencies.

Such a global survey probably would show that multinational mining companies have become more selective in deciding which developing countries should receive their exploration dollars. In addition, as some

writers have suggested, the share of private exploration expenditures in
developing countries may have declined (*Mining Annual Review*, 1978;
Mikesell, 1979; chapter 3 in this volume). This decline in part reflects
concerns of mining companies about the stability of investments in many
developing countries and a tendency of some governments to "change the
rules of the game" in their favor after investments are completed. However, this does not mean to imply that governments must maintain static
mineral policies or provide overly generous terms to attract investment.
Indeed, different policies can be successfully employed in different countries, providing they accurately reflect the relative competitive position of
the nation's mineral deposits in an international context and provide an
overall expectation of commercial success to the investor.

This chapter examines in detail two developing countries in different
parts of the world with different mineral policies and mineral possibilities—Botswana, located in southern Africa, and Papua New Guinea (PNG),
located in the southwestern Pacific (figure 4-1). Both countries have in
common: (1) honest, stable governments, (2) modern mineral legislation,
(3) substantial mineral exploration by major multinational mining companies, and (4) major mine developments in recent years. However, each
country has different policies pertaining to prospecting licenses and mining
agreements. In Botswana, prospecting license applications are reviewed at
the ministerial level and normally are issued within two months of submission. In PNG, several months to more than a year may pass before a
decision is made on a prospecting license application. In addition, the
probability of approval in Botswana is substantially higher than in PNG.
With respect to the specific fiscal terms of mining agreements, however,
Botswana follows an ad hoc approach of negotiating each mining agreement to fit specific commercial mineral deposit discoveries. In PNG, fiscal

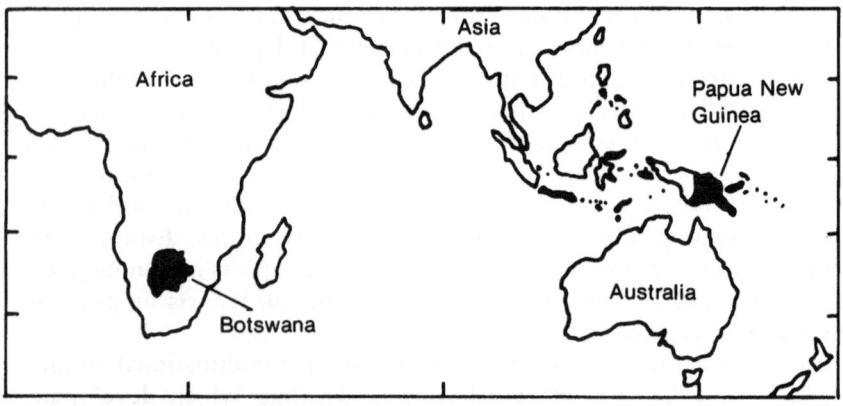

Figure 4-1. Location of Botswana and Papua New Guinea.

terms are established through legislation, which greatly reduces their uncertainty.

These two countries can provide useful insights into how exploration activity has responded to internal and external forces over time. Although neither Botswana nor PNG is a typical developing country, their experiences should be of value to other developing nations in formulating and implementing effective policies pertaining to mineral exploration and development.

Although petroleum exploration is excluded from the analysis, it should be noted that PNG has had active petroleum exploration—resulting in substantial natural gas and significant petroleum discoveries. Botswana has no major petroleum exploration programs.

The chapter is divided into four main sections. The first provides an overview and comparisons of mineral exploration in Botswana and Papua New Guinea, including selected comparisons with Australia and Canada. This is followed by two sections that describe the history of exploration in Botswana and PNG. The concluding section summarizes the findings.

Botswana and PNG in an International Context

As a beginning point in the examination of exploration trends in Botswana and PNG, these countries are compared with Australia and Canada. Australia and Canada are two less-explored developed countries with active exploration where major mineral discoveries continue to occur. These two countries provide an indication of exploration activity resulting from international market forces and represent global exploration activity within stable investment climates. Australia and PNG are located in the same geographic area, and most investors in PNG are from subsidiaries of Australian companies. Mineral exploration expenditure trends in PNG and Australia are expected to be broadly related. The most similar parallel to exploration trends in Botswana might be South Africa; however, due to the limited data available at the time this chapter was prepared, South Africa was not used for comparison. (The reader interested in South Africa is referred to chapter 5 in this volume.)

Figure 4-2 shows the trends in estimated mineral exploration expenditures in Australia, Botswana, Canada, and PNG in constant 1982 U.S. dollars.[1] Australia and Canada have much higher annual exploration expenditures, falling in the range of $200 million to $400 million, whereas Botswana and PNG have expenditures in the range of $2 million to $20 million.

---

[1]Except where otherwise noted, all figures in this chapter have been adjusted to constant 1982 U.S. dollars using the Consumer Price Index.

148

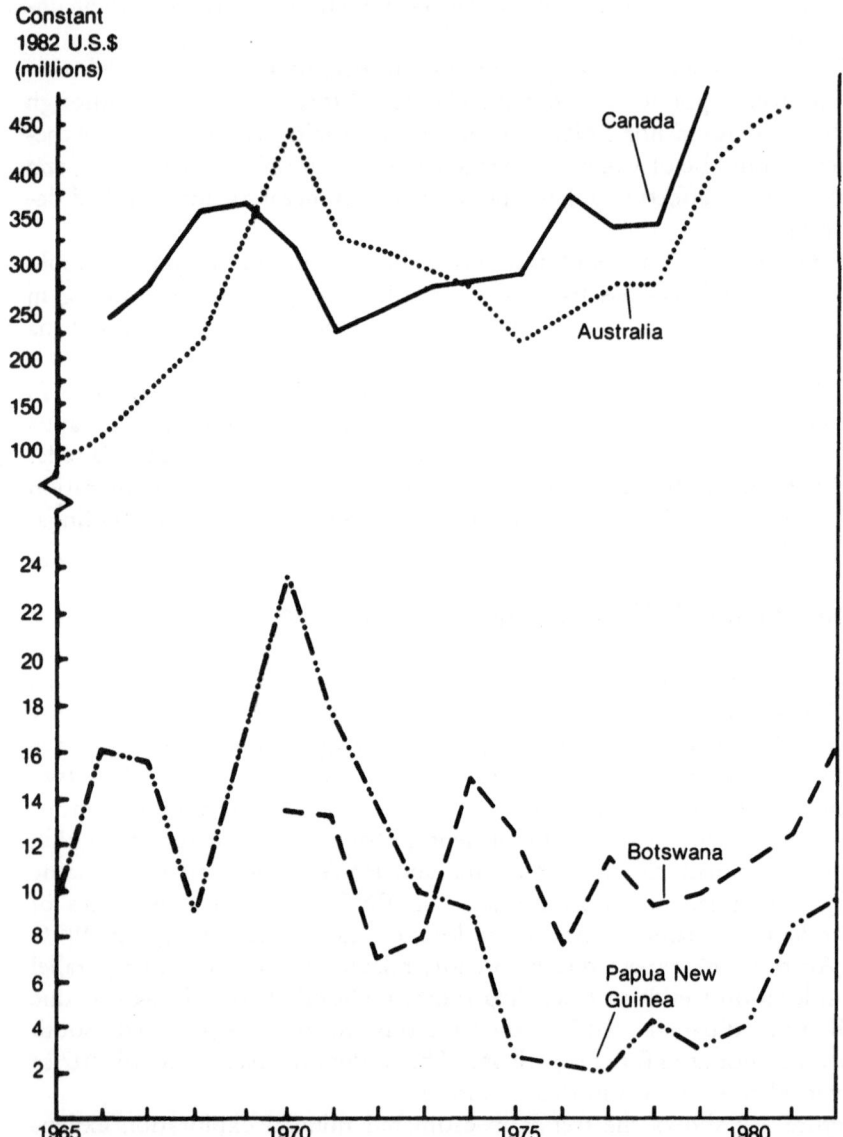

Figure 4-2. Mineral exploration expenditures (excluding those for oil and gas) in Australia, Botswana, Canada, and Papua New Guinea, 1965–82. The Australian data are for fiscal years ending in June. Note that two scales are shown on the expenditure axis; while this procedure is useful in illustrating relative trends and the correlation in changes in the direction of trends, it tends to produce closer apparent fits in trends than is actually the case. *Source:* Derived from unpublished government data.

The overall trends in exploration activity for all four countries show some similarities, most notably the high level of exploration expenditures in the period between 1969 and 1971, when worldwide active exploration for metals was taking place. Low levels in exploration expenditures occurred during 1976 and 1977 in Australia, Botswana, and PNG. All four countries experienced a resurgence in exploration after 1979, each for different reasons that will be discussed.

The four countries are quite different in size and are at different stages of their exploration history. To remove some of the distortion caused by the major size differences, exploration expenditures per square kilometer (km^2) are shown in figure 4-3. On a square-kilometer basis, the differences in expenditures among the four countries are reduced substantially. Expenditures in Australia and Canada, which are both at advanced stages of minerals exploration, have remained higher than in Botswana and PNG for much of the period examined. PNG had high expenditures paralleling Australia in the late 1960s, reflecting in part the high level of exploration and drilling activity for porphyry copper deposits in PNG following the major discovery on Bougainville Island. However, exploration in PNG was substantially below the other three countries between 1975 and 1980. This is attributable to a combination of factors, including depressed copper prices and hence low levels of exploration for porphyry copper deposits in the late 1970s, uncertainties surrounding the nation's independence from Australia in 1975, and the withdrawal of Kennecott from PNG after the failure to reach an agreement with respect to the large porphyry copper-gold deposit at Ok Tedi. In spite of major political instability in adjacent countries, exploration in Botswana was more than double the level in PNG during the late 1970s. This is the result of continued major investments in diamond and coal exploration, which offset major decreases in metals exploration. By the mid-1980s, however, exploration expenditures in PNG were exceeding those in Botswana.

Land Areas Covered by Prospecting Licenses

Figure 4-4 shows the trends in land areas covered by prospecting licenses in Botswana, Canada, and Papua New Guinea (similar information on Australia was not available). Both in Botswana and Papua New Guinea, large areas were covered by prospecting licenses in the early 1970s—a maximum of 38 percent (177,000 km^2) of the land area of PNG in 1970 and 57 percent (342,000 km^2) of Botswana's in 1973. By the late 1970s, however, exploration areas covered in Botswana and PNG had declined dramatically to the 20,000- to 45,000-km^2 range, slightly below the levels in Canada. The large amounts of land held under prospecting licenses in Botswana and PNG at the beginning of the 1970s reflect (1) the large areas available for prospecting licenses, (2) the reconnaissance level of explo-

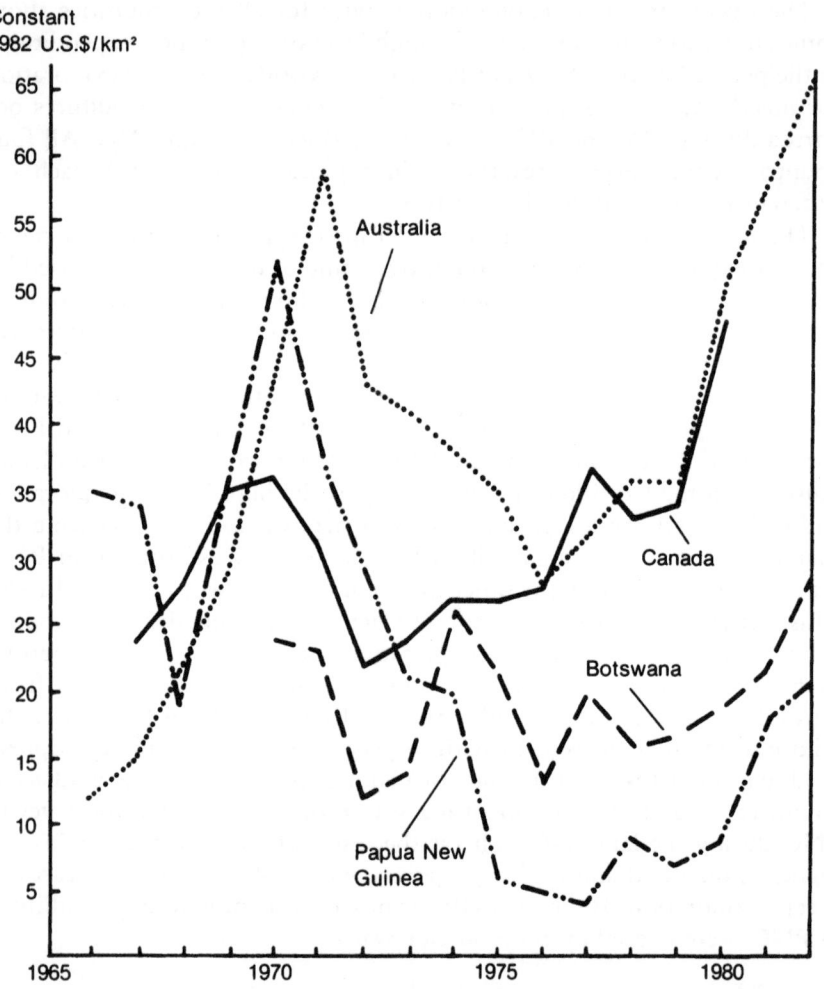

Constant
1982 U.S.$/km²

Figure 4-3. Exploration expenditures (excluding those for oil and gas) per square kilometer in Australia, Botswana, Canada, and Papua New Guinea, 1965–82. The Australian data are for fiscal years ending in June. *Source:* Derived from unpublished government data.

ration over much of the areas held (an indication of the early phases of exploration in both countries), and (3) expectations that both countries contained large undiscovered mineral deposits.

A comparison of figures 4-2 and 4-4 shows that trends in areas covered by prospecting licenses do not necessarily parallel trends in exploration expenditures. In PNG, there is moderate correlation between the trends

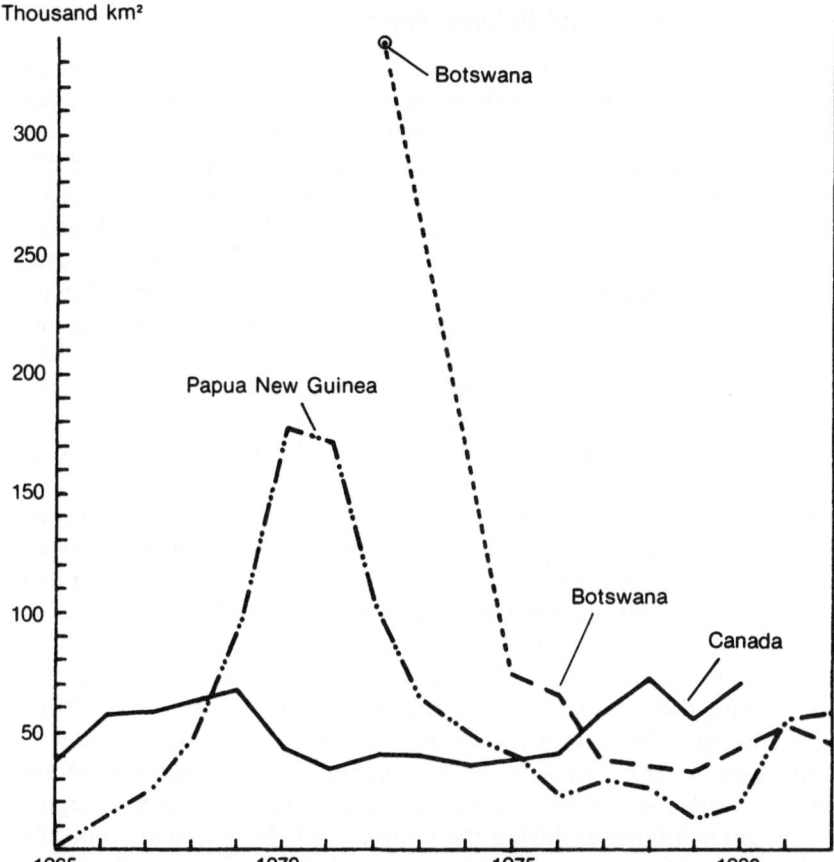

Thousand km²

Figure 4-4. Land area in Botswana, Canada, and Papua New Guinea covered by exploration licenses, 1965–82. Broken line for Botswana indicates that data are not available. *Sources:* Australian Bureau of Mineral Resources, Geology, and Geophysics, *Australian Mineral Industry* (Canberra, Australia, Bureau of Mineral Resources, various years); Energy, Mines and Resources Canada, *Canada Mines: Perspective from 1983*, MR 200 (Ottawa, Canadian Government Printing Center, 1984); Botswana and Papua New Guinea figures derived from unpublished government data.

in land areas covered by prospecting licenses and exploration expenditures. In Botswana, however, the trend in land areas covered by prospecting licenses has been downward while the trend in total exploration expenditures has been rising. The Botswana trends reflect the medium-term impact of legislative changes in 1976 and increased intensity of exploration for identified mineral deposits.

License Expenditures and Relinquishments

Two important concerns for the governments of developing countries are that companies (1) actively explore areas under license, and (2) do not retain licenses for long periods for speculative purposes.

An examination of 138 licenses issued in Botswana after 1975 and 378 licenses in PNG after 1959 indicates that most companies relinquish licenses rapidly. As shown in figure 4-5, between 40 and 50 percent of licenses are relinquished within one year, increasing to about 77 percent in Botswana and 84 percent in PNG within three years. The more rapid relinquishment of licenses in PNG may reflect a combination of factors, including (1) a shorter initial license period in PNG (two years versus three in Botswana), (2) generally higher minimum expenditure commitments in PNG, and (3) differences in exploration targets and exploration strategies (porphyry copper-gold deposits are usually detected early in exploration in PNG).

With respect to expenditures on land areas covered by licenses, in Botswana they have varied between about $250 and $350 per square kilometer subsequent to passage of more restrictive minerals legislation in December 1976. Excluding coal, exploration costs averaged about $200 per square kilometer over the 1979 to 1982 period.

Exploration expenditures in PNG averaged about $140 per square kilometer during the period from 1970 to 1982. Both the Botswana and PNG figures include substantial expenditures for drilling on known mineral deposits and therefore are substantially higher than expenditures for exploration at earlier stages. Grass-roots exploration in PNG averaged roughly $50 per square kilometer during the same period. New licenses typically have been issued for areas that average around 1,000 km^2 in both Botswana and PNG, and, assuming that PNG's expenditures are representative for grass-roots exploration, the average expenditure per year per license for grass-roots exploration is $50,000. This provides a general indication of what governments might expect at the grass-roots stage of exploration.

The cost per discovery in developing countries versus that in developed countries is not easily documented. With respect to Botswana and PNG, very large discoveries were made during the 1960s after an expenditure of a few million dollars. The total exploration expenditure in each of the countries is estimated to be between $175 million and $200 million from 1960 to 1982. If only proven commercial discoveries are considered, approximately $100 of in-place minerals have been discovered per dollar expended in Botswana (valued at 1982 commodity prices); in PNG, minerals valued at $150 have been discovered per exploration dollar spent.

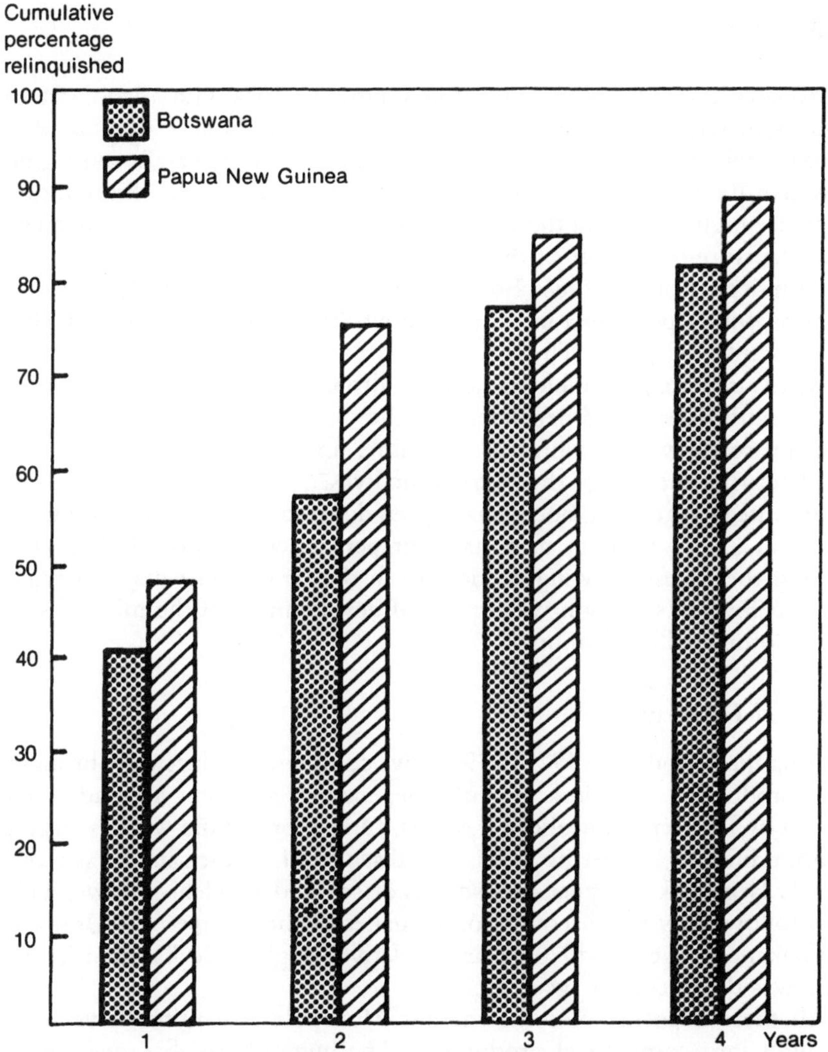

Cumulative
percentage
relinquished

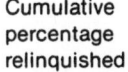

Figure 4-5. The average percentage of land held through prospecting licenses that was relinquished after one, two, three, and four years in Botswana and Papua New Guinea. Botswana includes 138 prospecting licenses issued between 1975 and 1982, and Papua New Guinea includes 378 prospecting licenses issued between 1959 and 1982. *Source:* Derived from unpublished government data.

Botswana

Botswana is a landlocked country in southern Africa with a land area of approximately 600,370 km² and a population of about 1 million. Approximately 80 percent of its land is covered by sand, generally averaging less than 100 meters in thickness (Reeves and Hutchins, 1982). Exploration costs are high because of the extensive sand cover and the lack of adequate infrastructure over most of the country.

At independence in 1966, Botswana ranked among the world's poorest nations with a per capita income of about $50 in current terms and $150 in constant 1982 terms. The nation's only export of significance was beef, which was both vulnerable to periodic droughts and outbreaks of hoof-and-mouth disease and dependent on the distant European market. The mineral sector was marginal at that time, and the existing small mines were nearing the end of their economic lives.

Seventeen years later Botswana's gross domestic product was estimated at about $1,000 million, or $1,000 per person, which placed Botswana among the fastest-growing economies in the world. The major driving force behind this growth was the rapid development of the mineral sector—dominated by diamonds.

Mineral History

During the period from 1971 to 1982, five commercial mines were brought into production. Within a year of independence and after a decade-long search by DeBeers Consolidated Mines, Orapa, one of the world's largest diamondiferous kimberlite pipes, was discovered. In the same year a relatively large nickel-copper deposit was discovered at Selebi-Phikwe. This was followed by the discovery of a smaller diamond pipe near Orapa at Letlhakane. In the mid-1970s, a large, relatively high grade diamondiferous pipe was discovered at Jwaneng.

All these deposits, plus the modest Morupule coal mine, have been brought into commercial production. The huge Orapa diamondiferous pipe and the associated town and infrastructure were developed in a mere two years, with production beginning in 1971. The Morupule coal mine began production in 1973, followed by the Selebi-Phikwe nickel-copper mine in 1974, the Letlhakane diamond mine in 1977, and the large Jwaneng diamond mine in 1982. Under active exploration by private companies in 1983 were numerous diamondiferous pipes, a modest copper deposit, major coal deposits, and small-lode gold deposits.

Mineral Policies and Legislation

Under Botswana's existing Mines and Minerals Act, enacted in December 1976, the government's policy has been to encourage active private-sector exploration, development, and operation of mines. Botswana's Geological Survey as well as its Department of Mines provides substantial assistance to private companies undertaking exploration and development of mineral properties. Prospecting licenses normally are approved within two months, providing that an acceptable exploration program is submitted with the application.

Private companies have pointed out two weaknesses in the Mines and Minerals Act. First, some have suggested that under the law the minerals minister has too much flexibility in setting the conditions for converting from a prospecting license to a mining lease. Second, the specific fiscal terms that will apply to a mining development usually have not been agreed upon before the discovery of a commercial mineral deposit. However, as of 1982, the government had been able to negotiate mutually acceptable terms with all companies wanting to develop a mineral deposit. Therefore, apparent weaknesses in the minerals legislation appear to have been overcome at the negotiating table.

For all minerals except diamonds, the Botswana government takes a minority equity position of 15 to 25 percent. In the special case of diamonds, the government and DeBeers are equal equity and voting partners. This reflects the government's recognition of the critical importance of diamonds to Botswana's economy as well as to DeBeers, which largely controls world rough-diamond sales to provide market stability. The fifty-fifty partnership between the government and DeBeers motivates them to find mutually acceptable solutions to their problems.

The impact of Botswana's ad hoc approach to the fiscal terms for mining developments is difficult to quantify. It can be argued that the absence of clearly specified fiscal terms has resulted in less exploration than otherwise would have occurred. It is not possible to evaluate this issue fully. It can be said, however, that because of Botswana's unique mineral situation and stable government, substantial levels of investment have been maintained under its existing system.

Both the Botswana government and foreign investors have initiated the renegotiation of mining agreements. In the early 1970s Botswana initiated the renegotiation of the Orapa mining agreement to increase both its profits and equity participation. This action was prompted because Orapa began producing much higher profits than had been discussed during negotiation of the original agreement.

The foreign investors in the BCL Limited copper-nickel mine have initiated renegotiations on three occasions. The renegotiations occurred as a result of technical start-up problems with the mine and smelter followed by falling metal prices that required financial restructuring of the agreements. A 1983 study of foreign mining projects concludes: "It should be noted that the financial difficulties encountered by BCL and the major shareholders were in no way attributable to the failure of the Botswana government to honor its agreements" (Mikesell, 1983).

Exploration Trends

Figure 4-6 shows the total annual expenditures on mineral exploration in Botswana in current and constant dollars from 1970 to 1982. The trend in current terms has been upward, from the low of $3 million in 1972 to approximately $16 million in 1982. The trend in constant 1982 dollars was slightly downward during the 1970s, turning sharply upward in the early 1980s as a result of a rapid expansion in coal exploration. Wide swings in the level of exploration are readily apparent in the figure. In an attempt

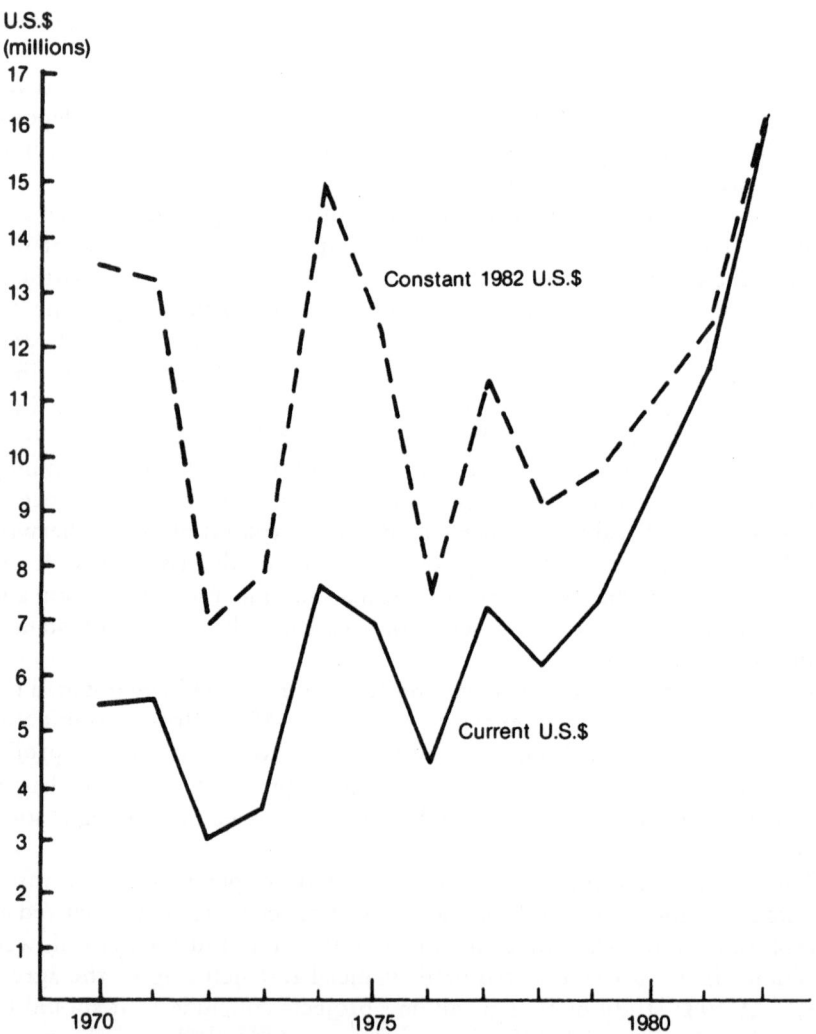

Figure 4-6. Mineral exploration expenditures in Botswana, 1970–82. *Source:* Derived from unpublished government data.

to interpret these swings, expenditures have been disaggregated into four main categories—diamonds, metals (primarily nickel and copper), coal, and radioactive minerals. These are shown in figures 4-7 through 4-10. The only significant industrial mineral exploration expenditure has been associated with the periodic evaluation of a large soda ash deposit at Sua Pan, which is not examined in this chapter.

Because of the confidentiality of some data from private companies, expenditure figures are not shown in dollars but are set to the base year 1980 = 1.0. Each year's exploration expenditure in 1982 dollars was divided by expenditures in 1980 to produce the ratio shown in the four figures.

Diamond exploration. Of all the countries in the world, Australia and Botswana had the best prospects during the 1970s for private-sector exploration for diamondiferous pipes. It is not surprising, then, that the

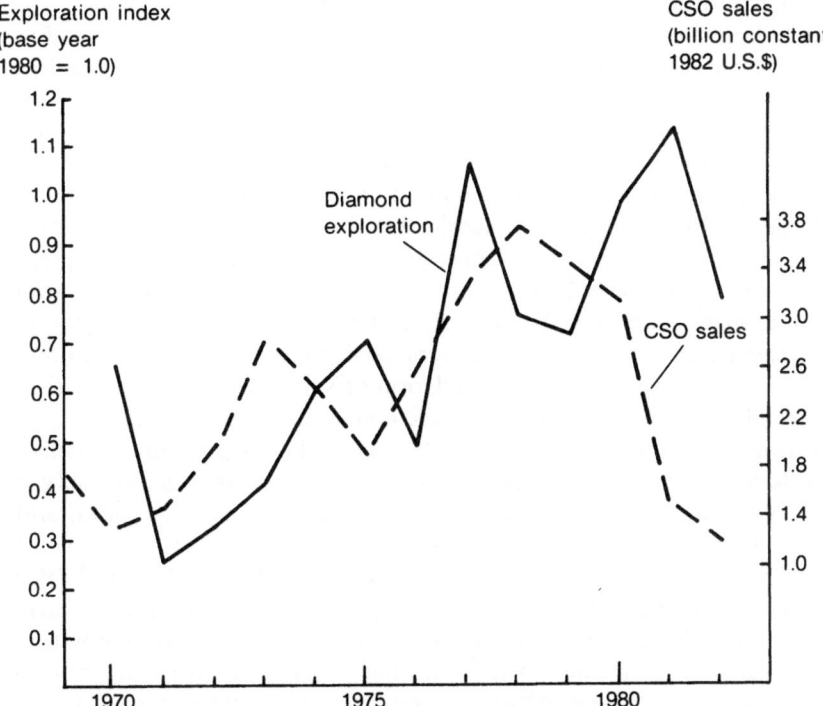

Figure 4-7.   Diamond exploration expenditures in Botswana, 1970–82, shown as the ratio of expenditures for each year divided by the expenditures in 1980 (constant 1982 U.S. dollars). The trend in Central Selling Organization (CSO) diamond sales is shown for comparison. *Source:* Derived from unpublished government data.

world's dominant diamond company, DeBeers, has maintained an active program in Botswana. In addition, active programs were carried out by Debswana, the fifty-fifty government/DeBeers mining company, to evaluate known deposits around existing mines. Beginning in the mid-1970s, Falconbridge, an important Canadian nickel-copper producer, began to explore actively for diamonds in Botswana. The Botswana government's view appears to have been that active competition in diamond exploration would ensure rapid discovery of the remaining undiscovered diamondiferous pipes. A large number of kimberlite pipes were found during this competitive period of exploration.

Figure 4-7 shows an upward trend with substantial annual swings in diamond exploration in Botswana during the period from 1970 to 1982. The peaks in 1970 and 1977 are related to detailed sampling of specific diamondiferous pipes. As 80 to 90 percent of world rough diamond production is sold through the DeBeers' subsidiary, the Central Selling Organization (CSO), it is assumed that CSO sales reflect world market conditions. Figure 4-7 shows the trend in CSO sales for comparison with the trend in diamond exploration in Botswana.

Excluding the exploration peaks in 1977 and 1981, the overall trend in exploration expenditures shows moderate correlation with the trend in CSO sales, allowing for a one- to two-year lag for exploration expenditures to respond to changes in CSO sales. Although not shown in figure 4-7, the trend in CSO sales in current dollars follows roughly the same trend as in constant dollars. The high exploration peak during 1980 and 1981 appears to be the result of accelerated diamond exploration by competing firms, prompted by the discovery of a substantial number of pipes in Botswana in the late 1970s, combined with the response to diamond shortages that occurred at the same time. The downturn in exploration in 1982 may be attributed partly to depressed conditions in the diamond market and, perhaps, to the consolidation of some exploration activities.

Before 1982, diamond exploration accounted for the largest share among the various exploration programs in Botswana. The overriding influences upon diamond exploration levels in Botswana are (1) that geologically it is a very favorable exploration area for large diamondiferous pipes, and (2) that DeBeers' long-term strategy has been to maintain effective control of rough diamond supplies to world markets. In 1983, 1984, and 1985 Botswana diamonds represented 19 to 20 percent of world supplies, clearly an important element in DeBeers' global strategy. Diamond exploration in Botswana has continued in spite of major changes in government diamond policies that substantially increased the government's share of profits and level of control in mining operations.

Coal exploration. For almost a century, major coal deposits have been known to exist in Botswana. Significant exploration by the Botswana

Geological Survey took place in the 1950s, and the first coal mine at Morupule began to produce in 1973. In the early 1980s, Morupule was producing about 400,000 metric tons per year of steam coal to meet domestic requirements.

Immediately after the 1973 petroleum energy crisis, Shell Coal commenced a major coal exploration program in Botswana. Shell remained the dominant company involved in coal exploration until the early 1980s when BP Coal Limited and Charbonnages de France actively entered the coal exploration scene in Botswana.

Although Botswana's total coal resources are not known, they may prove to be the largest of any developing country with the exception of China and India. Coal seams 7 to 10 meters thick are common, and resource estimates of up to 100 billion metric tons have been reported (World Bank, 1979), although Botswana government estimates for proven and indicated resources have been more modest—about 17 billion metric tons. The raw coal is relatively high in ash and sulfur content and does not have coking properties; however, washing produces a suitable export-quality steam coal with 0.5 to 1.0 percent sulfur and 11 to 13 percent ash.

Botswana's large coal resources are offset by its remote inland location. Only very large mining operations are likely to produce coal that is competitive in international markets. Therefore, coal exploration is dominated by very large companies such as Shell and BP that could develop projects having capital costs in excess of a billion dollars.

Most coal exploration during the period from 1970 to 1982 has focused on evaluating known occurrences of coal to determine the highest-quality deposits. Exploration has been dominated by drilling, resulting in high annual expenditures relative to exploration for other minerals in Botswana.

The driving force behind accelerated exploration in Botswana has been high energy prices, of which petroleum is the leading factor. Figure 4-8 shows the trends in both coal exploration in Botswana and world crude petroleum prices from 1970 to 1982. The year after the 1973 petroleum crisis and the associated large petroleum price rise, Shell Coal launched a major coal exploration program in Botswana. The second large petroleum price rise in 1979 brought three more coal exploration companies into Botswana. In 1982 there were three active exploration programs in Botswana aimed at evaluating the potential for large coal developments for export—on the order of 5 million metric tons per year. Major developments of coal are most likely to be associated with the development of a major railroad to the west through Namibia. The development of such a railroad might accelerate exploration for mineral commodities with a lower unit value such as soda ash, chromite, manganese, and asbestos.

Metal exploration.   Metal exploration in Botswana historically has been dominated by nickel and copper with much less interest in manganese,

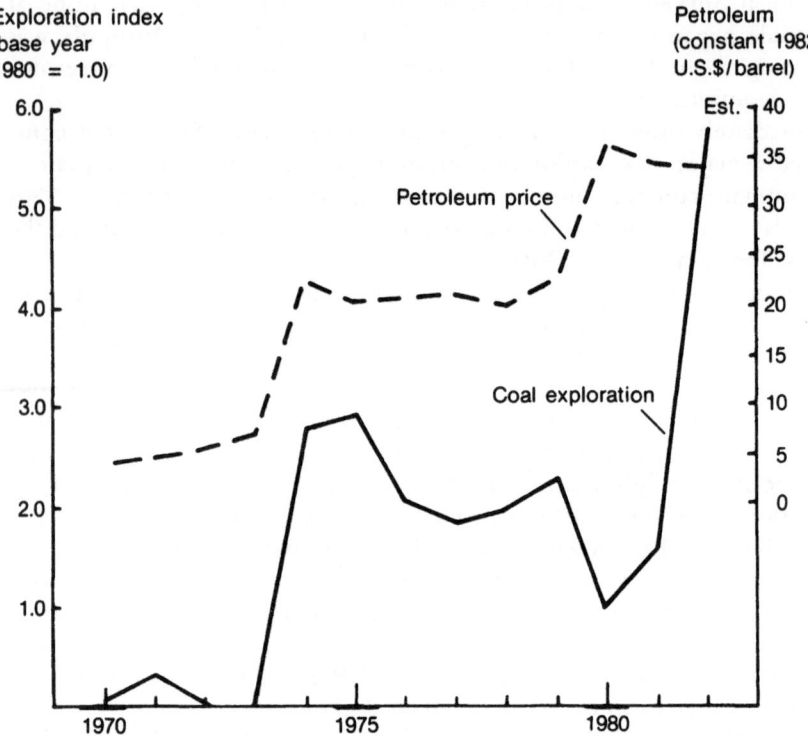

Figure 4-8. Coal exploration expenditures in Botswana, 1970–82, shown as the ratio of expenditures for each year divided by the expenditures in 1980 (constant 1982 U.S. dollars). The trend in petroleum prices is shown for comparison. *Source:* Derived from unpublished government data.

gold, and other metals. Figure 4-9 shows the trends in metal exploration and copper prices. The trend in exploration follows copper prices except in the period from 1970 to 1971 when exploration activity was high—a reflection of increased interest in Botswana following the decision to develop the nickel-copper deposit at Selebi-Phikwe and the high level of exploration worldwide for copper and nickel in the early 1970s.

Overall exploration activity in base metal exploration in Botswana has remained low for most of the post-1975 period. This is a result of the worldwide trend away from active exploration for nickel-copper sulfide deposits, due to depressed prices and low profitability, and a downgrading of the geologic potential for major nickel-copper deposits. Increased ex-

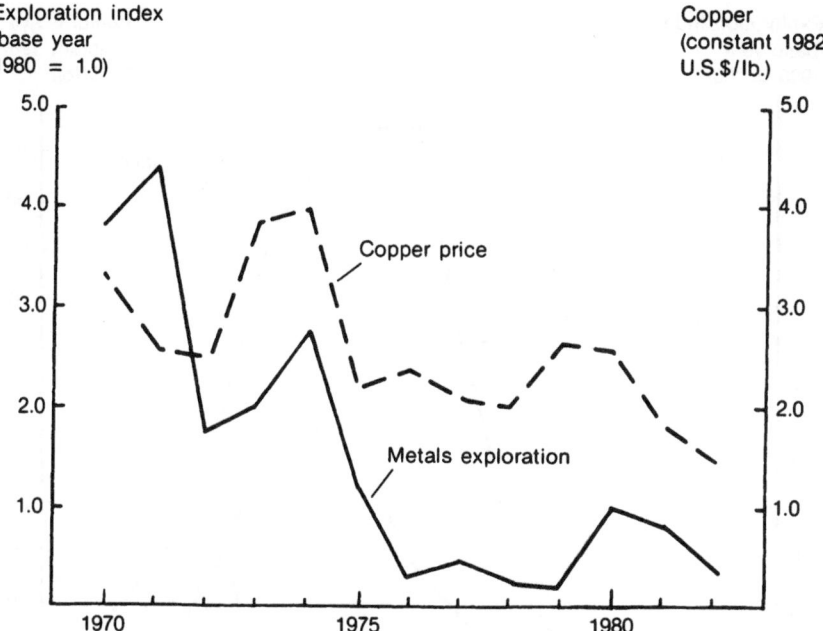

Figure 4-9.   Metals exploration expenditures in Botswana, 1970–82, shown as the ratio of expenditures for each year divided by the expenditures in 1980 (constant 1982 U.S. dollars). The trend in copper prices is shown for comparison. *Source:* Derived from unpublished government data.

ploration activity for gold in the early 1980s was undertaken by very small companies with a focus on reopening abandoned gold mines near Francistown.

Uranium exploration.   Botswana is not viewed by the mining industry as highly prospective for commercial uranium deposits, and only modest levels of exploration took place during the period from 1970 to 1982. During the late 1970s, Botswana attracted modest exploration activity in response to high uranium prices and increased worldwide exploration. Following the downturn in uranium prices in 1980, Botswana uranium exploration became a casualty of the worldwide contraction in exploration.

Figure 4-10 shows the trend in uranium exploration in Botswana and, for comparison, the trends in uranium prices and exploration levels in the United States. Uranium exploration in Botswana appears to be related to the rapid increase in prices and exploration activity in the late 1970s and the subsequent rapid fall in prices.

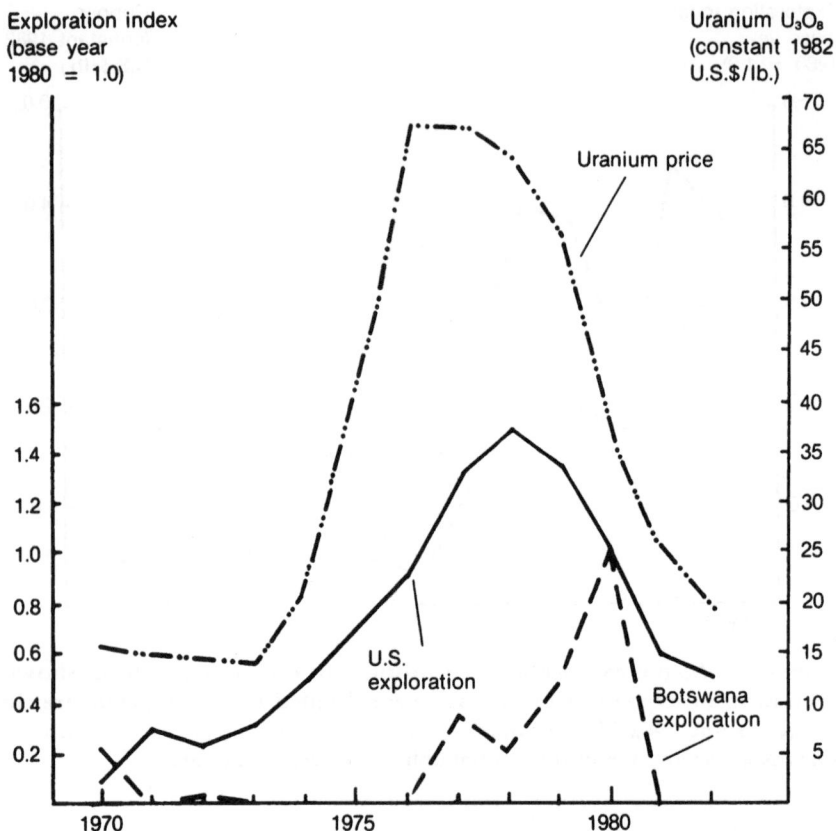

Exploration index
(base year
1980 = 1.0)

Uranium U₃O₈
(constant 1982
U.S.$/lb.)

Figure 4-10. Uranium exploration expenditures in Botswana, 1970–82, shown as the ratio of expenditures for each year divided by the expenditures in 1980 (constant 1982 U.S. dollars). The trends in uranium prices and the exploration index in the United States are shown for comparison with Botswana. *Source:* Derived from unpublished government data.

Profiles of Exploration Activities

Number of exploration companies.   Figure 4-11 shows that a small number of companies were engaged in exploration in Botswana during the period from 1970 to 1982. The average number was about twelve during the 1970s and increased to eighteen in the early 1980s. In comparison, when only those companies spending at least $25,000 per year are included, the total number decreases from a high of twelve in 1975 to five in 1979 before increasing to nine in 1981.

The jump in total number in the early 1980s reflects renewed interest on the part of small companies in abandoned gold mines in eastern Bot-

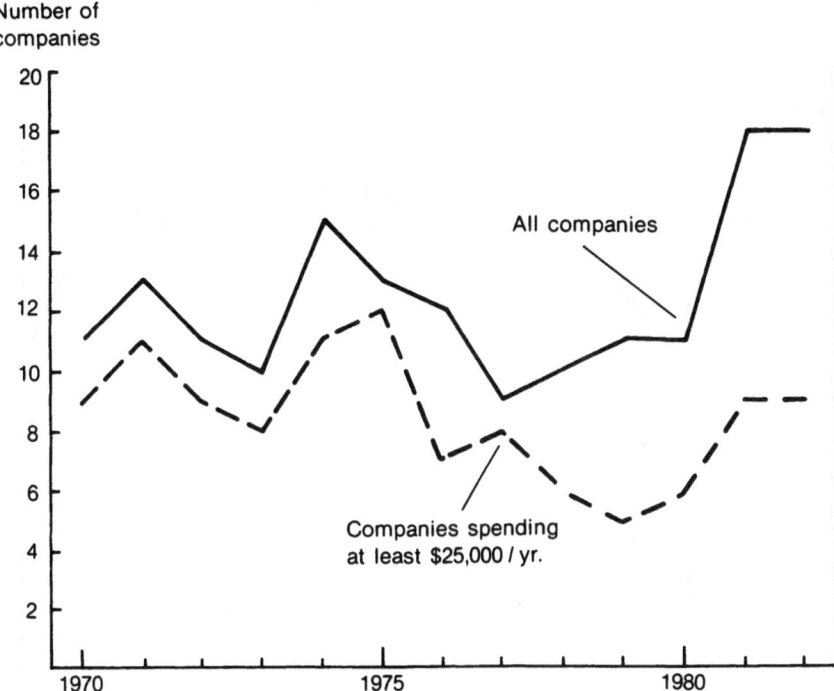

Figure 4-11. Total number of companies exploring in Botswana, and number spending \$25,000 (constant 1982 U.S. dollars) or more, 1970–82. *Source:* Derived from unpublished government data.

swana. Small-company activities in gold are not expected to increase significantly the level of mineral exploration or the total value of minerals produced in Botswana.

Figure 4-12 shows that in terms of total exploration, the largest three exploration programs account for 60 to 90 percent of total expenditures during the period from 1970 to 1982. Between 1976 and 1981 they accounted for more than 80 percent of exploration. The decline to about 60 percent in 1982 reflects the entry of two companies with relatively large coal exploration programs.

The three dominant companies have varied among seven major companies over the period, and no one company has led consistently in exploration expenditures. Botswana is not a country where small mineral deposits, with the exception of gold, are sufficiently attractive to support small mining companies. Because exploration targets are relatively large and costs are high, primary interest comes from large multinational companies.

Percent
of total
expenditures

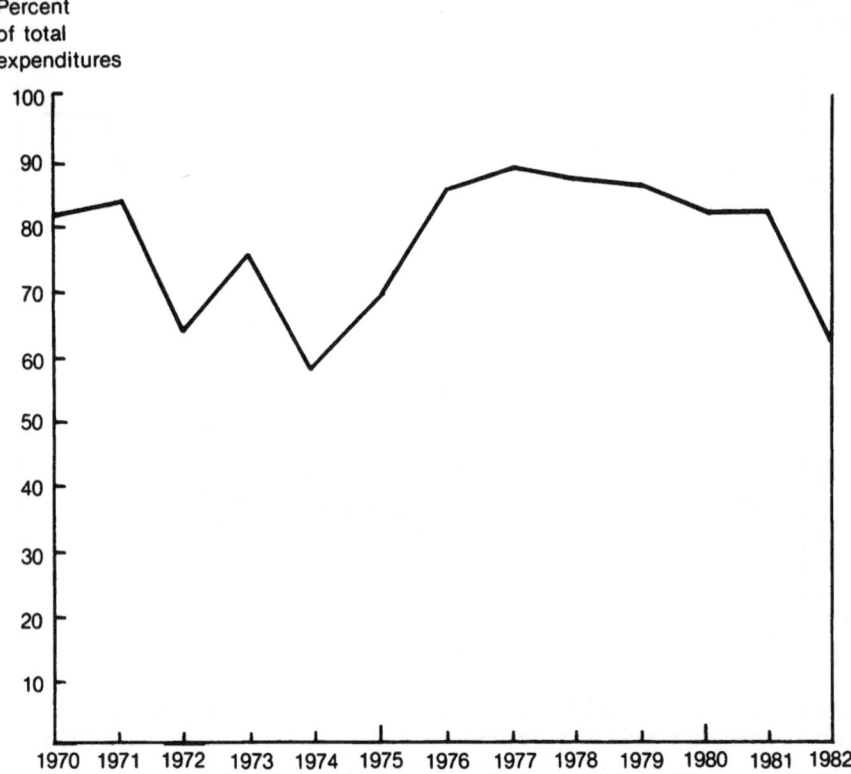

Figure 4-12.   Share of total exploration expenditures in Botswana of the three largest companies, 1970–82. *Source:* Derived from unpublished government data.

Prospecting licenses.   The number of prospecting licenses tripled from twenty-four in 1976 to seventy-two in 1982. This increase resulted from the intensity of exploration for diamonds and coal, and to a lesser extent for gold.

As shown in figure 4-13, the average size of licenses decreased from a high of about 2,800 km² in 1976 to about 640 km² in 1982. The major reason appears to be a provision of the mines and minerals legislation, enacted in December 1976, that limited all new prospecting licenses to a maximum size of 1,000 km². The figures for the early years reflect the large licenses issued prior to the 1976 legislation (*Botswana Government Gazette*, 1976). Most companies now apply for licenses of approximately 1,000 km². The smaller average license size shown for the later years is a result of license renewals that include a mandatory relinquishment of a minimum of 50 percent of the license area and small license areas issued

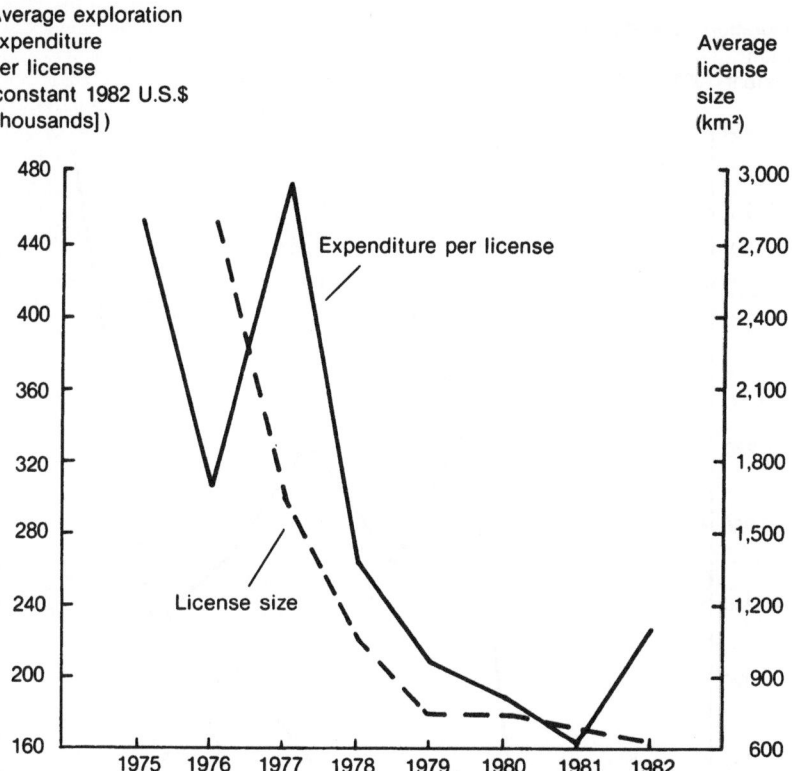

Figure 4-13. **Average exploration expenditure per license and average license size in Botswana, 1975–82.** *Source:* Derived from unpublished government data.

for gold to small companies. Licenses usually are issued for three years with an option for two renewals of up to two years each.

Figure 4-13 also shows that average expenditures per license have decreased in line with decreased average license sizes over most of the period. Average annual expenditures per license have varied from $453,000 in 1975 to $164,000 in 1981 and $227,000 in 1982.

To standardize the comparisons and remove the effects of large license areas, figure 4-14 shows expenditures per square kilometer; these have ranged from $233 to $356 over the period from 1977 to 1982. In 1976, however, expenditures were only $109 per square kilometer. The jump after 1976 is largely the result of the legislative reduction in license size. The major jump in 1982 reflects high expenditures in coal drilling and evaluation programs.

Expenditures in coal exploration are higher than for other minerals because of the higher share of drilling in coal programs. Figure 4-14 shows

Exploration expenditures
per km²
(constant 1982 U.S.$)

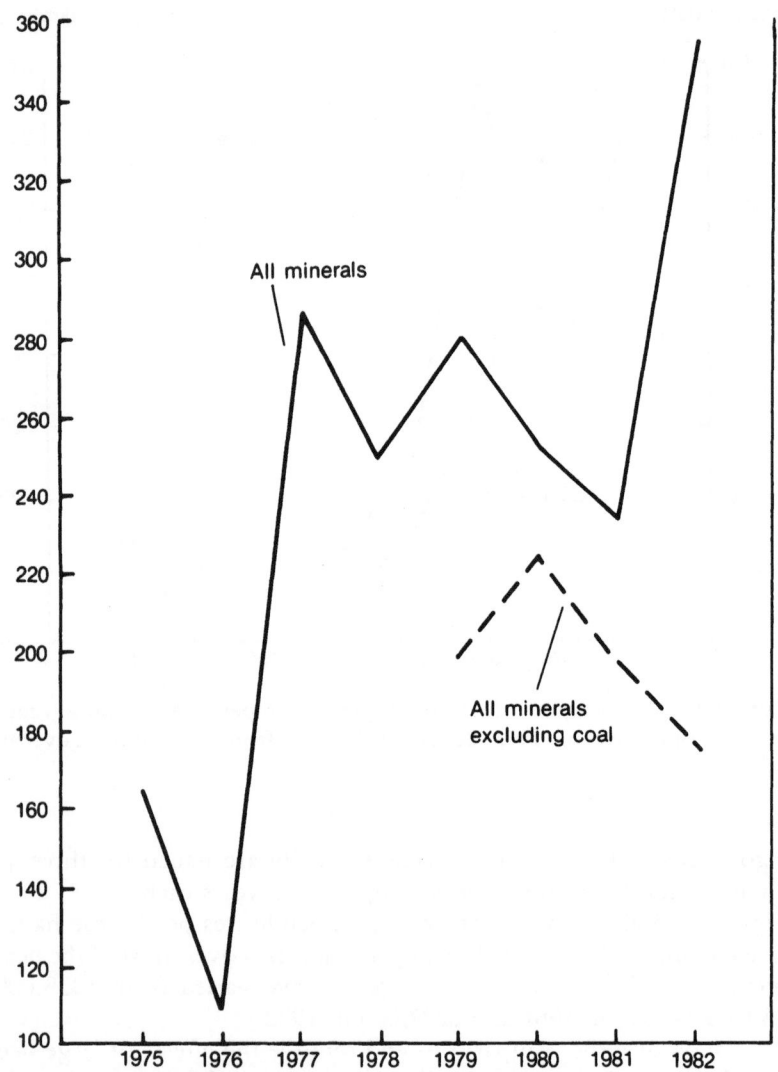

Figure 4-14. Exploration expenditures per square kilometer in Botswana for all minerals and for all minerals excluding coal, 1975–82. *Source:* Derived from unpublished government data.

expenditures per square kilometer—excluding coal. These average about $200 and have ranged from $174 to $224 between 1979 and 1982.

Botswana Exploration in Perspective

Botswana has approached the exploration and development of mineral resources largely from a commercial perspective with a minimum of the national rhetoric about control of multinational mining companies popular in many other developing countries in the 1970s. Botswana's goals have been to encourage efficient exploration and development of its mineral resources by multinational mining companies and to capture a substantial share of the economic rents resulting from commercial mine development.

Botswana encourages exploration through its straightforward system for submission and review of prospecting license applications. Potential investors have ready access to extensive files of geologic information at the Botswana Geological Survey Department, which is serviced by an efficient staff of professional geologists. License applications are reviewed and usually approved within two months of submission, providing that a reasonable exploration program is submitted by a reputable mining company. There are no significant fees to obtain a prospecting license, and no active exploration program had been terminated by the government as of 1982. Paralleling the private-sector exploration program have been a number of bilaterally funded programs for regional exploration, particularly in the least-explored areas of the country. The primary countries involved with bilateral aid in exploration are Canada, the Federal Republic of Germany, Japan, and the United Kingdom. The results of these programs normally have been made available to multinational mining companies.

Neither government promotion nor government policy appears to have attracted investors to Botswana. Rather, most appear to have selected Botswana based on regional or global exploration strategies for specific mineral commodities in countries considered to be politically stable and open to private investment.

Botswana's overall exploration trends for specific mineral commodities appear to have been influenced more by international commodity prices than by Botswana's mineral policies. Its ad hoc policies toward the fiscal terms in mining agreements may have had a significant negative impact on exploration for those mineral commodities where Botswana is less likely to have major mineral deposits (that is, copper, nickel, and uranium). However, for the major mineral commodities (diamonds and coal), long-term global strategies of major corporations appear to have overridden concerns about ad hoc fiscal arrangements. Investors willing to make a major exploration commitment (measured in millions of dollars) have obtained certain "good faith" assurances from the government that they would be treated fairly with respect to future negotiations. This strategy has sustained mineral developments in Botswana for more than a decade. Based on mineral projects under evaluation in 1982, it appeared likely that commercial development would continue—an uncommon record for much of the developing world.

Papua New Guinea

Papua New Guinea is located in the southwest Pacific, directly north of Australia and east of Indonesia. It has a land area of approximately 462,000 km² and a population of roughly 3 million. During the 1970s the nation took its place in the middle-income-country category with a GNP per capita of about $900.

For the most part PNG is a land of rugged, jungle-covered mountains as well as lowland swamps and meandering rivers, which, with the heavy rainfall and numerous offshore islands, create communication and logistical difficulties that greatly increase the cost of mineral exploration and development.

Prior to independence in 1975, PNG was administered for more than half a century by Australia. This has resulted in close and continuing economic ties with Australia.

Mineral History

Mining in PNG dates back to the 1870s when a minor gold rush took place near Port Moresby, the current capital of PNG. In the early 1900s, small-scale copper mining took place near Port Moresby. The first major placer gold discovery occurred near Wau in 1926, followed by a number of other gold discoveries including those at Edie Creek and Bougainville Island. By 1940, eight dredges were operating in the Wau-Bulolo area, plus several underground mines, producing a pre-World War II high of 8,447 kilograms (kg) of gold (worth approximately $122 million, assuming a 1982 gold price of $450 per ounce). Papua New Guinea clearly ranked as an important gold producer by world standards in the late 1930s.

Gold mining ceased in PNG during World War II and never recovered to prewar levels until the development of the major porphyry copper-gold deposit on Bougainville Island in the early 1970s.

Modern private sector exploration in PNG began about 1964 when Conzinc Riotinto of Australia (CRA) undertook a regional geochemical sampling program to locate porphyry copper deposits in the southwest Pacific. This exploration program resulted almost immediately in the discovery on Bougainville Island of the largest commercial porphyry copper-gold deposit in the Australia/Southeast Asia region.

This single discovery changed mining companies' perceptions of Papua New Guinea from that of an aging gold-mining province to a major

Table 4-1.   Important Mineral Discoveries in Papua New Guinea, 1960–83

Deposit	Size (million metric tons)	Date of discovery	Grade and metal	Gross in-place metal value (billion $)
Panguna (Bougainville)[a]	944	1964	0.48% Cu, 0.56 g/t Au, 3.0 g/t Ag	12.5
Mount Fubilan (Ok Tedi)[b]	34	1968	2.87 g/t Au (cap)	
	351		0.7% Cu, 0.6 g/t Au, 0.011% Mo	5.8
	25		1.17% Cu	
Lihir Island[c]	137	1983	2.66 g/t Au	3.6
Frieda River[b]	500	1966	0.5% Cu, 0.28 g/t Au	7.5
	260		0.4% Cu, 0.23 g/t Au	
Yanderra[a]	338	1965	0.42% Cu, 0.018% Mo	2.7
Porgera[b]	59	1964	3.56 g/t Au, 14.4 g/t Ag	2.3
Ramu River[b]	80	1968	1.14% Ni, 0.16% Co, 8% chromite	7.2
Total in-place metal value				41.6

Note: Includes only published data for deposits with an in-place metal value of at least $1 billion. The economically recoverable value of metals is smaller than the in-place values shown in the table. Assumed long-term prices in 1982 constant dollars are chromite, $65/metric ton; cobalt, $8/lb.; copper, $0.70/lb.; gold, $311/troy oz.; molybdenum, $3.75/lb.; nickel, $2.25/lb.; and silver, $7/troy oz.
[a]K. W. Doble, "Regional Geological Mapping and Mineral Exploration Activity in Papua New Guinea." Paper presented at conference on Resource Potential and Implications of Mineral Development in the South Pacific, Honolulu, Hawaii, September 1981 (Honolulu, East-West Center).
[b]R. J. Ward, "The Future Role of the Pacific Islands in Mineral Supply: The Papua New Guinea Case," *Materials and Society* vol. 10, no. 1, pp. 9–20.
[c]"Papua New Guinea," in *Mining Annual Review* (London, Mining Journal, 1986) p. 347.

porphyry copper-gold province. It is at this point that extensive exploration data become available and the current study begins.

As a result of the major Bougainville discovery and the general boom in exploration in Australia, exploration in PNG increased rapidly in the late 1960s reaching a peak in 1970 as previously shown in figure 4-2. Between 1964 and 1974 several major mineral deposits were discovered, the most important of which are shown in table 4-1.[2] During this time, a number of major porphyry copper-gold deposits were discovered, including the commercial Ok Tedi deposit in the Star Mountains in 1968. In rapid succession the porphyry deposits of Freida River, Yanderra, Misima Island, and Manus were also discovered. In addition to porphyry deposits, several nickel laterite prospects have been investigated, and particular attention has been given to a relatively large lateritic nickel-cobalt-

[2]The biggest gold discovery outside of South Africa was made on PNG's Lihir Island in 1983.

chromite deposit at Ramu River. Subeconomic bauxite deposits and chromite-rich beach sands have also been investigated.

Overall mineral exploration in PNG has resulted in discovery of the important mineral deposits shown in table 4-1 with gross in-place value of about $42 billion. It is important to note that only the Bougainville and Ok Tedi deposits had been reported as commercial discoveries; these two were estimated to have an in-place metal value of about $18 billion. The Bougainville deposit was discovered in 1964 and brought into commercial production within eight years. The Ok Tedi deposit was discovered in 1968 and brought into commercial production in 1984, sixteen years after discovery. Relatively large, low-grade, lode-gold deposits, under evaluation in the 1980s, are likely to be the next commercial mineral developments in Papua New Guinea.

Mineral Policies and Legislation

The PNG government has clearly articulated terms and conditions for mining and petroleum developments. Exploration and mineral development are left to the private sector; the PNG government, however, has the right to purchase a minority equity position in commercial mineral developments. Major mining enterprises are taxed at modest rates until they achieve a certain rate of return on the total investment (approximately 20 to 22 percent in current terms). Thereafter, an additional profit-tax formula applies, which substantially increases the government's revenue share from a mining operation.

The weakest link in PNG's mineral policies may be in the evaluation and issuance of mineral prospecting licenses, which have been slow and at times unpredictable, causing delays of several months to a year or more. As a result, more than a million dollars per year in exploration expenditures have been deferred. Grass-roots exploration expenditures in the early 1980s remained at about half the level that would have occurred with fewer constraints on issuing prospecting licenses.

Prospecting Authorities

Trend in Prospecting Authority applications. Mineral prospecting licenses in PNG, called Prospecting Authorities (PAs), are granted for up to two years with renewal subject to satisfactory proposals for additional exploration activities. Under certain provisions of the Mining Act (Amalgamated) 1977 (No. 8 of 1978), PAs are granted for areas not to exceed 25,000 km^2; however, individual PAs are normally limited to a maximum of 2,500 km^2. Most of the analysis of PNG's exploration activity is based on an examination of 378 PAs granted between 1959 and 1982. During this period, 303 PA applications also were denied or withdrawn and 52 were pending. There was a virtual moratorium on the issuing of PAs from 1979 to 1982.

The trends in PA applications are shown in figure 4-15. The peak year of active PAs, granted PAs, and refused or withdrawn PAs was 1970. From a high of 107 in that year, the trend in active PAs then fell to a low of 45 in 1977. Granted PAs followed approximately the same trend reaching a low of 5 in 1976 and then 4 in 1978, 1979, and 1982. The virtual moratorium on license applications between about 1979 and 1982 resulted from inadequate staffing in the Mines Department. Therefore, a total of 52 PA applications were awaiting government action in June 1982. The PA application fee was increased to $2,500 in the early 1980s, and this change was expected to reduce exploration interest by small companies.

Grass-roots versus total exploration.   As shown in figure 4-16, most exploration expenditures in PNG have not been for grass-roots work but have been related to proving and evaluating major discoveries. Whereas the peak in total exploration expenditures reached about $24 million in 1970, the peak in grass-roots expenditures was roughly $7 million in 1971. Typically, grass-roots exploration expenditures have remained at a mere $2 million or less—between 1964 and 1968, 1975 and 1980, and in 1982. This is quite modest considering the substantial remaining mineral potential indicated by exploration geologists. The pending PAs, if approved, were expected to double the level of grass-roots exploration in PNG. Because PNG was viewed primarily as a porphyry copper-gold province in the 1960s and 1970s, it suffered from the worldwide contraction in copper exploration. The much higher gold prices toward the end of the 1970s contributed to a resurgence of exploration interest for gold deposits.

PA size and expenditures.   Figure 4-17 shows the trend in average license sizes and expenditures for the period from 1966 to 1982. The peak in expenditures per license of $750,000 occurred in 1966 and 1967, when only a small number of licenses were active and intensive exploration was under way on the major Bougainville deposit. Expenditures declined to less than $200,000 per license from 1971 to 1974, before declining to less than $100,000 from 1975 to 1980—a reflection of the low level of exploration interest in PNG during this period. During the early 1980s, expenditures moved up into the $200,000 range, reflecting substantially higher exploration interest in PNG, which is now viewed as a favorable environment for medium-large, low-grade gold deposits.

Figure 4-17 shows that a peak in the range of 1,600 to 1,700 km^2 per license occurred in the period from 1969 to 1971 before steadily declining to less than 500 km^2 in 1976. The low of approximately 400 km^2 occurred in 1979 and 1980, before an abrupt increase to more than 1,100 to 1,200 km^2 in 1981. The large increase in the early 1980s reflects renewed interest in regional exploration for gold deposits by major corporations.

Number of
prospecting
authorities

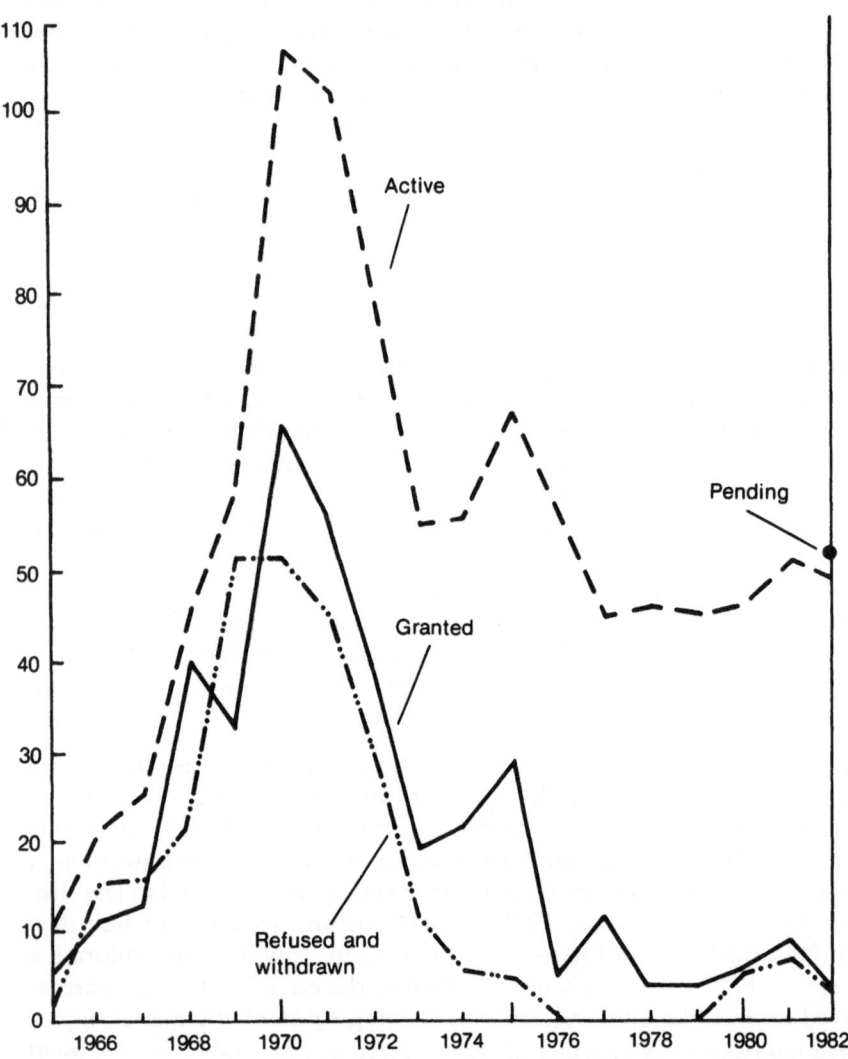

Figure 4-15. Number of Papua New Guinea Prospecting Authorities (PAs) granted, active, pending, refused and withdrawn, 1966–82. Granted PAs include those licenses issued in the year indicated; active licenses refer to all those issued over time that are still valid. The point marked "Pending" refers to the number of license applications under consideration by the government in 1982, when the moratorium was lifted. *Source:* Derived from unpublished government data.

Constant
1982 U.S.$
(millions)

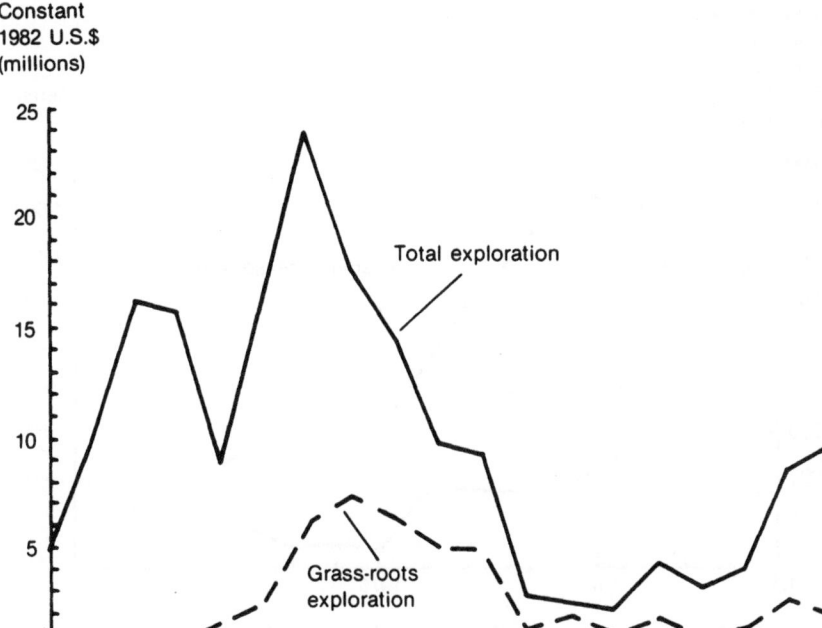

Figure 4-16.   Total exploration and grass-roots exploration expenditures in Papua New Guinea, 1964–82. *Source:* Derived from unpublished government data.

Major Versus Minor Exploration Companies

As shown in figure 4-18, major mining companies[3] have accounted for more than 80 percent of exploration expenditures except during the period from 1975 to 1978 when the majors greatly reduced exploration expenditures. As a consequence, a low was reached in 1976 when majors represented only 50 percent of expenditures. In 1982, majors accounted for 88 percent of exploration—a reflection of the continued dominant role of majors in PNG.

Figure 4-18 also shows the trend in the share of total exploration accounted for by the three companies with the largest exploration expenditures. These three programs represented more than 95 percent of exploration from 1964 to 1967, followed by a decreasing trend to 48 percent

---

[3]The following ten major mining companies have had active exploration programs in PNG: Broken Hill Proprietary, Conzinc Riotinto of Australia, Esso, ExOil, INCO, Kennecott, Mt. Isa Mines, Placer, Swiss Aluminum, and U.S. Steel.

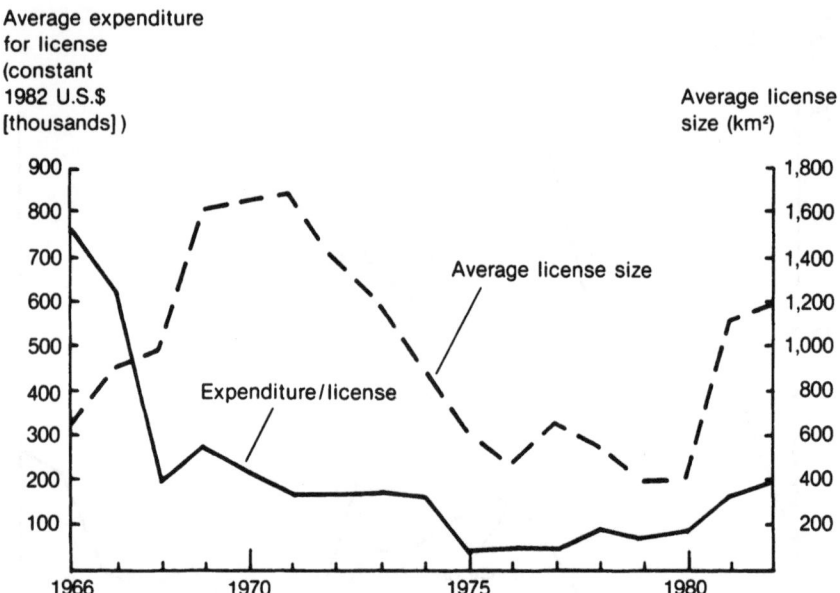

Figure 4-17. Average expenditure per license and average license size in Papua New Guinea, 1966–82. *Source:* Derived from unpublished government data.

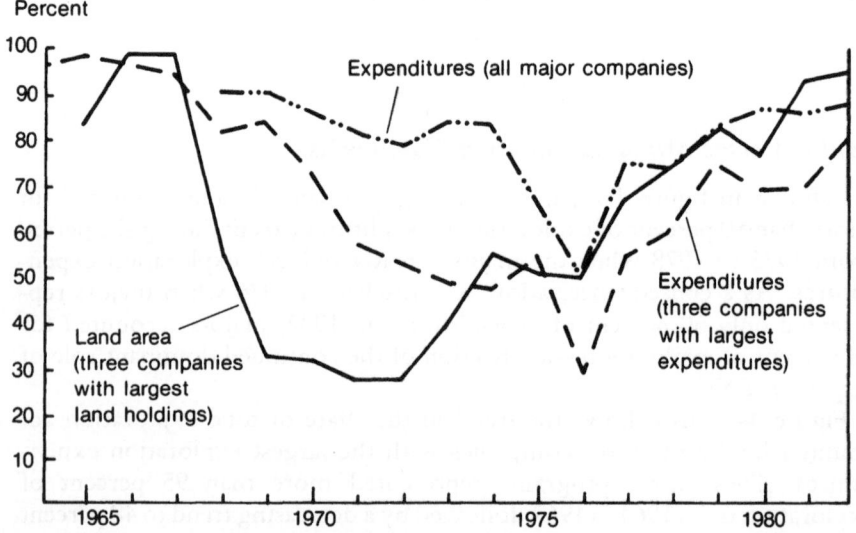

Figure 4-18. Share of expenditures on exploration programs by all major companies (1968–82) and by the three largest companies (1964–82), and the share of Prospecting Authority land areas held by the three companies with the largest land holdings (1965–82), Papua New Guinea. *Source:* Derived from unpublished government data.

in 1974 and a sharp temporary drop to 29 percent in 1976 when majors dramatically reduced exploration expenditures. From 1976 to 1982 the trend was generally upward, reaching a high of 81 percent in 1982—the highest level of concentration in more than a decade.

Figure 4-18 also shows the trend in total land held by the three companies holding the largest areas. The three companies with the largest exploration programs were not always the same as the three largest landholders. During the boom in exploration from 1969 to 1972 many companies held land, and this resulted in a decline in dominance by the largest three landholders to around 30 percent. However, by the early 1980s about 95 percent of all PA land areas was held by the largest three landholders. This reflects a combination of major programs by three companies and also the impact of the moratorium on licenses that kept most companies from obtaining PAs.

PNG Exploration in Perspective

Exploration in PNG from 1964 to 1982 can be divided into three different periods. After the major Bougainville discovery in 1964, there was a boom in exploration for porphyry copper-gold deposits that paralleled a boom in exploration in Australia. This boom peaked in PNG in 1970 and declined to a low in the mid-1970s when a second period of low levels of major company exploration began. Several factors appear to have contributed to this low level of exploration, including: (1) the failure of Kennecott Copper negotiations with respect to Ok Tedi and its withdrawal from PNG in 1975, (2) uncertainty after the government initiated renegotiation of the Bougainville agreement in the early 1970s, (3) PNG's independence in 1975, and (4) low metal prices and declining exploration activity for porphyry copper deposits. The third period began at the end of the 1970s with (1) the government's success in attracting an international mining consortium to develop the Ok Tedi project; (2) a combination of a substantial increase in gold prices and the recognition that PNG is an attractive geologic environment for medium-large, low-grade gold deposits; and (3) the assessment of some major mining companies that PNG's mineral policies and fiscal regime are among the most sophisticated in the developing world.

The high level of exploration interest in PNG that began in the early 1980s is not fully reflected in exploration figures because of the moratorium on issuing new PAs from 1980 to 1982. The cost of this moratorium measured in terms of lost grass-roots exploration is estimated to be at least $1 million per year. This estimate is based on estimated exploration expenditures proposed within applications that were frozen by the moratorium.

Conclusions

This chapter has examined exploration patterns in two developing countries from different regions of the world, with different mineral resources. Both countries are similar in that they have had major mineral developments financed by private capital during the 1970s and early 1980s. At a time when many developing countries were having great difficulty attracting private exploration investments, both countries had active exploration programs supported by a few major multinational mining companies. The mineral policies of the two countries differ on the important issue of fiscal terms. Botswana follows an ad hoc approach and negotiates special terms for each project, normally after a commercial discovery. Papua New Guinea spells out fiscal terms and environmental conditions in legislation, and it has a standard mining development agreement that is largely the result of the Ok Tedi negotiations (Papua New Guinea, 1981). However, with respect to the exploration phase, Botswana issues licenses rapidly with a minimum of uncertainty, whereas substantial delays are common in PNG.

Two countries with different mineral policies can each achieve high rates of mineral development because of the following four factors. First, mining industry representatives perceive of both governments as being honest, stable, and unlikely to expropriate major foreign mining investments. Second, both countries have an established record of successful major mineral developments by private companies. Third, the mineral potential of both is among the highest in the developing world for certain' minerals. Botswana's potential for diamonds and coal places it in the top five developing countries for both commodities. Likewise, PNG's potential for major porphyry copper-gold deposits and medium-large, low-grade gold deposits is among the top for developing countries. Finally, both countries are adjacent to rich, highly developed mineral economies and appear to receive a substantial exploration spillover effect from their neighbors. In the case of Botswana, most initial exploration interest has resulted from South African companies or South African subsidiaries of major multinational mining companies. Likewise, most investments in PNG exploration have come from Australian companies or Australian subsidiaries of major multinational mining companies. The mineral expertise available from South Africa and Australia probably has increased the level and quality of exploration programs in Botswana and PNG.

Both countries have modified provisions of prospecting licenses and mining agreements over time, and both have successfully renegotiated mineral agreements to increase the government share of profits. However, the effects of these renegotiations do not appear to have hurt either country. Both in PNG and in Botswana, the companies that had their agreements renegotiated continued to explore actively. Kennecott, which left PNG in 1975 after unsuccessful negotiations with respect to Ok Tedi, returned to

PNG in 1982 with a substantial exploration program. Conzinc Riotinto of Australia's Bougainville Copper mining agreement was renegotiated in 1974; nevertheless, the company maintained an active exploration program in PNG.

Apparently, then, changes in mineral policies that enhance the government's position—even the renegotiation of minerals agreements—will have only a short-term negative impact on exploration, providing the international mining community perceives that the net result leaves reasonable profit opportunities and an environment in which mineral projects can be operated efficiently.

Exploration levels appear to be the result of a multitude of factors that may vary considerably among countries and commodities and time. The ability of many governments to increase exploration levels by making modest policy changes may be more limited than generally assumed, and the impact of international trends in mineral commodity prices may be of more significance in stable developing countries.

Botswana and Papua New Guinea are among a small number of developing countries that continue to attract active private nonpetroleum mineral exploration. As previously discussed, the reasons are complex. However, the possession of a relatively high potential for discovery of very large, commercial mineral deposits in combination with a record of political and economic stability appear to provide the essential foundation from which successful mineral policies can be launched.

■

The governments of both Botswana and Papua New Guinea were generous in making data available for inclusion in this chapter. Throughout, confidential data have been aggregated to protect the interests of individual companies. The views and analyses in this chapter are the authors' and do not necessarily reflect those of either government.

References

Australian Bureau of Mineral Resources, Geology and Geophysics. Various years. *Australian Mineral Industry* (Canberra, Australia, Bureau of Mineral Resources).

Botswana Government Gazette, Supplement A. 1976. "Mines and Minerals Act," December 31.

Doble, K. W. 1981. "Regional Geological Mapping and Mineral Exploration Activity in Papua New Guinea." Paper presented at conference on Resource

Potential and Implications of Mineral Development in the South Pacific, Honolulu, Hawaii, September (Honolulu, East-West Center).

Energy, Mines and Resources Canada, 1984. *Canada Mines: Perspective from 1983*, MR 200 (Ottawa, Canadian Government Printing Center).

Kursten, M. O. C. 1983. "The Role of Metallic Mineral Resources for Countries of the Third World," *Natural Resources Forum* vol. 7, no. 1, pp. 71–79.

Mikesell, R. F. 1979. *New Patterns of World Mineral Development* (Washington, D.C., British-North American Committee).

————. 1983. *Foreign Investment in Mining Projects* (Cambridge, Massachusetts, Oelgeschlager, Gunn, & Hain).

Mining Annual Review. 1978. "Basic Policies Essential" (London, Mining Journal) pp. 7–12.

Mining Annual Review. 1986. "Papua New Guinea" (London, Mining Journal) p. 347.

Papua New Guinea, Department of Justice. 1981. *Standard Mining Development Agreement* (Waigani, Port Moresby).

Reeves, C. V., and D. G. Hutchins. 1982. "A Progress Report on the Geophysical Exploration of the Kalahari in Botswana," *Geoexploration* vol. 20, pp. 209–224.

Ward, R. J. 1986. "The Future Role of the Pacific Islands in Minerals Supply: The Papua New Guinea Case," *Materials and Society* vol. 10, no. 1, pp. 9–20.

World Bank. 1979. *Coal Development Potential and Prospects in the Developing Countries* (Washington, D.C., World Bank).

5

Mineral Exploration in South Africa

THEO E. BEUKES

South Africa is richly endowed with mineral resources. It produces more gold, platinum, manganese, and chromium than any other country in the world, and the overall value of its mineral production is second only to that of the United States in the Western world. (Figure 5-1 shows the distribution of mineral resources in South Africa.)

While the level of mineral exploration in much of Africa over the last ten to fifteen years has stagnated or declined, it has risen dramatically in South Africa. This chapter reviews this growth trend along with changes in the type of minerals sought through these exploration expenditures, and then identifies factors important to these shifts over time in the level and distribution of exploration expenditures. First, however, the chapter describes certain aspects of the South African mining industry, which are important for exploration and which distinguish South Africa from other mineral-producing countries.

Structure of the Mining Industry

Gold was first discovered on the Witwatersrand in 1886, and since that time it has been the major component of South African mining. Along with platinum and other precious metals, over the years it has accounted for about 70 percent of the total value of South African mineral production (see table 5-1). The remaining 30 percent has come in fairly even shares from (1) uranium, coal, and other energy minerals; (2) manganese, chromium, and other nonprecious metals; and (3) precious and semiprecious stones, industrial minerals, and building and ornamental stones. The overwhelming importance of precious metals, and in particular of gold, has

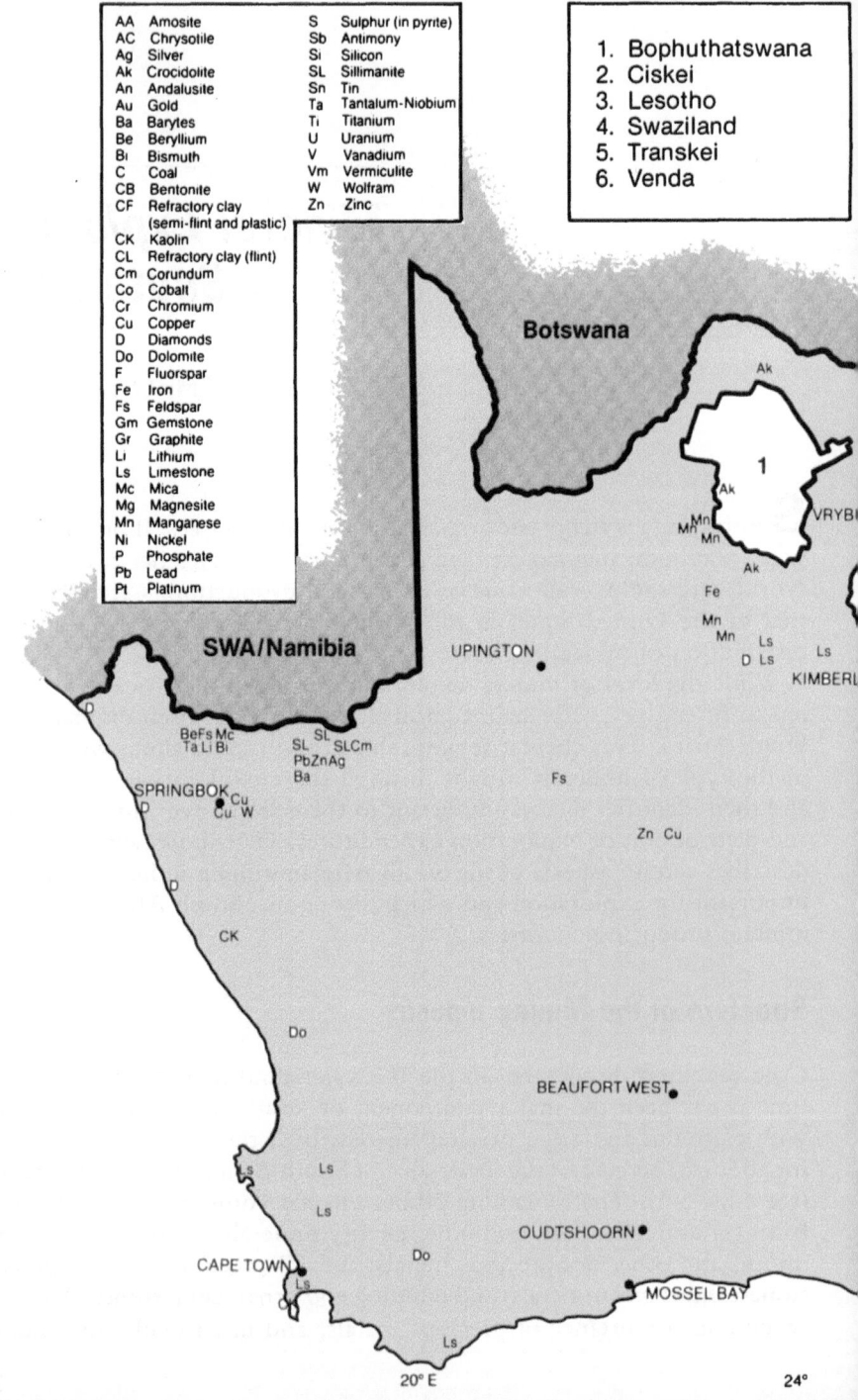

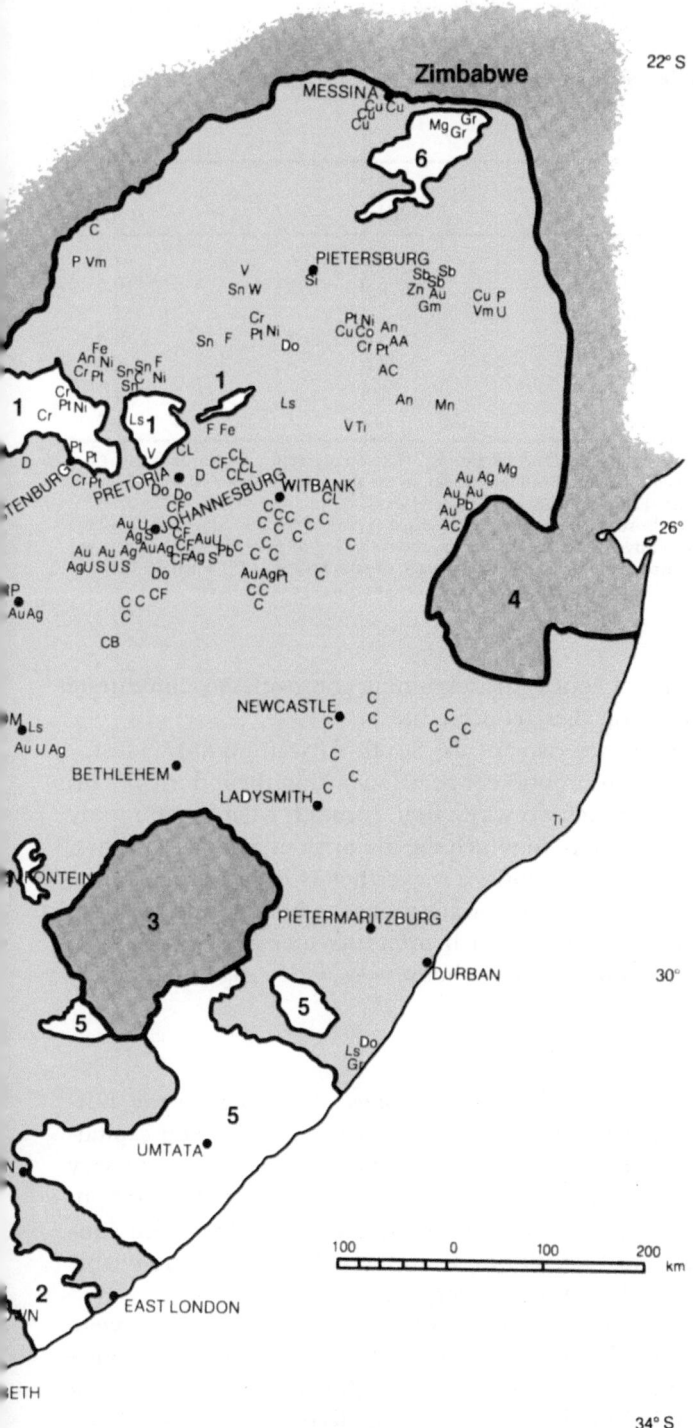

Figure 5-1. Distribution of mineral resources in South Africa. *Source:* adapted, with permission, from *South Africa Official Yearbook 1986*.

Table 5-1.   Total Sales of Mineral Commodities by Categories in South Africa, 1852–1973

(billions of U.S. dollars)

Commodity	Sales[a]	Percent of total
Precious metals	40.5	69.3
Energy minerals	6.2	10.5
Base metals (including manganese and chromium)	5.5	9.4
Precious and semiprecious stones	3.7	6.3
Industrial minerals	1.8	3.1
Building and ornamental stones	0.8	1.3

Source: Adapted, with permission, from D. A. Pretorius, "The Stratigraphic, Geochronologic, Ore-type, and Geologic Environment Sources of Mineral Wealth in the Republic of South Africa," *Economic Geology* vol. 71, no. 1 (January–February 1976) p. 12.

[a]Dollar values shown reflect the summation of receipts realized at time of sale. No adjustment has been made for the effects of inflation over time. Consequently, the real value of precious metal sales is underestimated, as these metals accounted for a higher percentage of sales prior to 1950 when the dollar in real terms was worth more.

encouraged the established South African mining companies to concentrate their exploration efforts on these commodities.

Another important characteristic of the South African mining industry is its cohesiveness, which promotes cooperation within the industry. This cooperation manifests itself in two ways: first, through what is commonly described as the Group System, in which the majority of individual mines, though completely independent entities, nevertheless have administrative and financial ties with certain major mining finance houses; and, second, through the Chamber of Mines, which offers its members—individual mining companies and mining finance houses—a wide range of services.

Group System

The Group System originated in the early days of mining on the Witwatersrand, when large sums of money were needed to develop capital-intensive, deep-level mines. Mining finance houses with the necessary financial resources and technical expertise undertook geologic exploration, development, and management of new mining ventures. They also consolidated, by merger and acquisition, a number of small and nonviable mining ventures into viable mining enterprises.

Of the six major mining finance houses that exist today, four were involved in the original development of the Witwatersrand goldfield—General Mining Union Corporation, or Gencor (formerly General Mining and Finance Company); Rand Mines, the mining division of Barlow Rand; Gold Fields of South Africa; and the Johannesburg Consolidated Investment Company, known as J.C.I. The Anglo American Corporation was established later, in 1917, but it rapidly became the biggest mining group in South Africa. The Anglo-Transvaal Consolidated Investment Company (Anglovaal) was established during the 1930s.

Each of these six major mining finance houses constitutes a group and each continues to finance exploration and to develop as well as manage mines. Mines so developed, however, normally are not divisions or subsidiaries of the mining houses, but rather they are separate companies with shareholders and boards of directors. A mining house typically will provide or arrange a significant portion of the capital for a new venture. Later, it will reduce its equity holding but will continue to provide under contract managerial, technical, and other services for the mine.

Historically, South African mining houses financed projects with internal capital. Joint ventures were negotiated between the groups. This traditional financing method to some extent has been replaced during the past decade by greater use of debt capital and joint ventures with foreign multinational mining corporations. Joint exploration programs also have become accepted practice.

Overall, the Group System links more than 100 mines financially and administratively to at least one of the six major mining houses. These six groups collectively account for more than 85 percent of the total value of mining production in South Africa.

Chamber of Mines

The mines and the major groups are linked closer still through their membership in the Chamber of Mines of South Africa, which evolved from a diggers' committee formed soon after gold was discovered on the Witwatersrand. A nonprofit organization, the Chamber provides a variety of services to its members, including labor recruitment, employee training, and labor negotiations. It processes and markets all of South Africa's uranium. It sells Krugerrand around the world and promotes the use of gold in jewelry and for industrial and investment purposes. It runs the miners' training colleges and hospitals and operates the Rand Gold Refinery.

The Chamber also acts as the industry's spokesman in dealings with the government and provides accounting, statistical, legal, economic, and other services to its members. Its numerous special committees focus upon a wide range of mine-related matters, such as pollution control, taxation, and financial assistance to universities and other educational institutions.

To promote new approaches to exploration in hopes of maintaining the discovery rate of new gold deposits, the Chamber supports research. It has financed, for example, the Economic Geology Research Unit of the University of the Witwatersrand, the Pre-Cambrian Research Unit at the University of Cape Town, and the Department of Mineral Economics at the Rand Afrikaans University. The mining finance groups, such as Jo-

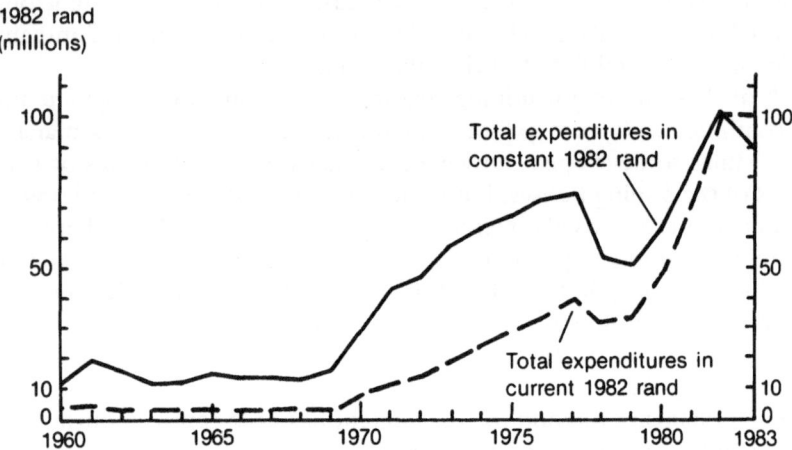

Figure 5-2.   Total South African exploration expenditures, 1960–83. *Source:* The exact expenditure figures from which these curves were drawn are included in tables 5-A-1 and 5-A-2 in the appendix to this chapter.

hannesburg Consolidated Investments, Anglo American Corporation, and Gold Fields of South Africa, also maintain individual research units that investigate new exploration techniques.

Trends in Exploration Expenditures

Annual exploration expenditures for South Africa are shown in millions of current and constant (1982) rand in figure 5-2 for the period from 1960 to 1983.[1] The data include exploration for nonfuel minerals as well as two mineral fuels—uranium and coal. Exploration for oil and gas essentially is conducted by the Southern Oil Exploration Corporation (Soekor).[2] Oil and gas exploration expenditures, except for those relatively insignificant amounts expended by South African mining corporations, are not included in the statistics of this chapter.

During the 1960s, exploration expenditures remained flat or stagnant, both in constant and current terms. This period was followed by a rapid increase during the early and middle 1970s. After a dip toward the end of this decade, exploration activity began to rise again during the early 1980s.

[1]Unless otherwise noted, all figures in this chapter are in constant 1982 rand. Equivalent values in constant 1982 U.S. dollars can be approximated by multiplying the constant 1982 rand figures by 1.15, the average exchange rate prevailing over 1982 between the rand and the U.S. dollar.

[2]Soekor is a state corporation established during 1965 by South Africa to encourage, coordinate, and conduct exploration for oil and gas on the subcontinent.

During the 1980s, however, South Africa experienced for the first time substantial inflation, so the increase in real terms was much more modest than the rise in current rand.

Real exploration expenditures, again measured in millions of constant (1982) rand, are broken down in figure 5-3 for precious metals, coal and uranium, and other metals and minerals. This figure indicates that exploration in South Africa during the years examined has passed through four distinct periods.

During the 1960s, it remained relatively stagnant, at about 16 million rand per year. Most exploration activity focused on the discovery of precious metals, and in the early years almost no effort was made to find coal and uranium. By the end of the decade, however, these commodities were attracting between 15 and 20 percent of the country's total exploration funds. Interest in nonprecious metals and minerals varied but on average consumed about a third of the available funds.

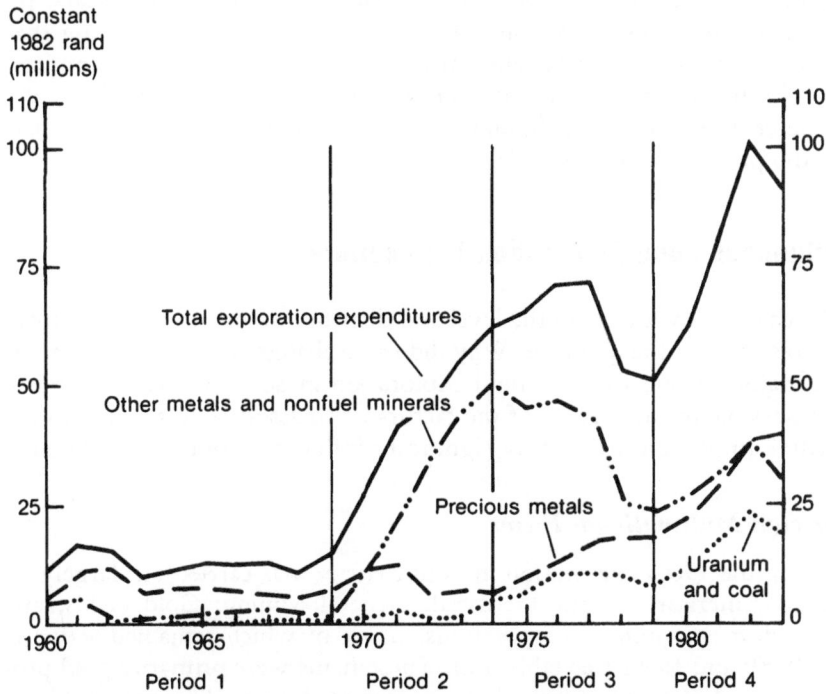

Figure 5-3.   South African exploration expenditures for precious metals (including gold and the platinum-group metals), other metals and nonfuel minerals (including copper, lead, zinc, tin, nickel, molybdenum, chromium, manganese, and fluorspar), and uranium and coal, 1960–83. *Source:* The exact expenditure figures from which these curves were drawn are included in tables 5-A-1 and 5-A-2 in the appendix to this chapter.

The first half of the next decade, the period from 1969 to 1974, is marked by a sharp rise in the search for nonprecious metals and minerals. Expenditures for these commodities jumped from less than 4 million to more than 45 million rand. This produced a substantial increase in the country's total exploration expenditures and a major shift in the allocation of these funds. As funding for coal and uranium increased much more modestly and actually declined for precious metals, nonprecious metals and minerals by 1974 accounted for 80 percent of all money being spent on exploration.

The third period, the years 1974 to 1979, experienced a decline of more than 50 percent in exploration for nonprecious metals and minerals. To some extent, however, this drop was offset by relatively modest but fairly persistent increases in exploration for precious metals and uranium and coal.

The last period, the years 1979 to 1983, recorded a doubling in the real expenditures for precious metal exploration, with significant though less dramatic increases in exploration for uranium and coal and for nonprecious metals and minerals. By 1983, after a hiatus of more than ten years, the search for precious metals once again was receiving more funding than either of the other two groups. Almost all of this effort was committed to gold. Indeed, during the entire period from 1960 to 1983, gold received 95 percent or more of the funding for precious metals. The rest was spent on the platinum-group metals.

Influences upon Exploration Expenditures

The changes over time in the level and distribution of exploration expenditures prompt the question, Why did these changes occur? And what are the major driving forces behind exploration in South Africa? While it is not possible to identify all of the factors or to assess with precision their relative importance, the more significant influences are fairly apparent.

Foreign Multinational Firms

During the 1960s exploration in South Africa was carried out largely by Anglo American, Anglo Transvaal, General Mining, Gold Fields, and other domestic mining corporations, many of which remained active in the 1970s and 1980s (see table 5-2). These firms were primarily gold producers, and in South Africa they concentrated on finding new gold deposits. While some were also looking for copper, nickel, and other base metal deposits, these efforts were largely in other African countries.

Toward the end of the decade foreign multinational mining corporations began to show an interest in South Africa, and during the period from 1968 to 1977 they initiated a number of exploration programs there. As table 5-3 indicates, more than thirty foreign firms were exploring in South

Table 5-2. South African Multinational Exploration Companies, 1977 and 1983

1977	1983
Anglo American Corp. of South Africa Ltd.[a]	Anglo American Corp. of South Africa Ltd.
Anglo Transvaal Consolidated Investment Co. Ltd.	Anglovaal Ltd.[b]
General Mining and Finance Corp. Ltd.	Gencor[c]
Gold Fields of South Africa Ltd.	Gold Fields of South Africa Ltd.
Johannesburg Consolidated Investment Co. Ltd.	Johanesburg Consolidated Investment Co. Ltd.
Rand Mines Ltd.	Barlow[d]
Union Corp. Ltd.	
Cape Asbestos South Africa (Pty[c]) Ltd.	
Industrial Development Corp.	
Minerts Development (Pty) Ltd.	
Messina Transvaal Development Co. Ltd.	
S.A. Manganese Amcor Ltd.	

[a]"Ltd." designates a registered South African company with limited liability. The stocks of such a company may or may not be quoted on the Johannesburg Stock Exchange.
[b]Anglo Transvaal changed its name to Anglovaal in 1981.
[c]General Mining and Finance Corporation Ltd. and Union Corporation Ltd. merged during 1980 to form General Mining Union Corporation Ltd. This name was changed to Gencor during 1981.
[d]Rand Mines Ltd. merged with the Barlow Group of companies during the fourth quarter of 1971 and effectively became the mining division of Barlow.

[e]"Pty" stands for proprietory, which in this context means that the limited-liability company has fewer than fifty shareholders. The stocks of a "Pty company" will not be quoted on the Johannesburg Stock Exchange.

Africa by 1977. The majority were American, but British, French, Dutch, German, and Canadian firms were also active.

This entry of foreign firms had two direct effects. It contributed to the severalfold increase in exploration activity that occurred during the early and middle 1970s (see figure 5-2). In addition, since most of the new firms were interested in base and ferrous metals, it stimulated a surge in exploration for nonprecious metals and minerals, reducing the historical emphasis on gold exploration (see figure 5-3).

The foreign mining companies also had an indirect effect on South African corporations, as their exploration led to the discovery of copper, lead, zinc, and nickel deposits. Among the more significant discoveries were Palabora by Newmont and Rio Tinto, Prieska by U.S. Steel and Anglovaal, Black Mountain and Broken Hill by Phelps Dodge, and Gamsberg by Newmont.

These successes prompted South African corporations to reappraise the country's base metal and mineral potential and to increase their base metal exploration. Concomitant with this reappraisal, there occurred an apparent shift in philosophy toward an increase in joint venturing between South African and foreign multinationals. The subsequent discoveries of Rozynenbosch (lead, silver, and zinc) by Phelps Dodge and Gold Fields, Van

Table 5-3. Foreign Multinational Exploration Corporations Operating in South Africa, 1977 and 1983

1977	1983
African Selection Trust Exploration (Pty)[a] Ltd. (B)	African Selection Trust Exploration (Pty) Ltd.[b] (B)
Aquitaine (F)	—
Armco Bronne (US)	—
Bethlehem Steel Exploration Corp. (US)	—
Billiton Exploration SA (N)	Shell Metals (Pty) Ltd.[c] (N)
BP Coal SA (Pty) Ltd. (B)	BP Coal SA (Pty) Ltd. (B)
Charter Consolidated Ltd. (B)	—
Chrome Chemicals (Pty) Ltd. (G)	Chrome Chemicals (Pty) Ltd. (G)
Dresser Minerals Intl. (US)	Dresser Minerals Intl. (US)
Eland Exploration (Pty) Ltd. (C)	Eland Exploration (Pty) Ltd. (C)
Essex Minerals Co. (US)	—
Esso Minerals (Pty) Ltd. (US)	—
Falconbridge Exploration Ltd. (C)	Falconbridge Exploration Ltd. (C)
Fluor-Genrec SA (Pty) Ltd. (US)	—
Hanna Minerals Holdings Ltd. (US)	—
Lonrho SA Ltd. (B)	Lonrho SA Ltd. (B)
Metallgesellschaft SA (Pty) Ltd. (G)	Metallgesellschaft SA (Pty) Ltd. (G)
Metallurg (US)	Metallurg (US)
Mission Exploration Co. (US)	—
Newmont SA Ltd. (US)	Newmont SA Ltd. (US)
Pandora Mining (Pty) Ltd. (US)	—
Phelps Dodge of Africa Ltd. (US)	Phelps Dodge of Africa Ltd. (US)
Placer Devel. (Pty) Ltd. (US)	—
Placid Oil Company SA (US)	—
Rio Tinto Exploration (Pty) Ltd. (B)	Rio Tinto Exploration (Pty) Ltd. (B)
Shell Coal SA (Pty) Ltd. (N)	Shell Coal SA (Pty) Ltd. (N)
Sorepmas (Pty) Ltd. (F)	—
Southern Sphere Mining and Development Co. (Pty) Ltd. (US)	Southern Sphere Mining and Development Co. (Pty) Ltd. (US)
TG Exploration Ltd. (US)	TG Exploration Ltd. (US)
Total SA (Pty) Ltd. (F)	Total SA (Pty) Ltd. (F)
Union Carbide SA (Pty) Ltd. (US)	—

Note: Capital letter(s) in parentheses after corporation name indicates country in which firm is based; B = Britain, F = France, US = United States, C = Canada, N = The Netherlands, G = Federal Republic of Germany.

[a]"Pty" stands for proprietory, which in this context means that the limited-liability company has fewer than fifty shareholders. The stocks of a "Pty company" will not be quoted on the Johannesburg Stock Exchange.

[b]Since 1980, part of BP International corporate group.

[c]Billiton was absorbed into the Shell Minerals and Metals Division.

Rooi's Vlei (tungsten and tin) by Phelps Dodge and Shell Metals, Geelvloer (zinc, lead, and copper) by Phelps Dodge and Gencor are only three of several discoveries of complex base metal ore bodies resulting from this shift in exploration goals and new cooperative philosophy.

The rise and redirection of exploration caused by the entry of foreign firms from the 1960s to the mid-1970s were not confined to base metals.

Both foreign and South African companies increased their efforts to find iron ore, manganese, and chromite. As a result, major deposits of manganese and iron ore were discovered, and numerous deposits of chromite were delineated.

While foreign mining companies clearly reshaped the level and distribution of exploration in South Africa, particularly during the 1970s, just why these companies chose this time to come to the country is less clear. The changing political situation on the African continent south of the equator, public policies in South Africa itself, and the development of new geologic approaches—all considered next—appear to be part of the explanation.

Political Environment

From 1950 until the mid-1960s, South Africa-based companies actively explored the African continent as far afield as the southern border of Sudan. Grass-roots exploration was carried out in western, central, and eastern Africa. South African multinationals were particularly active in Zimbabwe (then Southern Rhodesia), Botswana, Angola, Mozambique, and Namibia.

As British, French, and Portuguese colonies gained their independence, this situation began to change. From 1965 to 1974 the geographic range of South African firms contracted as first Zaire, then Zambia, Angola, Mozambique, and finally Zimbabwe became sovereign states—and more or less hostile toward South Africa. In some instances, the new governments barred South African firms, and in other cases, these firms simply found the new political environment too volatile and risky. By 1974, aside from some very modest efforts in Botswana, Namibia, and Swaziland, South African firms were not conducting any exploration in Africa outside their own country. This, coupled with the discoveries by the new foreign companies in South Africa, encouraged domestic firms to expand their search for minerals in South Africa and to devote more of their efforts to metals other than gold.

The changing political situation on the continent was, of course, not all beneficial to South Africa's minerals industry. Many of the newly independent states were openly hostile toward the South African government. This raised the possibility of conflict between the country and its neighbors and presumably reduced the incentives of domestic and particularly foreign companies to explore in South Africa.

The political hostility emanating increasingly from governments around the world, and particularly from African countries, toward South Africa has its origin in the country's domestic apartheid policies. Detailed discussion of these policies lies beyond the scope of this chapter. It is, however, appropriate to consider the impact of these policies upon mineral exploration in South Africa.

Historically it has been the declared policy of the South African government to encourage exploration and to promote foreign investment in the South African mining industry. No estimates of foreign ownership of South African mining exist, nor are there references in the media to a concern on the part of the government regarding foreign ownership. In fact, while countries such as Canada and Australia have established Foreign Investment Review Boards to prescribe levels of domestic control and equity participation, the South African government traditionally has encouraged multinational mining corporations to invest.

The positive impact of this tradition, however, has been substantially diminished by the existence of a political risk perceived by the potential foreign investor. The perceived political risk will be influenced by the availability and interpretation of information about South Africa, as well as the expectations about future political and economic stability in the subcontinent. This risk is expected to increase in intensity proportional to an increased intensity of anti-South African information, which will negatively influence expectations concerning political stability. This is true, regardless of the quality of the information. The perceived political risk will be effectively discounted by superior expectations concerning exploration successes.

Another type of political risk associated with exploration in South Africa is that imposed on the foreign multinational mining corporation within the country of origin. Specifically, this political risk can be identified as the pressure increasingly exerted on the corporation to withhold investment or to disinvest from South Africa. The multinational corporation's vulnerability to this political pressure will be influenced significantly by its domestic political exposure and sensitivity in its country of origin, or by its sensitivity to political actions of host governments in third countries, in which it may have established affiliated corporations. This type of political risk is proportional to the intensity of anti-South African sentiment within the government of the country concerned. It is anticipated that this political risk may increase in intensity in the future, if the growth in anti-South African sentiment, as witnessed during the past decade, is maintained.

As an aside, another type of political risk resulting from the international reaction to South Africa's domestic politics is that South African multinational mining corporations, regardless of their well-documented mining expertise, are experiencing increasing animosity to their exploration for minerals in other countries.

As part of its domestic political policy, South Africa has created independent homelands or states through the drawing of political boundaries within greater South Africa. These boundaries have ignored the geological boundaries of existing and yet to be discovered metallogenic provinces. At the very least, this action has diminished the potential for successful mineral exploration within the boundaries of South Africa, assuming that a relationship exists between the geographic size of a country and the number of mineral deposits to be discovered within it. It is acknowledged that the conflict between political boundaries and geological boundaries is not unique to South Africa. To the extent that a future political scenario for South Africa may involve a federation of states, such as exists in Australia, Canada, and the United States, it can be argued that the political boundaries created by the South African authorities are no different from the provincial boundaries that exist in these countries. It follows from this example, then, that the political risk associated with mineral exploration in South Africa is increased proportionately to the number of political boundaries created. At best, the mineral explorer or mining corporation will be exposed to a fiscal and mineral law disharmony between the federal and provincial governments of the same country. At worst, the mineral explorer will be exposed to the vagaries of political differences between independent countries, which in the South African case will be compounded by a lack of international recognition of the independence of the intra-South African countries. The increased risk applies with equal severity to the foreign multinational corporation and the South African corporation (the latter having obtained the changed status from domestic to foreign corporation, upon the date of declared independence of the newly independent state).

The newly independent state has no history with regard to its attitude toward foreign investment and mineral exploration and consequently the mining corporation has no guide in assessing the risks associated with the general policies of the government.

During the 1960s and early 1970s multinational mining corporations seemingly decreased their exploration effort in post-independence Africa. Factors that influenced this decision included the diminished security of tenure, changed tax regime, and increased probability of mineral property expropriation. It was suggested, then, that exploration activity in South Africa had increased substantially while exploration in post-independence Africa declined. Regrettably, the perception that political instability in South Africa may increase in the future has contributed to an apparent decline in exploration by multinational corporations (see table 5-2).

The net result of the political risk associated with exploration in South Africa, compounded by domestic politics, is that exploration will be undertaken only if the probability of success is considered great. And, the mineral deposit, once discovered, will be exploited only if it is considered a unique investment opportunity. The significant mineral economic and geologic costs of domestic political decisions are abundantly clear.

South African Mineral Policies

Despite any negative effect from domestic political policies, the mineral policies of South Africa encourage exploration. The country's tax laws, for example, allow companies to deduct from income all exploration and development expenditures undertaken to increase mineral reserves. The country also permits private individuals and companies to own mineral rights, and these rights can be separated from the surface or land ownership. The exceptions are gold and other precious metals, diamonds, uranium, and oil; the right to explore for and to exploit these minerals must be sanctioned by South Africa's Ministry of Mineral and Energy Affairs. The right to mine them requires a payment equivalent to a royalty payment. Private ownership encourages exploration by providing security of tenure, although it can be a detriment if it leads to the fragmentation of mineral rights over time, making it more difficult and expensive to secure the necessary rights before exploration can proceed.

South Africa also has a well-established infrastructure. The government is prepared to provide the roads and railways as well as the water and electricity supplies needed to undertake a new mineral project, provided only that the user is willing to guarantee a minimum level of use.

An important change in monetary policy, which clearly influences the mineral industry, also may help explain changes in levels of exploration expenditures and the shift away from gold and then back again. The change was the progressive relaxation of exchange controls after the institution of the "blocked rand" in 1961, which operated to prevent foreign capital from being withdrawn from the country except under the most exceptional circumstances. The pool of blocked foreign investment balances was mobilized by the institution of a two-tier exchange mechanism (a "financial rand" as well as the commercial rand), which was operated from 1979 until early 1983 and again from late 1985 to the present (1987). Exchange control on foreign investment was further relaxed in about 1981 when South African subsidiaries of foreign companies were permitted to borrow in rands up to 50 percent of their total investment requirements, compared to a maximum of 25 percent before that time. The financial rand, available to foreign investors, has invariably traded at a discount against the commercial rand. When the discount was 20 percent, for example, foreign investors could obtain assets worth 1 million rand for the equivalent of 800,000 rand (at the commercial rate) in foreign currency. Consequently, a foreign firm spending 1 million rand on exploration in South Africa received a sizable discount in foreign currency terms and could expect to gear up the venture more extensively. These changes made it more financially attractive to foreign firms to explore in South Africa, explaining in part their growing number during the 1970s.

New Geologic Approaches

During the 1960s the discovery of nickel deposits in Western Australia and of porphyry copper deposits in the Americas and Papua New Guinea led to new approaches or ways of interpreting geologic information with important applications for exploration. In South Africa, the use of these approaches led to major mineral discoveries in a region that had been considered promising beginning in the 1920s and that had been heavily— but unsuccessfully—explored. Located in the northwestern Cape Province, the four recent discoveries—at Black Mountain, Broken Hill, Big Syncline, and Gamsberg—all are complex ore bodies with main components of copper, lead, and zinc. The development of these new approaches enhanced the incentives of both domestic and foreign firms to explore in South Africa.

Mineral Demand and Prices

As shown in chapter 3 of this volume, the greater the demand and the higher the prices for mineral commodities, the more likely that exploration for them will continue. Consequently, a rise in mineral prices increases the expected returns from exploration and at the same time the incentives for firms and others to search for new mineral deposits. Mineral prices also can influence the direction of exploration in terms of the types of minerals sought for similar reasons. When the price of gold rises while the price of copper falls, for example, as was the case during the latter half of the 1970s, firms exploring for copper have a strong incentive to shift some or all of their exploration activities into the search for gold.

The influence of mineral prices on the level and direction of exploration is quite evident in South Africa. During the 1960s, when the real prices for energy and many metal commodities were declining, the country's exploration was concentrated on precious metals, particularly gold. The world price for gold, however, was set by the international monetary system at $35 (U.S.) an ounce and so actually was declining over the decade in real terms. As a result, gold exploration and in turn total exploration stagnated in South Africa.

The surge in exploration that occurred during the early and middle 1970s, as indicated in figures 5-2 and 5-3, also can be attributed in part to mineral prices. Copper and other nonprecious metals experienced strong booms and high prices in 1970 and again in 1973 to 1974. Widespread shortages during the latter period produced record high prices for a number of metal commodities and raised fears of longer-term supply deficiencies. These fears evaporated with the collapse of metal prices in 1975 and the prolonged recession in real metal prices that plagued producers for much of the following decade. Exploration for nonprecious metals responded by leveling off for three years and then dropping sharply in 1978 and 1979.

The severalfold increase in uranium prices that occurred during the 1970s and the sharp rise in coal prices that followed the world oil price increases in 1973 and 1979 stimulated the search for these energy raw materials. In constant (1982) rand, exploration expenditures for uranium and coal, which totaled only 2.7 million in 1973, reached 9.9 million by 1979 and peaked at more than 24.9 million in 1982.

A rise in price also clearly stimulates exploration for gold and other precious metals. Following the creation of the two-tier pricing system for gold in 1968, which allowed the price in open-market transactions to fluctuate freely (while the price for monetary transactions between central banks remained pegged at $35 [U.S.] an ounce), the nominal price for gold rose quite modestly through 1971. The real price, however, actually continued to fall over most of this period. Beginning about 1972, the price of gold began a strong upward climb until it peaked at more than $800 (U.S.) an ounce in 1980. Although the price since then has fallen to less than half this figure, the dramatic rise sparked a boom in gold exploration in the mid-1970s that has continued into the 1980s.

As an aside, it should be noted that a fluctuating rand/U.S. dollar exchange rate (particularly after February 1983) has often resulted in increased rand earnings for South African metal and mineral exporters. To the extent that export prices are quoted in U.S. dollars and domestic payments received in rand, the net effect of a weaker rand would be to compensate gold producers for a lower dollar price for gold. Against a background of enhanced rand earnings, exploration budgets at the very least would be maintained. Another influence related to a stronger U.S. currency would be the relatively lower cost of exploration in South Africa, all other things being equal.

Conclusions

Over the last quarter century, exploration in South Africa in one sense has come full circle. Back in the 1950s gold attracted many a prospector. Later in the 1970s, particularly during the first half of the decade, the search for other metals expanded rapidly, and the share of the country's total exploration effort devoted to gold declined markedly. During the 1980s, however, interest in gold rebounded, and once again this commodity became the most sought-after mineral commodity.

Yet exploration in South Africa has changed in other ways. The sheer magnitude of the country's exploration effort has exploded. In the 1960s, expenditures fluctuated between 12 million and 16 million (constant 1982) rand. By the 1980s, this figure had exceeded 80 million rand. Consequently, despite shifts over time in the distribution of funding, exploration for each of the commodity groups examined—precious metals, other metals and minerals, and uranium and coal—was considerably greater in the early 1980s than it had been two decades earlier.

These changes in the overall level and direction of exploration activity occurred for a number of reasons—some political, others purely economic.

It was somewhat surprising that the changing political environment in southern Africa caused by the independence of many former British, French, and Portuguese colonies did not particularly discourage exploration in South Africa, despite the open hostility displayed by many of these new states toward the country. This does not mean, of course, that South African exploration can continue to flourish in the midst of political instability and turmoil, both within its own national boundaries and those of its continent. Exploration and the development of newly discovered ore bodies are investments that take years to pay off and occur only in areas that offer stability over the very long run.

■

Through personal interviews and in their answers to a questionnaire, exploration managers, consulting geologists, and senior executives from all the major mining corporations identified in tables 5-2 and 5-3, as well as fellow academicians, provided the information on annual exploration expenditures and exploration strategy included in this chapter. The author gratefully acknowledges the assistance of those individuals who generously contributed of their time to this study.

References

Anderson, Anne M., and W. J. van Biljon, eds. 1979. *Some Sedimentary Basins and Associated Ore Deposits of South Africa*. Special publication no. 6 (Johannesburg, Geological Society of South Africa).

Anglo American Corporation of South Africa. 1970–83. *Annual Report* (Johannesburg).

Anglovaal Limited. 1970–83. *Annual Report* (Johannesburg). (Until 1981 this corporation was known as Anglo Transvaal.)

Chamber of Mines of South Africa. 1970–83. *Annual Report* (Johannesburg).

Gencor. 1976–83. *Annual Report* (Johannesburg). (General Mining and Finance Corporation Limited and Union Corporation Limited merged during 1980 and formed General Mining Union Corporation Limited. This name was changed to Gencor in 1981.)

Gold Fields of South Africa. 1970–83. *Annual Report* (Johannesburg).

Johannesburg Consolidated Investments. 1970–83. *Annual Report* (Johannesburg).

Pretorius, D. A. 1976. "The Stratigraphic, Geochronologic, Ore-type, and Geologic Environment Sources of Mineral Wealth in the Republic of South Africa," *Economic Geology* vol. 71, no. 1 (January–February).

Rand Mines of South Africa. 1970–83. *Annual Report* (Johannesburg).

South African Reserve Bank. 1960–83. *Annual Report* and *Quarterly Bulletin* (Johannesburg).

Appendix 5-A
Annual Exploration Expenditures, 1960–83

Table 5-A-1. Annual Exploration Expenditures in South Africa, 1960–83
(millions of current rand)

Year	Precious metals[a]	Coal and uranium	Other metals and nonfuel minerals[b]	Total
1960	1.485	0.010	1.072	2.567
1961	2.449	0.010	1.235	3.694
1962	2.684	0.035	0.580	3.299
1963	1.632	0.019	0.860	2.511
1964	1.773	0.311	0.665	2.749
1965	2.086	0.448	0.960	3.494
1966	1.892	0.590	1.010	3.492
1967	1.788	0.690	1.195	3.673
1968	1.727	0.682	1.076	3.485
1969	2.643	0.658	0.950	4.251
1970	3.758	0.716	3.500	7.974
1971	4.274	1.422	6.800	12.496
1972	2.479	0.875	11.000	14.354
1973	2.938	0.920	15.400	19.258
1974	2.812	1.996	19.500	24.308
1975	5.528	3.600	20.000	29.128
1976	6.844	5.400	23.000	34.888
1977	9.713	6.100	24.000	39.813
1978	10.893	6.300	15.000	32.193
1979	12.267	6.600	16.000	34.867
1980	17.871	9.950	20.900	48.721
1981	26.391	17.400	27.400	71.191
1982	38.245	24.900	38.500	101.645
1983	44.901	21.500	36.400	102.801

Source: Confidential company data.
[a]The category of precious metals includes gold and platinum-group metals.
[b]The category of other metals and nonfuel minerals includes copper, lead, zinc, tin, nickel, molybdenum, chromium, manganese, and fluorspar.

Table 5-A-2. Annual Exploration Expenditures in South Africa, 1960–83
(millions of constant 1982 rand)

Year	Precious metals[a]	Coal and uranium	Other metals and nonfuel minerals[b]	Total[c]	Consumer price index
1960	7.048	0.046	5.088	12.182	21.1
1961	11.390	0.046	5.744	17.180	21.5
1962	12.395	0.160	2.666	15.158	21.8
1963	7.397	0.086	3.898	11.381	22.1
1964	7.835	1.373	2.939	12.147	22.6
1965	8.866	1.904	4.081	14.848	23.5
1966	7.756	2.418	4.141	14.315	24.4
1967	7.089	2.735	4.739	14.563	25.2
1968	6.712	2.650	4.183	13.544	25.7
1969	9.954	2.478	3.578	16.008	26.6
1970	13.604	2.592	12.671	28.864	27.6
1971	14.621	4.864	23.262	42.746	29.2
1972	7.962	2.812	35.335	46.109	31.1
1973	8.620	2.700	45.190	56.510	34.1
1974	7.393	5.246	51.260	63.898	38.0
1975	12.803	8.338	46.320	67.460	43.2
1976	14.250	11.244	47.888	73.380	48.0
1977	18.199	11.429	44.967	74.596	53.4
1978	18.516	10.709	25.499	54.727	58.8
1979	18.438	9.919	24.049	52.409	66.5
1980	23.591	13.136	27.591	64.318	75.7
1981	30.256	19.948	31.412	81.616	87.2
1982	38.265	24.913	38.520	101.698	100.0
1983	39.932	19.121	32.373	91.426	112.3

Sources: This table is derived from the figures in table 5-A-1, using the consumer price index published in South African Reserve Bank, *Annual Report* and *Quarterly Bulletin* (Johannesburg, 1960–83).

[a]The category of precious metals includes gold and the platinum-group metals.

[b]The category of other metals and nonfuel minerals includes copper, lead, zinc, tin, nickel, molybdenum, chromium, manganese, and fluorspar.

[c]May not sum because of rounding.

6

Prospecting and Exploration in the Soviet Union

ALEXANDER S. ASTAKHOV
MICHAIL N. DENISOV
VLADIMIR K. PAVLOV

Prospecting and exploration in the Soviet Union are greatly influenced by both central planning and state ownership of mineral resources. For this and other reasons, the organization of prospecting and exploration differs significantly from that in countries where private companies are primarily responsible for these activities. Forecasts of future mineral demand and of the optimal options to meet this demand on a national scale make possible a description of prospecting and exploration in general. Projected long-term marginal costs of final mineral products are the best criteria for decision making.

This chapter begins by examining the role of prospecting and exploration in Soviet mineral development as a whole. Various stages in the search for new mineral deposits are first identified and then classified according to their tasks, inputs, and outputs. The organization of the agency primarily responsible for carrying out these steps—the USSR Ministry of Geology—is reviewed. The chapter then describes recent trends in Soviet prospecting and exploration with respect to the overall related expenditures, the geographic distribution of the deposits, and the types of minerals sought. The final section focuses on the procedures used in the planning and conduct of prospecting and exploration.

Stages and Organization

In the USSR, prospecting and exploration are considered to be a part—
and only a part—of the general process of mineral resource development.
Their effectiveness is evaluated solely in terms of their effect on overall
mineral supply.

The total mineral-supply system is illustrated in figure 6-1. The growing
need for mineral resources (block 1) encourages prospecting and exploration. During prospecting, geologists and other specialists search for and
discover mineral occurrences. Once the physical and chemical properties
of a deposit have been defined, technical means are selected to mine and
use the resources of the deposit (5). If there are no technological methods
appropriate for the efficient exploitation of the deposit (arrow 5-6), the
development of it is delayed temporarily (7), thereby providing a stimulus
to develop appropriate technology (6). After this becomes available (5-8
or 6-8), the potential social and environmental impacts are studied (8).

If the preliminary economic evaluation (11) is favorable, the principal
decision on whether to develop the deposit may be made (12). A favorable
decision leads to progressively more detailed exploration (13), providing
data for the detailed economic evaluation (14) and mine and mill design
(15). During construction (16) and subsequent mining (18), additional
exploration occurs at and near the mine (17). Mining helps meet national
mineral demand (18-1) and at the same time leads to depletion of known
mineral resources (18-19), which in turn stimulates additional exploration
(19-3-4), as well as technologic improvements in mining and processing
(19-6) and mineral substitution (19-2).

Stages of Prospecting and Exploration

In the Soviet Union, prospecting and exploration encompass six stages:
(1) regional geologic and geophysical surveys, (2) prospecting for mineral
resources, (3) preliminary exploration, (4) detailed exploration, (5) exploration
in the vicinity of an operating mine, and (6) operational exploration. These
stages are distinguished by their objectives, scale of activity, and exploration
methods. The initial stages are associated with the geosciences, the latter with
production engineering. New resources are discovered in the first two stages.
The last four stages then provide more detailed knowledge about these resources. The additional information generated at each stage helps decision
makers choose between advancing to the next stage or abandoning the project.

The first stage, regional geologic and geophysical surveys, covers four
activities. The first is regional geophysical surveys at scales of 1:200,000
or 1:100,000. The goal is a geophysical map that identifies the major
elements of a region's structure. The second is regional geologic surveys
at the same scales that produce a geologic map useful in locating areas for
subsequent prospecting and exploration. Third, more detailed geologic

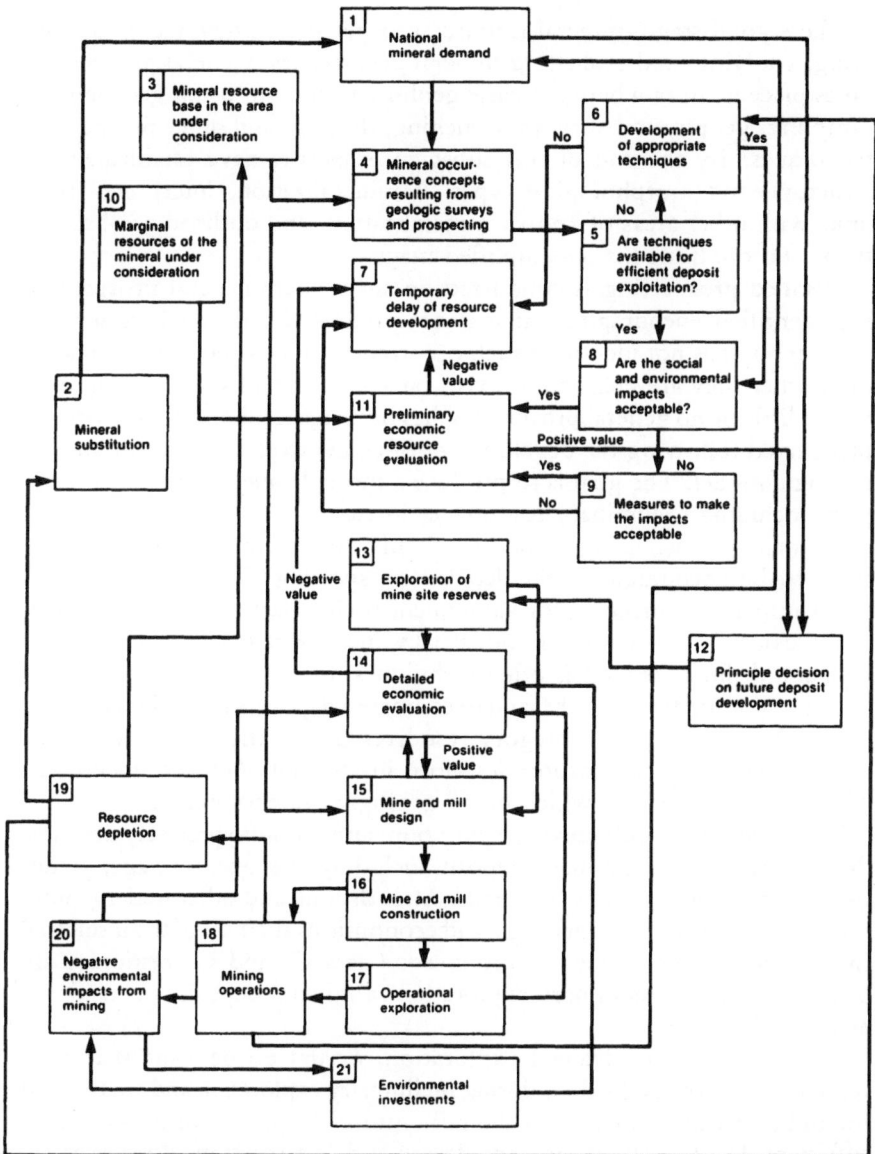

Figure 6-1. Major elements in the operation of the total system of mineral supply in the Soviet Union. (See discussion in text.)

surveys at a scale of 1:50,000 or 1:25,000 then permit identification of the ore-controlling structures and zones, which in turn will allow the production of more detailed mineral resource maps. The final activity is geologic mapping of deep fundamental structures in areas likely to hold mineral resources.

The second stage, mineral resource prospecting, is devoted to discovering, studying, and evaluating mineral occurrences. General prospecting takes place in areas where the basic geology is known, using geochemical sampling, geophysical surveys, trenching, drilling, and other prospecting techniques. By the end of this substage, scientists have established the structural and morphological types of mineralization, interpreted their links with other areas of known mineralization, and outlined ore-bearing fields and ore horizons on simplified maps.

Detailed prospecting is conducted on sites where general prospecting has identified encouraging mineral occurrences; on sites where general prospecting has not identified such occurrences, but where for one reason or another they are believed to exist; on sites in established mining areas where little or no general prospecting has occurred; and on sites previously prospected following the development of a new exploration technique or geologic model. The goal is to produce a more detailed report on a prospect, including preliminary resource estimates.

Estimative prospecting is conducted only on mineral deposits that continue to be encouraging after detailed prospecting. The goal here is to select mineral occurrences for subsequent preliminary exploration and to exclude occurrences with little industrial value. Scientists use surface workings and drilling, as well as other techniques, to construct geologic maps at scales of 1:10,000 to 1:1,000 and to estimate a deposit's possible economic value. At this point, C_2 category resources (under the Soviet system of mineral resource classification described in appendix 6-A) are identified. These are resources for which broad geologic evidence exists.

The third stage, preliminary exploration, aims at more precisely defining the characteristics of a mineral deposit, including its shape, tonnage, grade, ore quality, and mode of occurrence. After drilling and other underground workings, a preliminary geologic and economic evaluation of the industrial importance of the deposit is produced. Class C_2 and C_1 resources are estimated. After preliminary exploration, a mineral deposit is considered as part of reserves.

The fourth stage, detailed exploration, occurs on deposits that have received favorable evaluations during preliminary exploration and are planned for industrial development in the near future. Deposits are delineated in sufficient detail to allow for mine design and construction. Resources are partly reclassified into the higher categories of C_1, B, and A (special standards dictate the amounts of A and B necessary for design and mining activities to be undertaken).

The final two stages, exploration in the vicinity of an operating mine and operational exploration, occur after a mine is in operation, in order either to delineate better known reserves or to establish new reserves in the immediate vicinity of the operating mine.

Figure 6-2 depicts the general sequence of the first four of these six prospecting and exploration stages. The final two stages of exploration are omitted, because they occur at or near operating mines. The sequence flows from top to bottom, and the principal activities and information outputs of each stage appear in the left-hand column. On the right are the main information inputs that underlie and support the activities.

Table 6-1 compares the stages of prospecting and exploration in the USSR with those in certain Western countries. This comparison can be only approximate, because there is no official description or regulation of stages in Western countries as far as we know. Nevertheless, the table indicates that prospecting and exploration in these Western countries and the Soviet Union follow a similar sequence of activities.

Organization of the USSR Ministry of Geology

The Soviet state is the sole owner of in situ resources, and as such it controls the various activities affecting the management and use of mineral resources. While the system for mineral development encompasses a myriad of government agencies and departments concerned with geology, engineering, construction, mining, processing, and transportation, the Ministry of Geology is responsible for 85 percent of all prospecting jobs and has overall responsibility for geologic surveys and mineral discovery.

As shown in figure 6-3, the ministry has three levels of management. The first is the headquarters staff. The second encompasses the All-Union industrial and exploration corporations, the ministries of geology of the Russian, Ukrainian, Uzbek, and Kazakh republics (all either large or mineral-rich), and the boards of geology of the other Union republics. The third level includes industrial-geological corporations, research institutes, design bureaus, and factories for prospecting and exploration equipment.

In addition, the ministry is responsible for certain other activities, including transportation and other communication services, procurement and social services (such as training personnel and providing medical aid, health care, preschool child care, and housing). The ministry's main goal, however, is to identify the country's requirements for mineral resources. In practice, it is the geologic expeditions and crews of the industrial-geological corporations that actually carry out prospecting and exploration activities.

Trends

This section first examines shifts over time in the goals of Soviet prospecting and exploration. It then considers the role of mapping, deep drilling, and other efforts to enhance the effectiveness of the search for new mineral deposits. The section concludes with a review of recent trends in the overall level and distribution of Soviet prospecting and exploration.

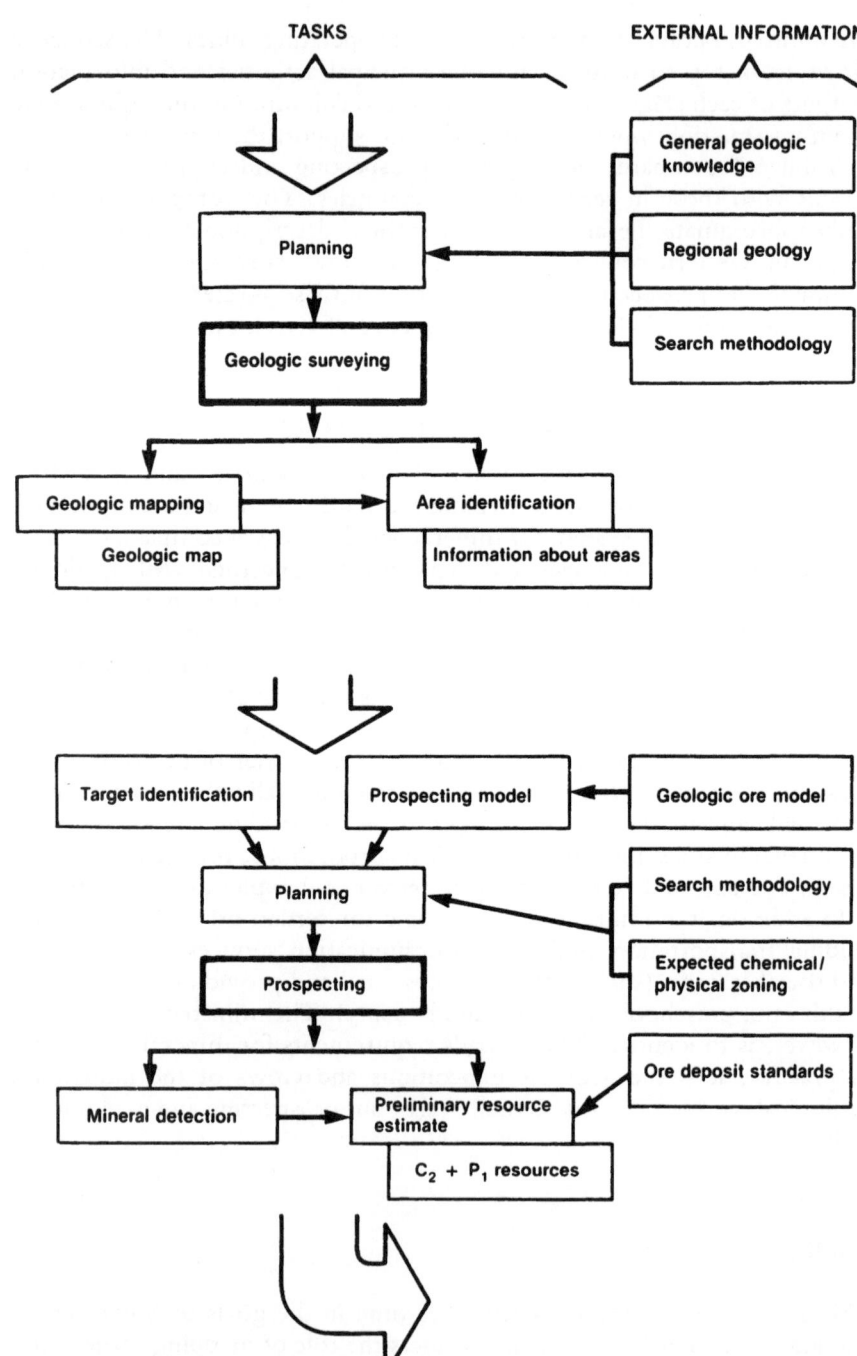

Figure 6-2. Sequence of prospecting and exploration activities in the Soviet Union. (See discussion in text.)

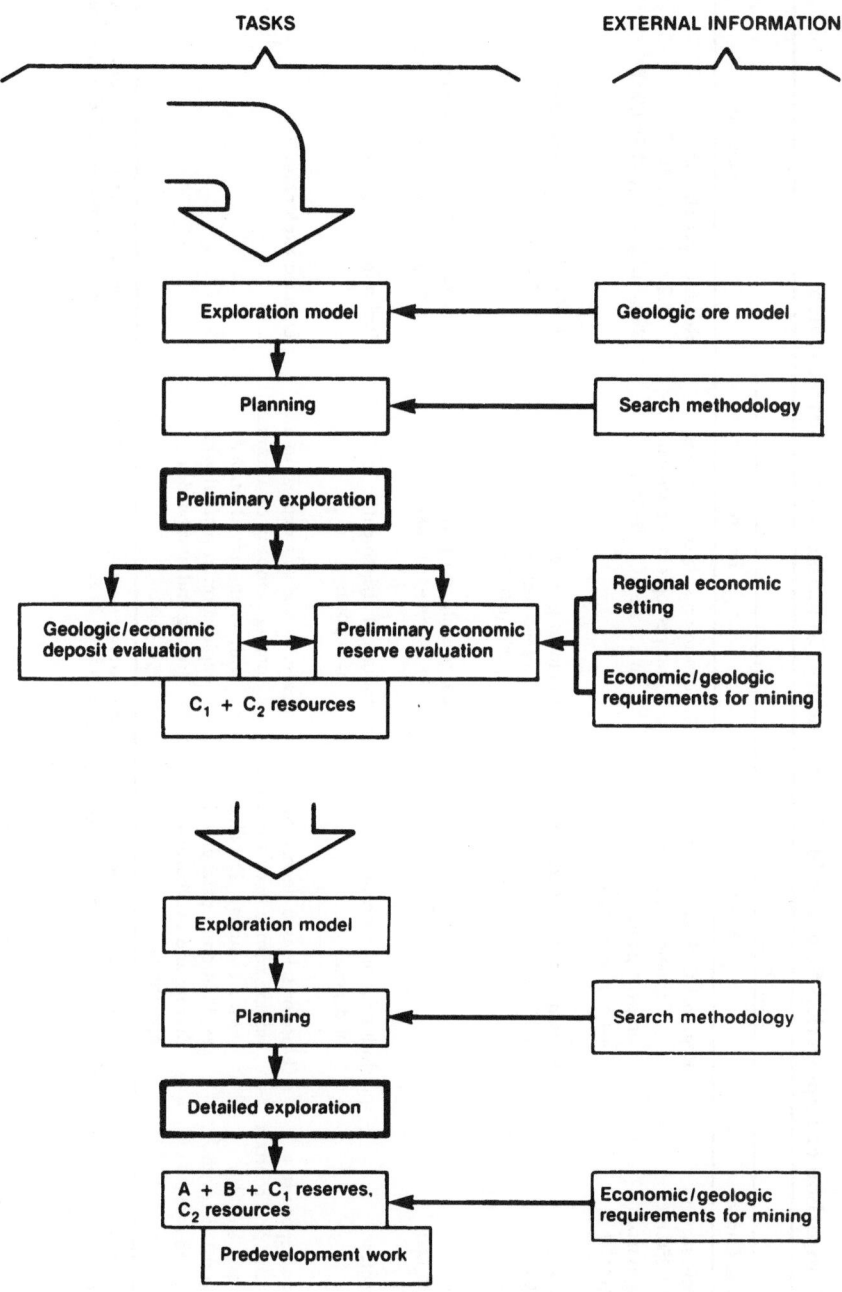

Table 6-1. Comparison of Exploration Stages in Market-Economy Countries and the Soviet Union

| United States and Canada | Market-economy countries | | Soviet Union |
	Australia	France	
Regional geologic surveys on a scale of 1:250,000; airborne geophysics; aerial photography	Regional geologic surveys on a scale of 1:250,000; airborne geophysics; aerial photography	Remote-sensing observations and geologic reconnaissance on a scale of 1:200,000	Regional geophysics (1:200,000–1:100,000) Regional geology (1:200,000)
Reconnaissance and regional geology (1:62,500)	Regional geology (1:63,360 and larger)	Regional geology and target selection (1:200,000–1:50,000)	Regional geology (1:50,000–1:25,000)
Large-scale prospecting (1:31,680–1:12,000)		Large-scale prospecting (1:20,000–1:10,000)	Mineral deposits prospecting (general, detailed, estimative)
Industrial assessment of deposits	Industrial assessment of deposits	Industrial assessment of deposits	Preliminary exploration
Lease or sale of a deposit to a mining company or a contract with a company. Detailed exploration in parallel with driving and overburden operations.	Lease or sale of a deposit to a mining company or a contract with a company. Detailed exploration in parallel with driving and overburden operations.	Lease or sale of a deposit to a mining company or a contract with a company. Detailed exploration in parallel with driving and overburden operations.	Detailed exploration

Sources: A. R. Sushon, *Stadiynost i stoimost geologo-razvedochnykh rabot na tryordye poleznye iskopayemye v capitalisticheskikh stravakh (Stages and Cost of Prospecting and Exploration in Capitalist Countries)* (Moscow, Viems, 1970); Sushon, *Organizacia i economika geologo-razvedochnykh za rubezhom (Organization and Economics of Prospecting and Exploration in Foreign Countries)* (Moscow, Niedra, 1979).

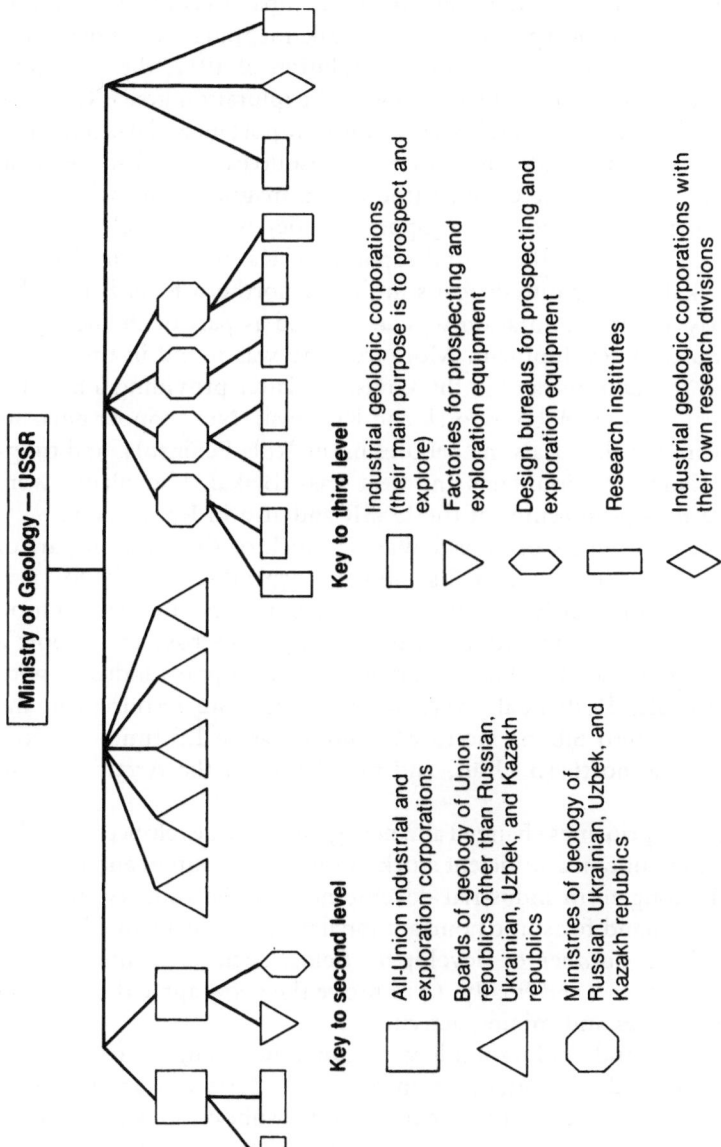

Ministry of Geology — USSR

Key to second level

All-Union industrial and exploration corporations

Boards of geology of Union republics other than Russian, Ukrainian, Uzbek, and Kazakh republics

Ministries of geology of Russian, Ukrainian, Uzbek, and Kazakh republics

Key to third level

Industrial geologic corporations (their main purpose is to prospect and explore)

Factories for prospecting and exploration equipment

Design bureaus for prospecting and exploration equipment

Research institutes

Industrial geologic corporations with their own research divisions

Figure 6-3. Organization of minerals prospecting and exploration in the Soviet Union.

Goals

In Czarist Russia mineral resources were little explored and known deposits limited in number. For the most part foreign companies controlled mining activities. No regular prospecting or large-scale mapping was undertaken. Within six months after the October Revolution of 1917, the new government formally announced prospecting and exploration goals. Of those goals, self-sufficiency in minerals was the most important and has remained so to this day. In the 1920s, huge iron ore resources were discovered in the European center of the country (the Kursk magnetic anomaly). Also discovered during this period were copper-iron deposits in the Ural Mountains, copper deposits in the central republic of Kazakhstan, tin deposits in Trans-Baikal, and apatite deposits in the far northern Kola Peninsula.

In the early 1930s, exploration was accelerated as part of the new goal of rapid industrial development. More iron ore was found in the Krivoi Rog basin in the Ukraine and in the Urals, the latter providing a base for the giant steel mills in Magnitogorsk and Kuznetsk. New copper deposits were discovered in the Urals, nickel ores in the Kola Peninsula, and tungsten in the Caucasus Mountains and near Lake Baikal. Phosphates were found in the European center of the USSR and also in Kazakhstan.

During the early years of World War II, much of the western part of the Soviet Union, upon which its industry depended heavily for raw materials, was temporarily devastated. The primary goal of exploration during this period was to provide a mineral resource base to support a shift in industry to the east. The second goal was to support industry with strategic minerals. Both goals were attained. Iron ore resources in the Urals and in western Siberia increased approximately 1.5 times, bauxite was found in the northern Urals, and nickel ores in the Arctic Norilsk region.

After the war, priorities changed and new goals were developed: namely, to achieve and maintain a sufficient stock of mineral resources and reserves to ensure the long-term industrial development of the country; to meet the growing demand for some nontraditional minerals, consumed in new applications by a number of developing fields, such as atomic power, electronics, and space; and finally, to improve the geographic distribution of known resources and mining activities.

As prospecting and exploration have become increasingly important in the post-war period, the organization of these efforts has grown more complex with the development of many regional subdivisions of the Ministry of Geology. Activities within the subdivisions are viewed and planned as interdependent elements of a global system, divided into stages beginning with geophysical mapping and ending with detailed operational exploration. The scientific base for the former has been highly developed, and modern methods for prospecting and exploration are used. Prospecting for shelf and ocean deposits has been accelerated. Collaborative programs with the Eastern European member-countries of the Council for Mutual Economic Assistance (COMECON) are also conducted.

In contrast with many other countries, the Soviet Union is nearly self-sufficient in mineral resources, importing only small quantities of a few mineral commodities. In spite of intense mining and associated depletion of some mineral deposits, reserves have increased in recent years due to successful prospecting and exploration.

A number of new prospecting districts and metallogenic provinces have been discovered, including the Badzhal tin district, the central Kazakhstan polymetallic belt, and the Charo-Tokkins iron district. New mineral deposits include the bauxite found at Vezhkhau-Vorykvinsk and Vyslovsk, the polymetallic ores at Kholodninsk, and the copper at Aktogai. Furthermore, the iron ore resources in the east—in the Urals, western and eastern Siberia, and the far east—have expanded considerably. Geologists have discovered the Charo-Tokkin, Sutam, and Larti ore basins with large explored and projected resources. The Aldan region in the eastern part of the country has been recognized as a large iron ore province. Exploration of the promising Porozhin manganese oxide deposit in the Yenisei Ridge was to be completed in the early to mid-1980s. New graphite discoveries in the Berdichev, Krivoi Rog, and Sub-Azov basins have been studied, reestimated, and prepared for prospecting and exploration in order to enlarge known resources of industrial graphite. A new apatite deposit has been discovered in the Zhytomir region, and complex apatite ores have been found in the Zaporozhye region, among others.

According to the statistics for the period 1970–80, 40 to 42 percent of the promising mineral occurrences identified during regional surveys and general and detailed prospecting were transferred to the estimative stage of prospecting. After estimative prospecting, 8 to 9 percent of the original promising mineral occurrences were designated as deposits, and 4 to 5 percent were transferred into the preliminary exploration stage as objects of primary interest. These average percentages vary considerably for different minerals; iron ore, phosphate, fluorite, and a few other minerals have the largest percentages.

Despite these successes, prospecting and exploration in the USSR face challenges. One is to expand the bauxite reserves in Siberia, where most of the country's aluminum production occurs. Similarly, the high-grade copper-nickel reserves in the Norilsk district and the Kola Peninsula need to be augmented. Finally, an intensive search is under way to add to the reserves of the lead and zinc mines in the Rudny Altai and Uzbekistan and the tin mines in the Primorje area.

We also face the important task of improving the geographic distribution of known resources, especially expanding resources in districts that have been mined for a long time. These are the southern Urals and northwestern Kazakhstan for iron ores, Rudny Altai for lead and zinc, and Primorje in northeastern USSR for tin. It is exceedingly important to ensure sufficient resources for the territorial-industrial complexes now being planned or developed (such as the region of the Baikal-Amur railway line).

With few exceptions, Soviet mineral reserves are sufficient to provide for the country's needs over the foreseeable future. Rather, it is the uneven distribution of these reserves that causes difficulties. It is thus not surprising that one of the principal goals of prospecting and exploration is rectifying this geographic imbalance. Indeed, this goal currently has priority over maximizing the increase in known resources.

Mapping and Other Efforts to Enhance Prospecting

At the same time, the Soviet Union is engaged in mapping at various levels. At the wide, regional level, data from traditional geologic surveying and remote sensing are being used to revise old surveys and to develop new methods of investigating large regions. Satellite and aerial geologic mapping along with ground checks of the remote-sensing data, all followed by interpretive mapping, have become compulsory practices in regional studies. Remote-sensing methods rather quickly provide considerable information about vast territories. All available geologic, geophysical, geochemical, geomorphological, and other maps of similar scales are used for interpretations. The country currently plans to use remote-sensing techniques to map all of its territories regardless of geologic potential at scales of 1:1,000,000 and 1:500,000. The program is to be accomplished during the next decade and a half.

The initial stage of prospecting proper, as pointed out in the preceding section, entails regional geologic and geophysical surveys at the 1:200,000 scale. Plans call for most of the territory of the USSR (22.4 million square kilometers) to be mapped at this level by the mid-1980s. Approximately 30 percent of the country is mapped at a scale of 1:50,000. This includes 80 to 100 percent of the most important mining areas in the European part of the country, the Urals, Kazakhstan, and central Asia; 60 to 70 percent of eastern Siberia and Primorje; and 30 to 35 percent of northeastern USSR. Over the last five years, the importance of large-scale geologic surveys increased considerably. These surveys enhance knowledge of the geologic structure and metallogeny of the USSR. They also have helped identify approximately 8,000 promising targets, of which, 1,700 have been recommended for estimative prospecting.

Deep drilling provides another means of increasing the effectiveness of prospecting. Most known ore fields have been drilled and studied down to depths of 0.5 to 1 kilometer. To advance our theories of ore formation and our understanding of the geographic distribution of ore deposits, we need a better understanding of ore sources, vertical zonation of ore-bearing structures, and metasomatic and geochemical aureoles. The intriguing possibilities opened up by deep drilling are demonstrated by the Kola super-deep borehole (SD-3), which has penetrated to a depth of more than

10,000 meters. Core samples and down-the-hole geophysical data from this borehole significantly altered understanding of the geologic structure of the Pechenga mining district, which previously had been based on surface geologic and geophysical evidence. We now know that structures favorable for hydrothermal mineralization extend to much deeper levels than previously assumed. Super-deep drilling is now under way in the Urals, western Siberia, and other important mining areas.

Another current effort to improve prospecting effectiveness in the Soviet Union involves searching for large, high-grade deposits in poorly prospected areas. These efforts, assisted by new exploration theories for tin, tungsten, gold, and porphyry copper deposits in eastern regions of the Soviet Union, have identified areas with significant ore concentrations, including several porphyry copper deposits.

Finally, the Soviet Union has undertaken research in three major areas to improve prospecting methods. The first seeks to develop theoretical and technical tools for use in underground prospecting. The second aims at advances in geophysical cybernetics to manage and optimize prospecting. The third focuses on developing new geophysical equipment for underground prospecting and exploration, drawing on scientific progress in associated areas such as magnetohydrodynamic generators and laser systems.

Level and Distribution of Prospecting and Exploration

Like most industrialized countries, the Soviet Union has increased prospecting and exploration expenditures in recent years. As shown in figure 6-4, expenditures increased at an annual rate of 6 to 7 percent from 1961 to 1980. From 1981 to 1985 expenditures continued that trend, rising at an annual rate of 6 percent. This growth reflects not only expanding geologic activities but also a rise in their costs, as prospecting and exploration have been conducted with greater frequency in more remote regions with poor access, at greater depths, and in more difficult geologic settings.

For example, between 1965 and 1980 the development of prospecting and exploration workings increased approximately 50 percent, while the cost per meter went up approximately 120 percent as workings in harder rocks and deeper boreholes became more prevalent. Over the same period the amount of mechanical column drilling increased by 36 percent, while the cost-per-meter went up approximately 60 percent. As a result, between the periods 1961–65 and 1971–75, prospecting expenditures per ton of extracted resources of commercial categories (A, B, and C_1) increased 2.2 to 2.4 times for tin, lead, and zinc, and 1.2 to 1.3 times for phosphates and bauxite.

The distribution of expenditures for nonfuel minerals by the USSR Ministry of Geology, which as noted earlier accounts for most of the prospecting and exploration in the USSR, is shown in table 6-2 for 1971

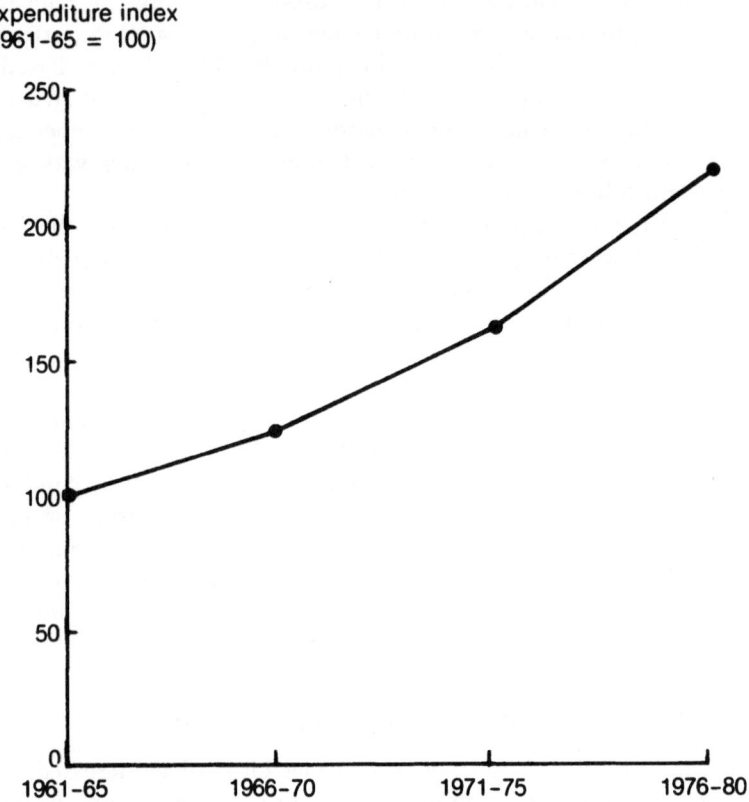

Figure 6-4. **Expenditure index for hard-mineral prospecting and exploration per five-year period.** *Source:* USSR Ministry of Geology.

and 1980. Prospecting expenditures have accounted for an ever-increasing
share of overall expenditures, as this activity has expanded in underde-
veloped and previously inaccessible regions and as the importance of deep
underground prospecting has increased in establishing mining areas. For
the period 1976–80, prospecting accounted for 52 percent of overall expenditures, compared with 40 percent during the period 1961–65. The
share of total expenditures devoted to prospecting varies among minerals:
it is 67 percent for tin; 53 to 58 percent for apatite, tungsten, and iron ore;
and only 35 to 42 percent for manganese and molybdenum.

Over time the allocation of prospecting expenditures has changed, reflecting shifts in mineral needs. Prospecting expenditures for antimony,
tin, and apatite increased by 200 percent between the periods 1961–65 and
1976–80. In contrast, expenditures rose only 100 percent for iron ore, lead,
and zinc; 50 percent for manganese and bauxite; and 20 to 30 percent for

Table 6-2. Distribution of Expenditures for Nonfuel Minerals by USSR Ministry of Geology, 1971 and 1980
(percent of total)

Stages and kinds of activities	1971	1980
Regional geologic and geophysical surveys	8.1	7.8
Prospecting	47.6	53.6
Preliminary exploration	10.2	8.5
Detailed exploration	16.2	15.6
Research activities	7.8	5.5
Other activities (e.g., construction, equipment purchase)	10.1	9.0
Total	100.0	100.0

Source: USSR Ministry of Geology.

copper and mercury. Expenditures for molybdenum, titanium, and potassium remained steady or even decreased. These changes reflect the industrial demand for previously discovered resources of each individual mineral material. For example, in the period 1986–90, prospecting budgets are expected to decrease for iron ore, copper, titanium, and potassium, for which explored resources have become sufficient.

Prospecting, as pointed out in the preceding section, encompasses three substages—namely, general, detailed, and estimative prospecting. Figure 6-5 shows that prospecting expenditures in the USSR are divided so that

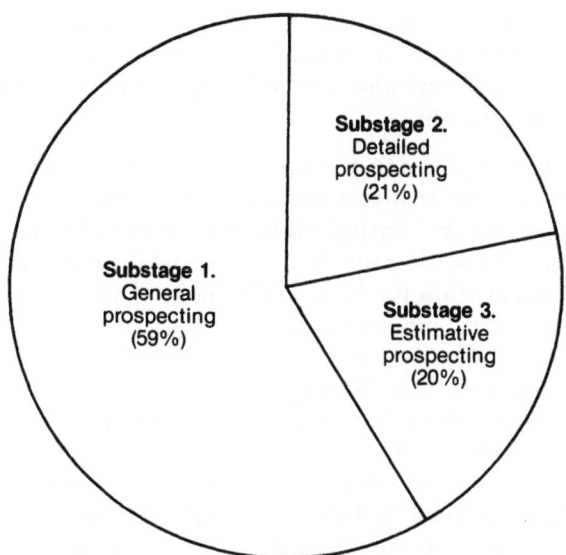

Figure 6-5. Expenditure shares devoted to the three substages of prospecting, 1978–82. *Source:* USSR Ministry of Geology.

more than half are devoted to general prospecting, with the rest split almost equally between detailed and estimative prospecting.

Planning and Decision Making

Decisions at all stages of prospecting and exploration are made in the context of their implications for national economic development as a whole. Prospecting and exploration activity together make up the initial element of the economy's overall mineral supply system.

The general goal of these activities is to meet the growing long-term demand for minerals in the national economy. The more specific economic goal is to manage efficiently the supply of mineral resources for the country. Some important points to be stressed are these:

- Expected mineral demand, not price, is viewed as the principal indicator of the need for exploration activities. Prices mirror demand, but no more. Thus no attempt is made to forecast prices; instead, national demand becomes the focus, and this can be estimated directly on the basis of the country's long-range plans and other factors.

- Long-term and current demands are taken into consideration. Current needs sometimes are sacrificed in the interest of future needs. An example is the development of mineral deposits in remote Siberian territories, which has been an expensive undertaking with payoffs expected only after intensive efforts.

- A multistaged, iterative, and adaptable procedure is used in planning prospecting and exploration specifically, and in developing mineral resources more generally.

The first step in this planning procedure involves the construction of a long-term (two decades or more) scenario of the national economy. Demand-supply matrices are constructed for this scenario by the State Planning Committee. Total demand for metals and financial and material resources are estimated on the basis of the national plan. The demand is then taken as an external input into the mining activities planning system.

In the second step, the mineral reserves and resources necessary to meet the projected demand for metals and other raw materials are estimated. Then, an assessment is made regarding which mineral reserves need to be increased and by what amounts.

In the third step, the Ministry of Geology allocates funds to various regions and corporations on the basis of the assessment made in the second stage and an analysis of specific projects prepared by regional geological surveys. In the final step, corporations select prospecting areas and sites to fund on the basis of the professional experience of geologists and other information. The decisions are usually made by the managers of regional corporations, although exceptions do occur. Decisions concerning a few important sites are made directly by the ministry headquarters staff.

Tools for Planning and Decision Making

A variety of models, many of which are computerized and highly mathematical, are used in planning mineral prospecting and exploration. The total system of models (being used or in preparation) is displayed in figure 6-6. The first of its three major sections consists of external-input models that describe the major links between mineral development and the economy as a whole. The second includes many specialized models of each prospecting and exploration stage, supplemented by auxiliary models for data collection, economic resource evaluation, and other purposes. The third section lists models for managing particular prospecting and exploration activities such as geophysical surveying and drilling.

For example, to identify areas favorable for prospecting and then to guide subsequent prospecting, geologists and other scientists are now using mathematical models that essentially compare characteristics of areas with known deposits with those of new areas under study. If the characteristics are similar, the new area may be recommended for further study. If not, the area will not be recommended. These models provide, for example, quantitative forecasts of ore tonnage, grade, and volume for various regions prior to extensive prospecting.

A computerized management system, called "Geology," is being developed to improve the management of prospecting and exploration. The system takes into account all the main participants, from the Ministry of Geology to the crews on geologic expeditions.

Organizations that now use computerized geologic and geophysical data-processing systems are compiling large data files in the course of prospecting and exploration. A standard system is being designed for geologic and economic estimates of ore resources. It provides for processing prospecting and exploration data, estimating resources under different economic and geologic assumptions, assessing the accuracy of the resource estimation under different scales of activities, and evaluating the possible economic damage in case the resources are not confirmed.

Assessment of Economic Value

One important goal of prospecting and exploration is to maximize potential economic value, that is, to maximize society's gain from these activities. This requires maximizing the difference between total revenues from the smelted ore and total costs. The latter consists of operating and capital expenditures for everything from prospecting through smelting,

216

EXTERNAL-INPUT MODELS (AIMS AND CONSTRAINTS)

Long-term plan for
prospecting and exploration

Optimal planning of mining activities
(national and regional models)

Resource base

Management of national
economy's mineral sector

MODELS OF PROSPECTING AND EXPLORATION ACTIVITIES

Auxiliary models

Managerial models

Stages

Regional geologic and geophysical
surveys (1:200,000 scale)

Geologic surveys
(1:50,000 scale)

General
prospecting

Resource evaluation of areas

Data collection

Resource evaluation of sites

Select areas

Prioritize prospecting areas

Optimize fund allocation

Simulation management model
for surveys and prospecting

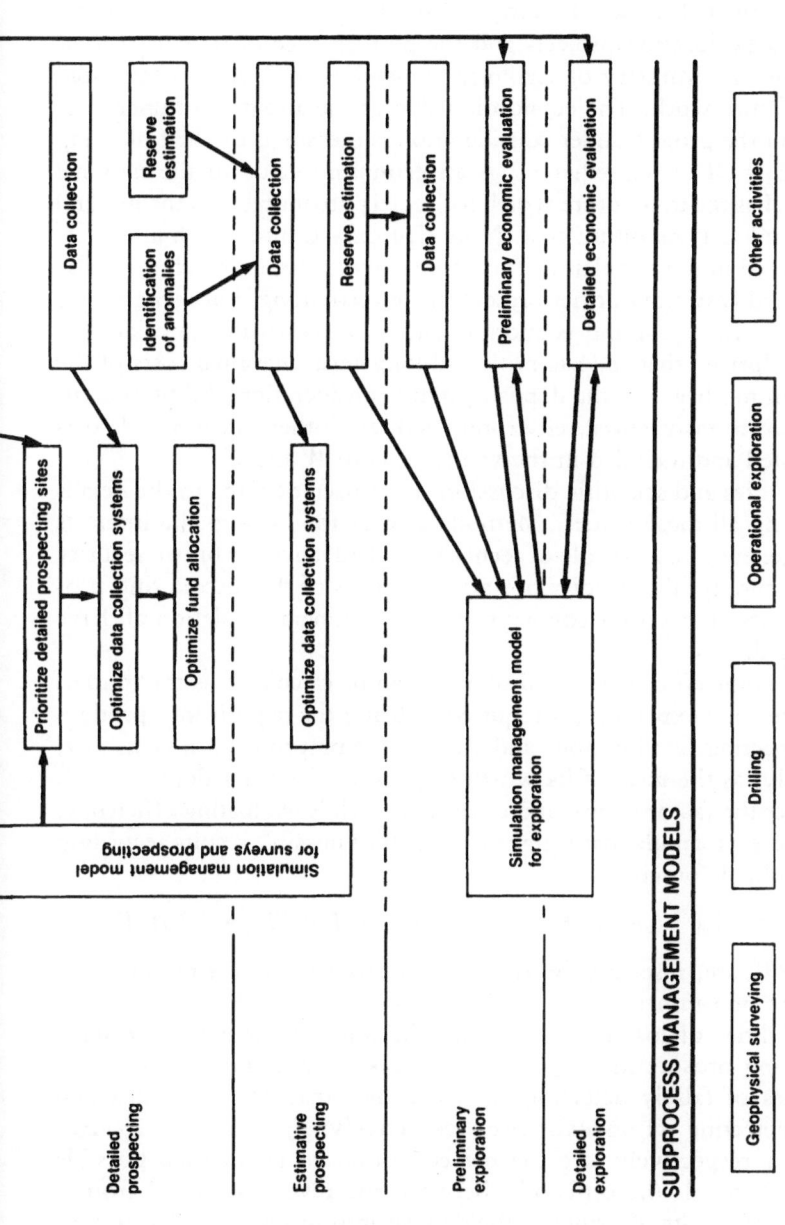

Figure 6-6. System of models for the prospecting and exploration management process. (Each box represents a model or series of models.) *Source:* A. Sushon, unpublished material.

including costs arising from environmental impacts. At each stage of mineral development, beginning with prospecting, estimates must be made of potential future benefits and costs, both direct and indirect.

Cost-benefit analyses are used to evaluate the quality of individual prospecting and exploration projects and the performance of structural divisions within the Ministry of Geology, as well as to develop centralized plans for future work. The economic value of the additional mineral resource from the project under consideration is measured by the following formula: $E = W - C$, where E is an economic value to the national economy; W is the total future worth for society attributable to the mineral resource; and C is its future cost. Note that only future worth and costs are considered in the calculation. To some extent, this approach is similar to discounted cash flow or net present value accounting, but some values are of a broader social and economic nature. W is usually calculated in terms of a "price" that reflects projected long-term marginal costs of the mineral commodity, not the deposit, under consideration. Additional indirect social, infrastructural, environmental, and other impacts and costs are forecasted and included in the values of both W and C.

Many studies and scientific discussions have been devoted to the details of estimating all these values. Methods of cost forecasting, the interests of future generations, the role of technological advance for future security of mineral supply, the validity of various aspects of the discounting procedure and its ability to account properly for mineral depletion all have been debated.

The economic efficiency of a discovery can be calculated as the ratio of its worth to total expenditures required to bring the deposit into production (prospecting, exploration, and subsequent preproduction costs); or, alternatively, as the ratio of its worth to prospecting costs alone.

To reflect the factor of risk and uncertainty while evaluating efficiency, specialists use a formula that weighs a number of possible results according to their probable outcomes:

$$\text{Expected value} = P(1)DE_d + P(1)D_oE_{do} + P(0)FE_f + P(0)F_oE_{fo}.$$

$P(1)$ and $P(0)$ are, respectively, the a priori probabilities for the presence and absence of a mineral occurrence; $P(0) = 1 - P(1)$. D and D_o reflect the probabilities of locating or missing the mineral occurrence, respectively, during prospecting; $D_o = 1 - D$. Likewise, F and F_o reflect the probabilities of falsely detecting an occurrence that does not exist and correctly detecting its nonexistence, respectively; $F_o = 1 - F$. E_d, E_{do}, E_f, and E_{fo}, respectively, are the net economic values of each possible outcome. E_d equals the difference between the revenues gained from a deposit and the costs of bringing the deposit into production. E_{do} and E_{fo} are the expenditures incurred while missing or failing to detect a mineral occurrence and while correctly detecting a nonoccurrence, respectively. E_f equals expenditures on prospecting while falsely identifying a mineral occurrence that does not exist, plus any costs incurred as a result of this misperception.

Expenditures related to prospecting and exploration make up a necessary component of the project evaluation. In order to simplify these estimation procedures, a system of standard expenditures is used by geologic organizations in the USSR. Prepared by special agencies and recommended by the Ministry of Geology, these standards provide a basis for planning expenditures and designing prospecting and exploration programs.

Depending on their purposes, the standards are highly aggregated or differentiated. At the differentiated end of the spectrum is the Single Rates Estimates Collection. Much more aggregated are the Special Aggregated Estimates and Regional Complex Estimates, which are used at the prospecting and early exploration stages, when little is known about the mineral occurrence under study. They incorporate only the most important components of overall expenditures (those related to prospecting, transportation, and temporary housing), differentiated according to the region of the country and type of mineral deposit.

In many cases, data collected during estimative prospecting are insufficient to be used during preliminary exploration in applying even Regional Complex Estimates. The best solution in these instances is to select a geologic analogue—a previously studied deposit that is similar to the deposit of interest. The exploration budget for the new deposit, in this case, is based on historical exploration expenditures for the old deposit. When the new deposit differs from the analogue in some recognizable manner, the exploration budget may be adjusted to reflect these differences.

When exploration projects are located in remote areas that lack an infrastructure, it is inappropriate to choose analogues simply on the basis of geologic information. Estimating accurately how much it will cost to build transportation systems, power plants, and other facilities may be considerably more important than estimating accurately how much it will cost simply to develop the deposit. In this case, several analogues may be used to determine the exploration budget. For example, one analogue might consider the costs of evaluating or proving up the deposit, a second the costs of processing, and a third the costs of transportation and other regional development.

Nevertheless, it is not a simple procedure to select analogues and to equate characteristics of the deposits under comparison. More than 20,000 deposits and occurrences have been discovered in the USSR. They have been evaluated separately, at different times, using different techniques. For each known deposit and mineral occurrence, scientists are collecting standardized information on geology, geography, ecology, mining engineering, and economics. Special state-run offices maintain the quality of the data, so that they can be used in standardized calculations, and coordinate data collection between regional offices and Moscow. These

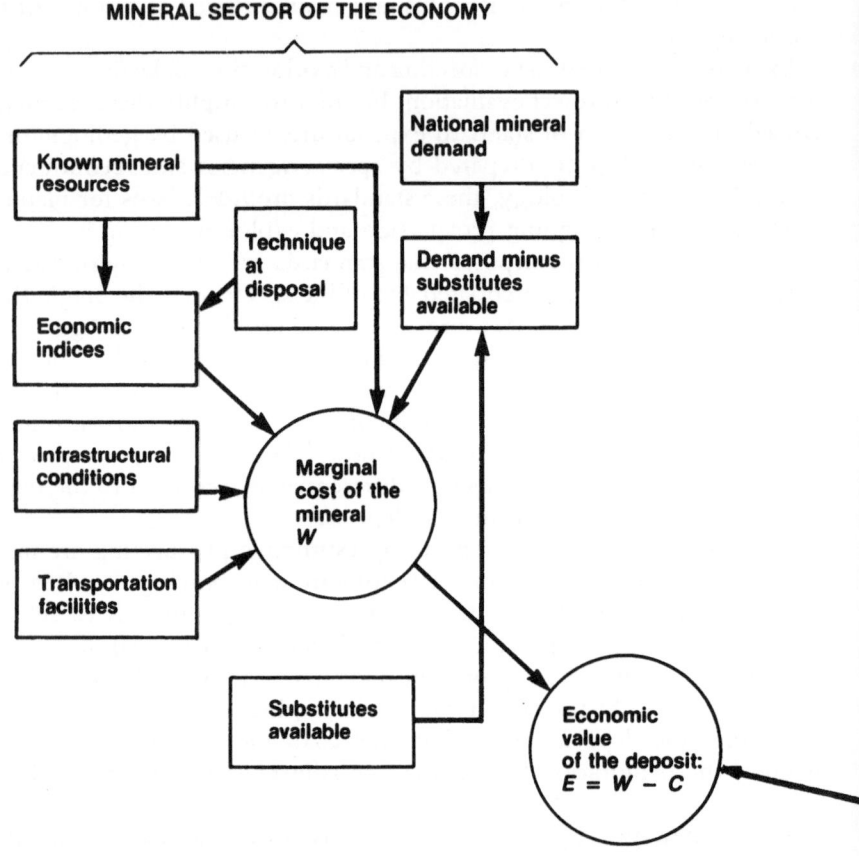

Figure 6-7. Factors influencing the potential economic value of a mineral deposit. (See discussion in text.)

data play an invaluable role in improving prospecting and exploration planning.

Several general principles are used when making decisions during prospecting and exploration:

• During early stages of prospecting, when potential benefits and costs are very uncertain, the decision-making procedure is adaptive and employs a number of simplified methods and rules of thumb.

• A decision is never permanent. When available data are scarce, broad decisions are made; when additional data become available, decisions become more detailed and more long-term in scope. The guiding principle is to make decisions that can evolve as conditions change with time and new information is gained.

DEPOSIT UNDER CONSIDERATION

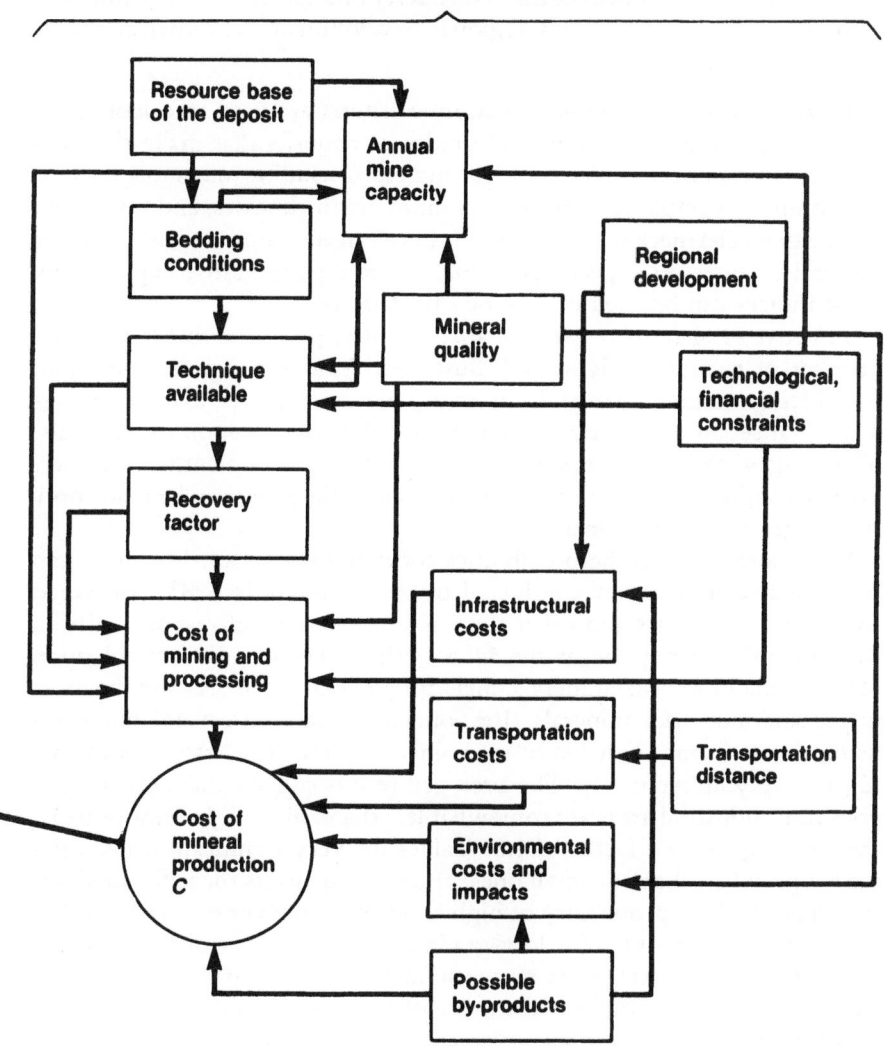

• The criteria for decision making are specialized at each stage according to the different objectives and data availability. At the earliest substages of prospecting, geologic considerations weigh heavily in the decision-making process. Economic criteria become more important at the more advanced stages of prospecting and exploration. During preliminary exploration, economic values are assumed to be constant over time, whereas during detailed exploration these values are carefully forecasted throughout the total life of the project.

• Additional noneconomic considerations (such as environmental, social, infrastructural, and regional development concerns) are taken into account.

In view of these principles, economic criteria are used to calculate the marginal geologic conditions and grades of deposits that make their extraction uneconomic. The geologic indices (including measures such as minimum acceptable ore grades, minimum ore thicknesses, and maximum impurity levels) mark the cutoff points between economic and uneconomic mineral occurrences. Special bulletins, updated periodically, explain how these indices can be used in specific calculations.

Effective long-term resource management requires the forecasting of a number of factors, as figure 6-7 illustrates. Among the more important variables are national demand, discovery of new deposits, and technological advances in exploration, mining, and the consumption of minerals. The demand forecasting system used in the USSR was mentioned earlier. Forecasts of new discoveries and advances in technology can be made only with a great degree of uncertainty.

Some insights into the possibilities for new discoveries, however, can be gained from a survey conducted by A. Astakhov in 1983. Answered by twenty geologists and other scientists in positions of responsibility in the mineral-rich regions of the USSR, the questionnaire focused upon nineteen major nonfuel minerals, including several metal ores and the most important industrial minerals. Respondents answered several questions using a scale from: "very likely" (4 points), "likely" (3), "rather unlikely" (2), to "very unlikely" (1). The averaged results suggest that new discoveries near industrial areas are somewhat less than "likely." In remote areas, they are higher by a factor of 1.3. Values for very significant discoveries are roughly half those of small ones, and three-quarters those for medium size deposits. The probability of high-grade ore discoveries is rather high, only three-quarters that for low-grade ores. No serious differences, according to the experts, exist between shallow and deep deposits, or between simple and complicated modes of occurrence. The probability of new findings in areas with inconvenient climatic conditions is rather high.

The experts were not greatly concerned about depletion, and they expected Soviet reserves to remain stable or even to increase over the foreseeable future. They also were asked to rank various ways to compensate for depletion of known mineral resources, and their rankings from most to least important follow: (1) mining lower-grade ores, (2) discovering new deposits of traditional resources, (3) mining at greater depths below the surface, (4) exploring for additional reserves at known deposits, (5) exploiting small deposits, (6) developing nontraditional mineral resources, (7) substituting other minerals, (8) improving recovery rates at known deposits, and (9) lowering per capita mineral consumption.

They saw technical advancement as the main factor that would keep the costs of resource development from rising in the future. They were also asked whether the positive influence of technical advancement on costs would overcome the negative impacts of lower ore grade, more remote mines, and other increasingly difficult natural conditions of mineral resources development in the future. The question was asked first with regard to new discoveries and second with respect to known but yet unexploited deposits. In both cases the responses were evenly divided between yes and no. As for the discovery costs per one ton of new metal, a related but narrower question, the prevailing view was that they would increase by at least 30 percent.

Certainly, expert opinion is not always the most reliable forecasting technique, but such opinions may provide useful insights. In any case, the views of experts, whether they are right or wrong, are likely to influence analysis concerning the overall level and distribution of prospecting and exploration expenditures.

References

Astakhov, A. S. 1981. "Principles of Systems Approach to Economic Evaluation of Mineral Resources." In 4 Pryrodnye resurcy i okruzhayushchaya sreda. *Dostizhenya i Perspektivy* vol. 23, no. 8, pp. 21–29.

Kozlovsky, Je. 1982. "USSR Geology and Future Research Goals," in *Soviet Geology* no. 12, pp. 3–18 (in Russian).

Sushon, A. R. 1970. *Stadiynost i stoimost geologo-razvedochnykh rabot na tryordye poleznye iskopayemye v capitalisticheskikh stravakh* (*Stages and Costs of Prospecting and Exploration in Capitalist Countries*) (Moscow, Viems).

———. 1979. *Organizacia i economika geologo-razvedochnykh za rubezhom* (*Organization and Economics of Prospecting and Exploration in Foreign Countries*) (Moscow, Niedra).

U.S. Bureau of Mines and U.S. Geological Survey. 1980. *Principles of a Resource/ Reserve Classification for Minerals.* U.S. Geological Survey Circular 831 (Washington, D.C., Government Printing Office).

Appendix 6-A
Mineral Resource Classification

A uniform resource classification system is used throughout the Soviet Union. The system covers mineral resources that have been explored, estimated, and projected.

Explored resources encompass Categories A, B, and C_1. **A-category** resources consist of the most highly proved part of explored resources. They have been thoroughly investigated, and the mode of occurrence, shape, and structure of an ore body are known. This information is derived from drilling and mine workings.

B-category resources include those deposits whose characteristics have not been studied quite so thoroughly, although some of their major characteristics have been delineated. The extent of these resources are determined with the help of information from drilling and mining operations, but their extrapolation is permitted to a limited extent only.

C_1-category resources are even less well delineated. Only their most general characteristics are known. This information comes from drilling and geologic and geophysical data.

Preliminary estimated resources, the **C_2 category**, are established on the basis of geologic, geophysical, and geochemical evidence, and measurements of the ore body in exploratory workings. Resources in this category also can be estimated by extrapolation of geologic data. The size,

Table 6-A-1. Approximate Correspondence of Resource Categories in the United States and the Soviet Union

Country	Total resources				
United States	Identified resources			Undiscovered resources	
	Demonstrated reserves		Inferred or possible reserves	Hypothetical resources	Speculative resources
	Measured or proved	Indicated or probable			
Soviet Union	Explored reserves			Projected resources	
	A + B	C_1	C_2	P_1	P_2	P_3

Note: A, B, C, and P indicate categories of resources, as discussed in the text of this appendix.
Sources: A. R. Sushon, *Stadiynost i stoimost geologo-razvedochnykh rabot na tryordye poleznye iskopayemye v capitalisticheskikh stravakh* (Stages and Cost of Prospecting and Exploration in Capitalist Countries) (Moscow, Viems, 1970); U.S. Bureau of Mines and U.S. Geological Survey, *Principles of a Resource/Reserve Classification for Minerals*, USGS Circular 831 (Washington, D.C., U.S. Government Printing Office, 1980).

shape, and structure of an individual ore body is only approximately defined. Its quality and technological properties are inferred by analysis of limited samples or by analogy with similar deposits.

Projected resources of basins, areas, and fields are estimated by analogy with similar and explored resources elsewhere. They are subdivided into three categories: projected resources of explored deposits or those under exploration (P_1); projected resources of undiscovered deposits thought to exist on the basis of evidence from geologic surveys, prospecting, and geophysical and geochemical tests (P_2); and projected resources of potentially promising areas where new deposits may be discovered (P_3).

Table 6-A-1 compares the Soviet resource classification system with that used by the U.S. Geological Survey and Bureau of Mines. Category A and B resources, for example, are comparable to proved reserves, and category C_1 to probable reserves. Categories C_2 and P_1 correspond to possible reserves, while categories P_2 and P_3 are undiscovered resources.

7

Multilateral Exploration Assistance: The United Nations Programs

MELINDA CRANE-ENGEL
ERICH SCHANZE

Among the most significant changes in the international mining industry during the past twenty-five years has been a trend toward greater heterogeneity. Whereas private mining firms formerly dominated all phases of mineral development, including exploration, public capital from both mineral-exporting and mineral-importing countries has become increasingly important to the industry. Public capital has flowed into mineral exploration through multilateral channels as well, under the auspices of development assistance programs. This chapter examines multilateral exploration assistance programs, particularly those connected with the United Nations (UN), and their influence on the worldwide search for minerals.

Objectives of Multilateral Exploration Assistance

Multilateral mineral exploration assistance programs have two general objectives. The first is to increase the international supply of minerals and the second is to promote the development of Third World countries. Of the two, however, economic development is generally considered to be the principal reason behind the major assistance programs.

Supply-Side Objectives

The supply-side view rests upon either of two assumptions concerning private foreign investment in mineral exploration.

227

The first and more prevalent assumption relates to a historical geographic distortion in worldwide mineral investment patterns. Beginning with the energy crisis in the mid-1970s, fears of import dependence and scarcity began to arise in the nonfuel resources realm (Bergsten, 1974; Mikesell, 1974). The U.S. proposal in 1976 for an International Resources Bank, like the Brandt Commission's similar call four years later, drew attention to the high concentration of exploration expenditures by market economies in a few developed countries (World Bank, 1977a; Brandt Commission, 1980). Given the historic concentration of mining in developed countries, observers predicted that exploration in developing countries stood a higher chance of success; consequently, the failure of the developing countries to attract a larger share of exploration funds was interpreted to mean that those funds had not been allocated according to purely geologic or commercial considerations (World Bank, 1977a, pp. 8–9). Political risk was held accountable for this "exploration gap."

Proponents of this view predicted that the exploration gap would lead to considerably higher prices, as firms explored for and brought into production relatively high-cost deposits in developed countries in preference to lower-cost but "riskier" properties in developing countries (British–North American Committee, 1976, pp. 14–19; Mikesell, 1979, pp. 11, 19). Moreover, if the former were exhausted and the latter remained unexplored (and therefore undiscovered), long-run shortages also were possible. According to this view, then, multilateral initiatives to increase capital flows into the Third World minerals sector were required to counter the threat of higher mineral prices and shortages and their deleterious effects upon mineral-consuming economies. It is not surprising that certain of these initiatives, such as the 1976 U.S. proposal, emanated from industrialized countries dependent on mineral imports.

While the exploration gap and its supply-side effects continue to be cited as reasons to support the existence of multilateral exploration assistance programs (United Nations Secretariat, 1980a, p. 4; Fozzard, Gurman, and Huhta, 1983, pp. 1–2), a second variant of the supply-side view has gained some measure of credence. This theory suggests that distortions in the overall level and timing of exploration expenditures are due to market imperfections. High interest rates and imperfect capital markets may limit the ability of companies to borrow and thereby discourage exploration during periods of slack demand and declining mineral prices, when internal funds are low (Stockmayer, 1982, p. 66). Exploration expenditures become "lumpy," highly vulnerable to shifts in the business cycle. A second type of market imperfection—an oligopolistic industry structure—makes investment patterns lumpy in a second respect: exploration decisions may be made on the basis of other firms' actions rather than solely in response to cost and price variables. It is argued that both phenomena may lead to a misallocation of resources if exploration is left in the hands of a few private firms.

Development Objectives

The second rationale for multilateral assistance focuses on the needs and welfare of the developing countries rather than the developed countries. This view also has two basic variants. The first is premised in part upon investment shortfalls. Resources represent a possible avenue toward industrialization, fiscal revenues, foreign exchange, and the transfer of skills and technology. This avenue, however, is blocked for many developing countries by limited capital and technology. These domestic problems are compounded by the decline in private foreign investment in exploration generally and in Third World exploration in particular. Multilateral assistance, according to this view, provides a crucial means of removing these impediments by stimulating the flow of capital into the Third World minerals sector.

A second variant of this view sees the barriers to resource development within the context of bargaining theory, rather than in connection with a presumed exploration gap. As expounded by Vernon (1967), Smith and Wells (1975), and others, bargaining theory draws attention to the role of information in altering the dynamics between a host country and a foreign investor. According to this theory, private investors may be willing to undertake exploration within developing countries but at a price varying inversely with the prevailing level of geologic uncertainty or political risk (Smith, 1977, p. 89; Kirchner and coauthors, 1979, pp. 159, 185). The purpose of multilateral assistance, therefore, is to provide the mineral-producing country with a neutral source of information concerning its resources (Smith, 1977, p. 89). No longer need it commit to mineral policies in a vacuum; possessing at the outset the data needed to assess its bargaining power and resource stock, the mineral-producing country is able to make informed decisions about the need for and acceptable terms of foreign investment.

Multilateral Exploration Assistance: The Development Banks and the European Community

Mineral explorers normally must rely on risk capital to finance the preliminary stages of exploration, as loans are seldom available for investments that do not promise definite returns within a certain time period (see chapter 2). Third World mineral exploration is no exception. Of the various multilateral exploration assistance efforts, only two have participated to a significant degree in the high-risk stages of mineral exploration—the United Nations Development Programme (UNDP) minerals program and the United Nations Revolving Fund for Natural Resources Exploration (UNRFNRE). Before examining these two UN programs in depth, the chapter provides a brief overview of multilateral exploration assistance provided by the development banks and the European Community.[1]

One outcome of the 1976 U.S. proposal for an International Resources Bank was a report recommending an expanded role for the World Bank in the nonfuel minerals sector (World Bank, 1977a). This study suggested that the bank (1) help negotiate concession agreements between foreign mineral investors and host countries, and (2) provide token loans to host countries for the financing of exploration. The latter was to be carried out with or by a foreign sponsor that would guarantee repayment in the case of an unsuccessful exploration.

The recommended extension of the bank's nonfuel mineral activities has not occurred. There has been some qualitative shift away from the development and operation of mines, upon which the bank had concentrated before 1977, toward advanced exploration and reexploration for the purpose of rehabilitating or expanding existing mines. The first category of advanced exploration includes engineering/feasibility study loans—some cofinanced by bilateral donors but not guaranteed by investors as recommended in 1977—to Peru (1980/phosphate feasibility study/$7.5 million); Burundi (1981/nickel feasibility study/$4 million); and Thailand (1981/potash feasibility study/$8.9 million) (World Bank, 1980, 1981). Loans to Bolivia (1977/$12 million) and Mexico (1980/$40 million) illustrate the second, reexploration category; both provided assistance to enhance the productivity of existing small-scale mines (World Bank, 1977b, 1980). Bolivia's National Mineral Exploration Fund received another loan of $7.5 million for purposes of re-lending to small miners for exploration work (World Bank, 1979a).

Two regional development banks—the Inter-American Development Bank (IADB) and Asian Development Bank (ADB)—also have expressed an interest in devoting more funds to the nonfuel minerals sector generally and to exploration in particular. Nevertheless, support from both banks for mineral-related projects remains relatively minor. IADB has accorded three loans for exploration and feasibility work: to Argentina (1964/iron ore feasibility study/$100,000); to Brazil (1974/regional exploration/$4.3 million); and to Colombia (1977/phosphate feasibility study/$1.7 million). ADB has financed three projects in its member states: Korea (1977/mineral resources survey/$200,000); Thailand (1979/aeromagnetic survey/$10 million); and Bangladesh (1980/improvement of the exploration capacity of the Geological Survey/$6.2 million) (Giraud, 1983, p. 266).

The Lomé Convention of 1979 between the European Community and the African, Caribbean, and Pacific States (ACP) allotted the equivalent of about $400 million (U.S.) in risk capital to the ACP states by the European Investment Bank (EIB). An additional $980 million (U.S.) has

---

[1]All dollar amounts throughout this chapter are given in current U.S. dollars for the year in question and have not been adjusted for inflation.

been allocated in loan capital. The first Lomé Convention in 1974 had simply declared such funds available for use in a variety of areas, but Lomé II specifically referred to mineral exploration work—preliminary as well as advanced—as a possible use. However, little of the allotted money has supported mineral-related work, with the exception of feasibility studies in Ghana (diamonds), Senegal (iron, phosphate), Zaire (tin), and Uganda (copper) (Cheysson, 1980; Europäische Investitionsbank, 1983; Giraud, 1983, p. 265).

United Nations Development Programme

The United Nations Development Programme (UNDP) was established in 1966, through the merger of the UN's Expanded Program of Technical Assistance and its Special Fund. The UNDP was founded to provide developing countries with financial and technical assistance in numerous fields. In the minerals sector, UNDP endeavors not only to provide technical exploration support but also to strengthen the capability of developing countries to carry out future exploration on their own.

Operational Structure

UNDP provides financing for technical assistance projects that for the most part are executed by other departments or agencies within the UN system. These include the United Nations Industrial Development Organization (UNIDO), the International Labor Organization (ILO), and the United Nations Secretariat Department of Technical Cooperation for Development (DTCD). Occasionally, the International Atomic Energy Agency (IAEA) or the World Bank will execute UNDP projects.

This section of the chapter focuses primarily on projects executed by DTCD, as these constitute more than 85 percent of all UNDP mineral exploration projects. As of 1983, DTCD had eleven New York-based technical advisers and seven other professional staff members, who supervised project teams operating in the field. In general, the technical advisers are drawn from a range of earth sciences and engineering disciplines, as are the consultants hired to serve on project teams. Technical advisers and consultants have come to DTCD from both private industry and from public service in national geological surveys, mining departments, and other agencies.

When the Special Fund began operations in 1959, it had $34 million at its disposal, of which approximately $5 million was allocated to mineral projects. This amount grew steadily through 1970, when the annual budget for mineral-related assistance through the UNDP reached $20 million (Bosson and Varon, 1977, p. 48). In 1971, however, UNDP adopted a new method of assistance programming under which the recipient state was given the primary responsibility for determining the distribution of funds among different sectors and for identifying projects within them. The low priority assigned to the mineral sector by most governments, combined with UNDP funding cutbacks during the 1970s and 1980s, have greatly decreased in real terms the UNDP mineral-related contribution.

Project Design and Selection

Prior to the 1971 reorganization, UNDP chose mineral projects on the basis of economic and geologic criteria. The country programming system leaves the selection and design of all projects to the initiative of recipient countries. UNDP is financed through the contributions of UN member states. These funds are allocated to individual developing countries on the basis of size and economic need (GNP per capita). Each recipient may draw upon different UN services, including those related to mineral exploration, up to the total amount of its respective allocation, or its Indicative Planning Figure (IPF). Thus, the IPF represents a pool of available funds to be apportioned among projects proposed by different ministries within the particular country. The country's central planning authorities select those project proposals considered meritorious and forward them to UNDP headquarters in New York for approval.

The influence that the UN can bring to bear in planning exploration projects depends on the country and project involved, varying with the recipient state's level of development and familiarity with the mining sector. The point of contact between the recipient and the organization is generally the UNDP resident representative in the country concerned, rather than the minerals experts in New York. To the extent that the representative is sufficiently well-versed in relevant technical and institutional areas, he or she presumably can suggest to the country those UNDP services most relevant to its needs, and thereby attempt to steer the selection and design of the project. In the last analysis, however, the recipient country has the decisive voice in determining how and where UNDP grants are to be allocated.

To the extent that UNDP and DTCD are able to influence project selection, they have attempted to encourage projects intended to improve the follow-up capabilities of the recipient country itself, as well as projects designed to attract foreign investment. A recently initiated investment promotion campaign has laid particular emphasis upon the latter. This campaign encompassed a three-pronged effort to attract follow-up investment through identification and formulation of projects intended to (1) generate information of a specific, commercially appropriable nature; (2) upgrade the transmission of information to potential investors; and (3) provide assistance and information designed to minimize uncertainty concerning property rights.

Budget Allocation

Total contributions to UNDP have averaged between $600 million and $700 million in recent years. Over $100 million are used to pay annual UNDP program and administrative costs. As a result of the country programming system, about 3 percent of the total budget, some $18 million a year, is allocated to mineral-related activities. Roughly 13 to 14 percent of this amount is used to cover the overhead costs of DTCD. Almost half of the rest is allocated to minerals projects in African countries, while Asian and Latin American projects account for approximately 25 percent and 10 percent, respectively (Lewis, 1983).

Between its inception (in 1966) and 1983, UNDP had undertaken more than 250 projects and had spent or committed almost $230 million (United Nations Secretariat, 1980a, appendix; United Nations Department of Technical Cooperation for Development, 1983). As of 1979, recipient governments had contributed the equivalent of $115 million in their respective local currencies; they presumably committed an additional $44 million to projects begun in the four years thereafter.[2] The following sections describe in greater detail the activities supported with these funds.

Institution-Strengthening Projects

About 30 percent of the projects undertaken by UNDP have been designed to strengthen mineral-related institutions ranging from geological surveys and ministries of mines, to analytic laboratories and the geology departments of national universities. Both consultants and equipment may be employed to accomplish this objective.

Institution-strengthening projects generally emphasize training. Each consultant supervises the on-the-job training of a national counterpart. In addition, nationals may be sent abroad for fellowships or group study tours including visits to similar mineral-related institutions in neighboring countries. Training may also be provided by the manufacturer of equipment supplied by UNDP.

The institution-building projects supported by UNDP have included the restructuring of the Greek ministry of mines, in which competition and overlap among departments hampered efficient surveying and basic exploration. To enhance expertise in the analytic and engineering areas important to later phases of the mining cycle, UNDP has helped two Latin

---

[2]As of 1979, UNDP had spent about $165 million on approximately 200 projects. As of 1983, close to $64 million had been spent or budgeted for projects begun after 1979. Data for government contributions to projects begun after 1979 are unavailable. Our estimate of $44 million is based upon the fact that the government contribution generally totals about 41 percent of the total project budget.

American countries to establish centers for mining and metallurgical research (United Nations Secretariat, 1980c, pp. 3–4). UNDP also contracted with a mining consulting firm to train personnel of the Turkish Mineral Research and Exploration Institute to carry out feasibility studies at standards acceptable to international lending institutions. The contractor supervised Turkish nationals preparing studies on previously known but unevaluated copper and iron ore deposits and also sent trainees on fellowships to its home country (United Nations Secretariat, 1980b, p. 5).

Another project in a country with a relatively high level of geologic expertise provided the staff of the national institute for geology and mining with specialized practical and theoretical training in areas such as mapping and plate tectonics, economic geology, geostatistics, and electronic data processing. In addition, UNDP funded a consultancy in labor economics and industrial relations, as well as fellowships in business administration and accounting.

As the preceding examples demonstrate, most UNDP institution-building projects involve cooperation with public agencies rather than commercial organizations. However, UNDP on occasion has worked with state-owned enterprises. In Ghana, for instance, a $2 million project to strengthen the management and supervisory capacity of the state gold-mining corporation provided consultants in geology, metallurgy, mining, and mechanical and electrical engineering. A second project supplied the corporation with a raise-borer and personnel trained in its use and maintenance (United Nations Secretariat, 1980b, p. 4).

Direct Exploration Projects

The projects described in the preceding section were aimed primarily at improving the ability of recipient countries to carry out autonomously various phases of the mining cycle. Projects involving direct exploration, either exclusively or in conjunction with institution-building activities, may perform a similar function in that they also emphasize training and provision of equipment that in most cases remains the property of the recipient country. However, these projects, which account for almost 70 percent of all UNDP mineral-related endeavors, often involve exploration that the recipient country can follow up only with access to external capital or expertise or both.

The majority of successful UNDP projects have led to the discovery of metallic minerals. The hope that such minerals could be developed to provide foreign-exchange earning exports, as well as the fact that demand for industrial minerals in many developing countries is limited, accounts for this focus. UNDP projects have located precious metal, nonferrous heavy metal, iron, and ferroalloy prospects, whereas industrial minerals and uranium have accounted for relatively fewer finds (United Nations Secretariat, 1980a, appendix) (see table 7-2 in the section below on Results of UNDP Exploration).

DTCD has carried out a great deal of grass-roots exploration on behalf of UNDP. Although the data generated may be applied to the nonmineral sectors, DTCD prefers to limit its role to mineral exploration rather than to the production of more general geologic information. In the view of DTCD, the budget and time constraints faced by many developing countries favor an approach that emphasizes from the beginning mineral discovery (Carman, 1979, p. 17).

UNDP-financed work in the large-scale reconnaissance stage has involved techniques such as airborne geophysical or geochemical surveying and geologic or geochemical stream sediment sampling. DTCD has flown over 1 million kilometers in airborne geophysical surveys and has undertaken the production of numerous large-scale geologic maps (Lewis, 1983). Projects undertaken within this stage have been distinguished by the size of the areas explored, which individually have covered as much as 250,000 square kilometers (Carman, 1979, p. 31). So long as these efforts are systematic and comprehensive, the chance of finding targets meriting further detailed work will improve as the size of the project area increases. However, a very large scale also may increase the risk of "butterflying": exploration activities that, by virtue of an overambitious work plan or inadequate budget, produce too little information to identify targets for detailed follow-up work. As the project descriptions below reveal, butterflying problems occasionally have decreased the quality of information produced by UNDP exploration.

In some cases, UNDP assistance has ended after the first phase of the mining cycle, with the provision to recipient-country governments of maps and other data obtained. When projects have resulted in the delineation of target areas, final project reports have included recommendations for detailed reconnaissance. In other cases, DTCD has gone on to perform follow-up work directly. In recent years, DTCD has begun to emphasize smaller project areas and detailed work on prospects previously identified through its own work or that of others. This shift reflects in part the department's policy that stresses participation in the downstream phases of the mining cycle, but it also may be attributable in part to UNDP budget cuts that tend to render large-scale projects less practicable.

Case Histories of Exploration Projects

The following case studies describe projects in which UNDP has supported activities ranging from large-scale or detailed reconnaissance to prefeasibility work.

African Case History

In one African country UNDP has expended more than $3 million for eight years of regional and detailed reconnaissance work, including technical analyses of promising findings. The project began in the mid-1970s with regional stream sediment surveys and geophysical reconnaissance in one promising region in the country. In the project's second phase, this work was followed by semidetailed geochemical reconnaissance, which in turn prompted the sampling of almost fifty anomalies. These samples resulted in the discovery and preliminary evaluation of promising nickel sulfide mineralization. In another region, preliminary geochemical reconnaissance and airborne geophysical surveys indicated a potential for certain precious and heavy metals and alloys, including gold, copper, cobalt, lead, and zinc. The project's third phase, scheduled to run from 1980 to 1983, concentrated on further mapping and surveying of this latter area, with detailed follow-up on promising occurrences.

UNDP contributed over $3.2 million to the project, covering the costs of foreign consultants, travel, equipment, and training. Seventy-two percent of the UNDP share financed the services of several exploration geologists, a geochemist, a geophysicist, an analytical chemist, a spectroscopist, and an administrative staff. Another 14 percent supplied analytic, mineralogical, and geochemical equipment such as an atomic absorption spectrophotometer, an ion meter, a polarizing microscope, spectrometers, and proton magnetometers. The government contributed over $1 million to pay for the services of the national project team, composed of a project director and counterparts to the foreign experts, as well as for equipment and office and laboratory premises.

The government and its project team worked well for the most part with the UNDP and DTCD staff members. This cooperative working relationship facilitated the transfer and efficient use of information; in formulating and implementing a systematic work plan, the project team was able to rely upon existing photographs, maps, and data provided by the government. At one point the government sought to expand the project into yet another unexplored region, but it was convinced subsequently that the limited time and resources available would be better spent on the narrowly defined work plan that originally had been approved. In this particular case, DTCD and UNDP appear to have possessed significant influence over decisions about project design or adaptation.

Along with a comprehensive terminal report, the project members produced semiannual progress reports, technical data, maps and diagrams, and various technical reports. The final report contains a detailed account of project activities, describing, among other things, sampling methods and densities, analytic and ground follow-up techniques, and the number, angle, and depth of drill holes completed. One section summarizes the grade and characteristics of uranium deposits discovered, the length and thickness of nickel mineralization encountered, and the location and merits of anomalies delineated. The report concludes with specific follow-up recommendations on methods to be used in further investigation. This final section also contains general suggestions, based upon project experience, for the use of existing maps and diagrams in inferring the geologic characteristics of other regions in the country. Along with the other reports issued during the project, this final report was turned over to the government at the end of the project.

Although on the whole this information looked promising, a breakdown in the DTCD-government working relationship marred the quality of the data generated in one respect. As the project neared its conclusion, DTCD's influence was overridden by government officials who were anxious to begin to develop one of the deposits discovered. DTCD was pressed to cease its operations when only one half of the deposit had been fully prospected. As a result, the tonnage figure that was reported was lower than it would have been with additional drilling. The investment-dampening effects of this were aggravated by the reluctance of the government officials to seek private investment by disseminating the necessary information to interested companies. The government later changed its approach, but the case illustrates the difficulties that may arise because technical reports and data are supplied to the government, which alone possesses the authority to decide whether and to whom the information will be revealed.

Asian Case History

A long-running project in Asia illustrates the problem of butterflying, referred to earlier in this chapter. With a total budget of more than $2.5 million, composed of a UNDP contribution of $2.1 million and a host-country share of $400,000, the project was to have conducted both an integrated survey of a region covering 16,000 square kilometers as well as specific investigations on the feasibility of known gold, iron ore, and industrial mineral deposits.

Although the team of experts supplied by DTCD was similar in composition to that of the African project described earlier, its effectiveness appears to have been hampered by frequent changes in the host-country staff. These uncertain working relationships may have exacerbated the risks inherent in an ambitious and complicated work plan.

A low priority was assigned to one of the most important objectives of the project: the completion of the regional reconnaissance step in the exploration process. The basic mapping and surveying of the region was especially important because the mineral targets to be investigated had been identified on the basis of incomplete data. In spite of this, the regional reconnaissance was undertaken in a fragmented fashion. Of three planned stages—(1) reconnaissance, geochemical, and mineralogical surveys, (2) followed by detailed geochemical work and geologic mapping, and (3) ending in a geophysical survey, drilling, and trenching—only portions of the first and the second were completed. Project resources and available data were inadequate to support the intended compilation of geologic maps; as a result, the project did not remedy the dearth of comprehensive geologic data concerning the region.

The main focus of the project constituted prefeasibility work—geophysical surveying, drilling, trenching, and beneficiation testing of samples to estimate the reserves, grades, and tonnages of two graphite prospects. Similar methods were used to investigate and evaluate gold-bearing gravels and gemstones. Only in one case did the project team decide not to carry through a specific investigation on the ground that to do so would represent a waste of resources, given the inconclusiveness of available data.

This project also produced periodic progress reports, as well as those focusing on the specific prospects under investigation, for review by DTCD, UNDP, and the government. The terminal report describes project activities and findings in detail and contains both technical and institutional recommendations concerning development of the gold, graphite, and gemstone deposits that were investigated. In recognizing the problems to which the project was subject, the final report calls attention to the importance of carrying out basic exploration before proceeding with additional specific work on known deposits.

Two aspects of the terminal report merit special consideration, especially because they reflect an attempt to minimize follow-through problems, such as those encountered in the African case history described. First, the report stresses the importance of public policy in creating an attractive investment climate and refers to the government's commitment, undertaken at the outset of the project, to set up public corporations or arrange for private investment. Second, the report weighs the prospects of local as opposed to foreign private investment and focuses upon measures permitting small-scale operations within the capacity of local public or private enterprises. Since the project's conclusion, further exploration work or development of the deposits has been undertaken by state-owned corporations, with prospects of early mineral production.

Burundi[3]

The mineral exploration project in the central African country of Burundi has been one of the UNDP's longest running—and its most expensive—to date. Upon completion of the project's final phase UNDP was expected to have spent more than $11.5 million, or about $450 per square kilometer—more than the program has allocated to mineral exploration in any other country. In the course of over fifteen years most of Burundi's explorable territory has been covered by reconnaissance campaigns, and

---

[3]The following two case studies are based upon research conducted by Melinda Crane-Engel in Burundi and Burkina Faso (formerly Upper Volta) between September and December 1984.

a number of minerals have been subjected to detailed examination. UNDP's work has increased the knowledge of Burundi's resource base significantly. However, substantial outlays for advanced exploration work, including prefeasibility studies, failed to uncover a commercially exploitable find.

The project began in 1969 with large-scale reconnaissance. Photogeological and airborne geophysical surveys were delayed and then terminated when only partially complete, in order to pursue detailed work on lateritic nickel deposits. The second phase of the project also concentrated on these and was extended beyond schedule to allow time to assemble data for prefeasibility analysis on metallurgical processes for recovery of the nickel and on means of producing peat as an energy source. In 1977, a third phase was launched to resume regional prospecting, which had been abandoned earlier, and to finance the prefeasibility analyses on the Musongati lateritic nickel deposit. The analyses were used in a preliminary feasibility study, which concluded that exploitation of the deposit would be potentially economic, contingent upon solution of certain problems associated with processing the peat and upon achievement of major improvements in energy and transportation infrastructure (United Nations Secretariat, 1980b, p. 5, and 1980c, pp. 5–6).

That conclusion appears to have been overly optimistic. A follow-up feasibility study, financed by the World Bank, has indicated that while the Musongati deposit's geological characteristics (estimated nickel reserves of more than 700 million tons and associated values in copper, cobalt, platinum-group metals, and chromium) are attractive, infrastructural barriers are overwhelming. Burundi lies more than 1,000 kilometers from the sea in a region with a relatively low level of industrialization and correspondingly low demand for minerals. Given the lengthy delays and high costs incurred in local transport, only high-unit-value (hence exportable) minerals, or low-unit-value products with a local market, hold any promise of commercial feasibility. Nickel, however, has a relatively low unit value.

The infrastructural constraint should have been abundantly clear by 1973 or 1974, when rising energy costs further tipped the cost-benefit calculus. Continued work after this point may represent an inefficient allocation of resources partly attributable to the government's desire to pursue development of what it believed to be a bonanza. To the extent that this belief was based upon overly optimistic reports by DTCD, the blame must also be shared by that organization for relying upon experts who lacked the experience or incentives to apply commercially relevant criteria. Because review and evaluation of the project activities were lax, UNDP officials in New York also failed to realize that the exploration team had underestimated the nongeologic influences upon the nickel's economic viability.

The fourth and final phase of the project was sought to correct a number of the shortcomings experienced in previous phases. A detailed work program was drawn up in an attempt to prevent the butterflying that had occurred earlier. The revised project aims are to complete regional prospecting and to undertake detailed work on certain anomalies. The project's documents stress the need to concentrate on minerals that are of high unit value, marketable, and economically exploitable. Indeed, gold has become the chief exploration target.

Burkina Faso

UNDP's mineral exploration work in another landlocked African country dates even further back. The first project in Burkina Faso commenced in 1965 with regional prospection. The second phase was devoted to detailed exploration of a large manganese deposit at Tambao, which is located in the northern part of the country. As in the case of Musongati, considerable infrastructural barriers stand in the way of developing this deposit. Although UNDP sponsored various studies on mining operations and infrastructure requirements in the hope of proving economically exploitable reserves, it did not lose sight of the broader goal of improving knowledge of the country's resource base. The project's third phase returned to regional prospecting, covering the northeast and southwest of the country in the search for targets located close to existing infrastructure. During the project's third phase, a UNDP exploration team found a highly promising lead–zinc–silver anomaly 35 kilometers from Ouagadougou. After carrying out a limited preliminary evaluation that included drilling twelve boreholes, the project turned the deposit over to the government and resumed general reconnaissance.

Although the known geologic characteristics of the Perkoa deposit appeared encouraging, its grade, tonnage, and geometry remained uncertain. The government initially hoped to find a private company to carry out follow-up work and did elicit some interest through notices in the mining trade press. However, investors may have been deterred by the geologic uncertainty, compounded by Burkina Faso's unclear mining and fiscal legislation. The government eventually reconsidered and decided instead to obtain a World Bank loan for the purpose of completing a feasibility study. Had the government remained unwilling to go to the capital markets or had it been unable to do so, the decision to terminate DTCD work on the deposit could have been a mistake. DTCD would bear partial responsibility for that decision in that competent members of the project team did not try to dissuade the government from its course. Yet the actual outcome may represent precisely the correct allocation of resources in the sense that loan capital was used for relatively risky prefeasibility work, leaving UNDP grant funds free to support high-risk reconnaissance.

Results of UNDP Exploration

As of the end of 1982, UNDP had expended approximately $130 million and recipient countries the equivalent of about $70 million on a total of 150 projects that had been completed. As table 7-1 indicates, these projects resulted in eighty-three mineral discoveries, primarily in Africa, Asia, the Middle East, and Latin America. Of these, eight were in production and a ninth was under development. Five more appeared to be promising discoveries, bringing the number of actual or prospective mines discovered with UNDP assistance to a total of fourteen. Table 7-2 provides more details about these ventures.

These successful projects can be divided into two groups. The first includes discoveries resulting from basic exploration initiated without prior knowledge of existing mineralization (Carman, 1977, p. 321; Ventura, 1980, p. 209). These include porphyry copper finds in Ecuador, Panama, and Malaysia, as well as a bauxite deposit in the Solomon Islands and marble and limestone reserves in Togo.

In Ecuador, UNDP-sponsored geochemical exploration and subsequent surveys provided sufficient information on a porphyry copper deposit to spur investment by a Japanese company. Although the firm withdrew after spending $1.5 million on additional drilling, the deposit was under continuing evaluation in 1982 (*Mining Annual Review*, 1983).

Japanese investors also were attracted by a UNDP exploration program at Petaquilla in Panama. The project team carried out regional reconnaissance over an area of 17,000 square kilometers and pursued significant anomalies through detailed geologic, geochemical, and geophysical surveys. On the basis of the terminal reports, the government obtained and accepted a bid for follow-up work from a Japanese joint venture (Carman, 1977, pp. 323–324; Ventura, 1980, p. 199; *Mining Annual Review*, 1983).

The Malaysian discovery probably constitutes UNDP's greatest bonanza to date. In the course of a general geochemical survey and a soil-sampling program, the DTCD team located significant copper anomalies. Further work by the local geological survey induced the Overseas Mineral Development Corporation of Japan to develop the Mamut mine, which opened in 1975 (Carman, 1977, p. 322).

The second group of successful projects covers deposits where the project teams helped establish the commercial viability of existing deposits. In Mexico and Chile, for example, UNDP assisted the government in evaluating its porphyry copper mine sites and thereby enhanced their prospects for development. Mexico's La Caridad mine began to operate in 1979. Anaconda made plans to develop Chile's Los Pelambres deposit but postponed these plans following declines in copper prices. Three of the other discoveries in this class include porphyry copper at Sar Chesmeh in Iran and two offshore tin deposits in Indonesia and Burma; all were in

Table 7-1. UNDP Direct Exploration: Number of Mineral Discoveries

Mineral	Africa	Asia, Middle East	Europe	Latin America	Total
Precious metals (Au, Ag, Pt)	10 (Au)	1 (Au)	—	1 (Ag) 2 (Au)	14
Nonferrous metals (Cu, Pb, Zn, Sn, Sb, Hg, Cd, In, Bi, Tl)	3 (Cu) 1 (Pb–Zn)	1 (Pb–Zn) 1 (Sn–W) 6 (Cu) 4 (Sn)	1 (Pb–Zn)	10 (Cu)	27
Iron and alloys (Fe, Mn, Cr, Ni, Co, Mo, V, W)	2 (Ni) 1 (Mo) 2 (Fe) 1 (Ni–Cu sulfide) 1 (Mn)	1 (Fe)	—	1 (Fe, Mn) 1 (Ni-laterite) 2 (Fe)	12
Light metals (Al, Li, Mg, Be, Cs, Sr)	—	1 (bauxite-Al)	1 (bauxite-Al)	—	2

Fertilizers and industrial minerals (asbestos, diamonds, fine and kaolin clay, phosphate, salts, feldspar, potash, limestone, marble, gemstones, building materials)	2 (P) 1 (limestone) 3 (diamonds) 1 (potash and rock salts) 1 (asbestos) 1 (limestone and marble) 1 (P, potash, rock salts)	1 (limestone) 1 (potash) 2 (P) 1 (gemstones) 2 (building materials)	1 (potassium salts)	1 (emeralds)	19
Special metals (Ti, Zr, Hf, Nb, Ta, rare earths, beach sands)	1 (rare earths) 1 (beach sands)	—	—	—	2
Uranium	3	1	3	—	7

Note: Deposits are classified according to their main constituent mineral.

Sources: United Nations Secretariat, "Two Decades in Mineral Resources Development," paper presented at the Institution of Mining and Metallurgy Conference (New York, United Nations, 1980); United Nations Department of Technical Cooperation for Development, "Natural Resources and Energy Division Minerals Branch: Operational Projects," (New York, United Nations, 1983); W. Gocht, *Handbuch der Metallmärkte* (*Handbook of Metal Markets*) (Berlin, Springer Verlag, 1974); Diercke, *Weltwirtschaftsatlas* (*World Commercial Atlas*) (Munich, Deutscher Taschenbuch Verlag, 1981).

Table 7-2. Results of UNDP Direct Exploration

	Country	Deposit type	Project expenditures (U.S.$, in thousands)		Exploration dates	Status (as of 1983)	Geostatistics (metric tons in thousands, unless otherwise noted)	Investment cost (U.S.$, in millions)[a]
1. Deposits discovered by UNDP/DTCID that have secured finance for development	Malaysia (Mamut deposit)	Prophyry copper	UNDP DTCID	770 <u>790</u> 1,560	Began 1965	Developed by Overseas Mineral Development Corp., Japan; came into production 1975	Reserves = 178,000 Grade = 0.48% Recoverable mineral = 692	82
	Togo	Marble	UNDP Govt.	1,327 <u>513</u> 1,840	Began 1962	Into production 1969	—	6
		Limestone			Began 1962	Into production 1972	—	43
2. Deposits with development potential discovered by UNDP/DTCID	Ecuador (Chaucha)	Prophyry copper	UNDP Govt.	4,015 <u>836</u> 4,851	Began 1969	Investment by Japanese company that subsequently withdrew; deposits under evaluation	Reserves = 60,000 Grade = 0.7% Recoverable mineral = 390	—
	Panama (Petaquilla)	Porphyry copper	UNDP Govt.	1,548 <u>1,071</u> 2,619	Began 1966	Joint venture by Japanese investors proved reserves and granted 10 year option for 10% share	Reserves = 300,000 Grade = 0.7% Recoverable mineral = 1,701	300 (estimated)

3. Known mineralizations further explored by UNDP/DTCD that have secured finance for development

Country	Mineral	Source	Amount	Exploration	Development status	Reserves	Value
Solomon Islands (Rennell Island)	Bauxite	UNDP Govt.	985 586 1,571	1964–68	Development plans being formulated	—	300 (1979 estimate)
Burma (Heinze Basin)	Offshore tin	UNDP Govt.	5,618 1,491 7,059	Began 1971	Into production 1980	Reserves = 31 million m³ Grade = 0.22 kg/m³ Recoverable mineral = 6	28
Indonesia	Offshore tin	UNDP Govt.	1,080 2,381 3,461	Began 1969	Into production 1975	Reserves = 63 million m³ Grade = 0.22 kg/m³ Recoverable mineral = 20	15
Iran (Sar Chesmeh)	Porphyry copper	UNDP Govt.	1,566 2,000 3,566	1960–68	Into production 1981	Reserves = 800,000 Grade = 1.2%	1,400
Mexico (La Caridad)	Porphyry copper	UNDP Govt.	897 1,850 2,747	Began 1966	Into production 1979	Reserves = 739,000 Grade = 0.7% Recoverable mineral = 4,162	600

(Continued)

Table 7-2 (continued—see p. 244 for units of measure)

	Country	Deposit type	Project expenditures		Exploration dates	Status (as of 1983)	Geostatistics	Investment cost[a]
3. (cont.)	Morocco	Rock salt	UNDP	1,100	1967–72	Under development	Reserves = 3,000,000 Grade = 98.6% Recoverable mineral = 2,430,000	182
			Govt.	2,537				
				3,637				
	Tanzania (Buck Reef)	Gold	UNDP	626	1964–68	Into production 1982	—	2
			Govt.	395				
				1,011				
4. Known mineralizations with development potential further explored by UNDP/DTCD	Chile (Los Pelambres)	Porphyry copper	UNDP	5,377	Began 1969	Purchased by Anaconda; development plans postponed 1982	Reserves = 428,000 Grade = 1.8%	400 (1979 estimate)
			Govt.	3,064				
				8,441				
	Guinea (Mount Nimba)	Iron ore	UNDP	806	Began 1970	Development plans awaiting approval of financing institutions	Reserves = 600,000 Grade = 56% Recoverable mineral = 486,000	760 (estimated)
			Govt.	410				
				1,216				

Sources: United Nations Secretariat, "Two Decades in Mineral Resources Development," paper presented at the Institution of Mining and Metallurgy Conference, London (New York, United Nations, 1980); United Nations Department of Technical Cooperation for Development, "Natural Resources and Energy Division Minerals Branch: Operational Projects" (New York, United Nations, 1983); J. Carman, *Obstacles to Mineral Development: A Pragmatic View* (New York, Pergamon Press, 1979); United Nations Secretariat, "Two Decades of Mineral Resources Development: The Role of the United Nations," *Natural Resources Forum* vol. V (January 1981) pp. 15–31; *Mining Annual Review* (London, Mining Journal, 1983).

[a]Dollars are in current years and are not adjusted for inflation.

production in 1982 (Carman, 1977, pp. 322–324, and 1979, pp. 48–59; *Mining Annual Review*, 1983).

Even if UNDP work on known deposits is ignored, its record in locating viable deposits is favorable. Private companies generally estimate the chance of finding a deposit that will result in a profitable mine as one in two hundred at best (Arne, 1982, p. 320). Considering only those discoveries within the first group of cases that were in production in 1982, UNDP has done well, with two operational mines out of a total of 150 projects.

Comparing the UNDP success rate with private industry's, however, is misleading because of the different objectives of each. UNDP expenditures may include substantial outlays for basic mapping and training work that a private company would not undertake. Indeed, the transfer of skills is as important to UNDP as the identification of potentially viable deposits. At the same time that UNDP encourages development, it also fosters world mineral supply: future mineral discoveries represent one intended spin-off of mapping and training activities. For example, training carried out in the course of the Panama's Petaquilla project enabled nationals to find later the San Blas mineralized belt. Moreover, Petaquilla heightened interest in the area at large, spurring private and public exploration in neighboring countries and regions, resulting in the discovery of one of the world's largest porphyry copper bodies at Cerro Colorado, Panama. Once copper prices recover, this deposit will be developed further (Harkin, 1976, pp. 34–36; *Mining Annual Review*, 1983).

The United Nations Revolving Fund for Natural Resources Exploration

The United Nations Revolving Fund for Natural Resources Exploration (UNRFNRE) was established in 1973 and came into operation in 1975 with approximately $5.5 million donated by Japan and The Netherlands. The concern over world mineral supply in the early and mid-1970s helps to explain the creation of the Fund and the generous support provided by Japan, whose import dependence gave it reason to fear possible price increases or shortages. As a program devoted exclusively to natural resources, the Fund was designed in part to offset the reduction in UN mineral sector funding that had followed the adoption by UNDP of the country programming system. The two programs, in fact, are connected at both policy and operational levels. UNDP, DTCD, and World Bank officers participate in the Joint Operations Group, which advises the Revolving Fund's administrator on major decisions concerning the selection and implementation of projects. The Fund possesses its own technical and managerial staff but also employs the services of UNDP technical advisers.

Operational Structure

The Revolving Fund differs fundamentally from UNDP in that it is intended to be self-supporting: after initially relying on contributions, the Fund is supposed to sustain itself through payments, called replenishment contributions, from recipient countries derived from the proceeds of successful projects. Before the Fund begins work, the host-country government in theory agrees to turn over 2 percent of the annual value of the minerals discovered, or, if the country is among those designated by the UN as least developed, only 1 percent. The Fund may modify the timing of repayment for an economically marginal project (United Nations Development Programme, 1983, p. 6).

The project agreement defines a geographical area where the Revolving Fund possesses exclusive exploration rights. The replenishment contribution applies to all resources within this area that the project team identifies as "reported minerals" in its final report. Potential economic viability is critical, so reported minerals must be investigated sufficiently with respect to grade and tonnage (United Nations Revolving Fund for Natural Resources Exploration, 1982b, pp. 9–10; United Nations Development Programme, 1983, p. 13).

The replenishment contribution is payable for fifteen years from the beginning of commercial production, up to a ceiling of ten times the total project costs at constant prices. Contributions are suspended when the deposit is not producing, and the fifteen-year period extended accordingly. Moreover, no replenishment contribution is required if commercial production does not begin within thirty years after the beginning of the project, since the project agreement is effective only for that length of time (United Nations Development Programme, 1983, pp. 6–7).

In calculating the level of replenishment contributions needed to allow the Fund to revolve, its founders evaluated the past exploration expenditures and success rates of UNDP and two members of the UN. On this basis, it was estimated that a program of twenty new projects per year and a replenishment rate of 2 percent would achieve revolving status in about twenty years (Ventura, 1980, p. 240; United Nations Secretary General, 1974, p. 9). This estimate, however, is contestable; most mineral industry observers assert that the Fund never will be self-supporting at the current contribution level (Carman, 1979, pp. 66, 71–72; Ventura, 1980, p. 240).[4]

As the following sections indicate, the replenishment contribution concept affects both project selection and budgetary allocation.

---

[4]Ventura (1980) estimates that even without a ceiling limit, a contribution rate of 2.4 percent to 4.7 percent would be needed to make the Fund revolve. Carman (1979) puts the figure much higher, at 16.4 percent.

Project Design and Selection

Under the UNDP system, economic feasibility enters into the design of a project only after the overall level of Indicative Planning Figure funds available to particular recipient countries has been established based on size and need. In contrast, commercial feasibility plays a role in decisions at the Revolving Fund from the very beginning; equitable distribution among the various developing countries is just one issue considered in selecting projects. According to its mandate, the UNRFNRE was established to explore for natural resources that have a potential market, either domestic or foreign. The dual objectives of contributing to the recipient country's economic development and enabling the Fund to become self-supporting lie behind the injunction to choose projects according to their technical and economic viability. Within these criteria, the Fund is to give "due consideration" to equitable distribution and to the special situation of the least-developed, landlocked, and island countries (United Nations Development Programme, 1983, p. 3).

Unlike UNDP, the Revolving Fund authorities bear sole responsibility for the ultimate formulation of project activities. Project requests normally are presented by the would-be recipient government to the local UNDP representative, who forwards the request, with comments and supporting data, to the Fund. Other suggestions for projects may come from regional bodies such as the Economic and Social Commission for Asia and the Pacific, which has been particularly active in the mineral sector (United Nations Revolving Fund for Natural Resources Exploration, 1982a, p. 10).

In some cases, the Revolving Fund has undertaken follow-up work on targets prospected under UNDP auspices.[5] The Fund seeks to coordinate closely with UNDP, for example, especially so that IPF funds may be used to support additional investigation when the exploration data are insufficient. If IPF funds are unavailable, the Revolving Fund in "exceptional circumstances" will carry out a limited survey to determine the project area and its potential (United Nations Development Programme, 1983, p. 15, note 1).

The information obtained from the UNDP, the Revolving Fund, or, in some cases, third-party reviews is considered in light of the selection criteria mentioned above. Some requests are rejected, some accepted, still others may be restructured. The economic viability and potential market criteria require an orientation toward discovery; broad geologic surveys, for example, lie outside the scope of the Revolving Fund's mandate. However, the potential market criterion has been interpreted with a fair degree of latitude to include projects in early stages of the exploration cycle. Exploration may have proceeded to the stage of target identification, but

---

[5]In Burkina Faso, for example, the Revolving Fund is carrying out follow-up work on a site partially explored by UNDP.

this is not always a requirement. Promising mineralization in one area may fulfill the potential market criterion for a project that includes other areas of higher risk as well. Particularly where equitable distribution factors weigh heavily in a country's favor, the Fund may be willing to take on a project at a stage close to grass roots. Thus, in a small, infrastructure-deficient country with no existing mineral industry and little prior exploration, the Fund may agree to undertake broad-based exploration aimed at locating mineralization that, by virtue of high unit value and low processing requirements, would be feasible to export.

During its first few years, the Revolving Fund had trouble finding projects that met its technical and economic criteria. Only four proposals were approved by the administrator in 1975 and 1976, and in the following year not one was accepted. The source of the difficulty appears to have been government resistance to the replenishment contributions. More than a third of the countries contacted in the Fund's first three years either disavowed interest in assistance or withdrew their initial acceptance (United Nations Revolving Fund for Natural Resources Exploration, 1979, pp. 11–12). The 2 percent reimbursement requirement presumably was perceived by such countries as unreasonable (Ventura, 1980, p. 239).

In an attempt to offset this governmental reluctance, the Fund in 1981 adopted the ceiling on replenishment contributions and the 1 percent alternative for least-developed countries. These measures seem to have helped (United Nations Revolving Fund for Natural Resources Exploration, 1982a, 1983). As of mid-1983, seven projects stood on the Fund's operational roster, while another twenty were at various stages of review (Fozzard, Gurman, and Huhta, 1983, p. 2; and table 7-1). Some observers consider this an inadequate pool, asserting that an increase in the quantity of applications does not represent an improvement in their quality. Governments, they argue, submit only projects of dubious viability or those that have been rejected by other possible funding sources, preferring not to subject projects with more potential for profit to the repayment requirement. From this premise, some observers conclude that the replenishment principle has hampered the selection of projects and, ultimately, the efficacy of the fund.[6]

But what constitutes efficacy? If, in this case, it is perceived as the fulfillment of the Fund's mandates to become self-supporting and to make the maximum contribution possible to world mineral supply, then efficacy requires a low-risk strategy stressing commercial viability. To the extent

[6]This view was put forward by members of the German mining industry and by officials of the Federal German Ministry for Economics in interviews with the authors. For a slightly different elaboration of the argument, see Nooten (1979, p. 46). The premise can be challenged, however, on several grounds. Insofar as proposals submitted to UNRFNRE relate to projects at an early stage of exploration, they would be characterized by a level of geologic risk rendering highly subjective and uncertain any judgment regarding their viability. Secondly, exploration philosophies differ, so that a project—even one in a more advanced stage of exploration—perceived as marginal by one explorer might be attractive to another.

that the replenishment principle has deterred the submission of potentially economic projects, it then could be said to have reduced efficacy. On the other hand, the Revolving Fund can be considered a development program with an implicit mandate to provide financing that contains a certain grant or subsidy component. In this case, the Fund should focus on precisely those projects that have been rejected by other sources of capital but still stand some chance of attracting investment in the future. Under that view, the replenishment principle actually might promote efficacy.

Government resistance to the replenishment principle has affected project selection in another way. Under the best of circumstances, the process of negotiating an agreement between the Revolving Fund and a recipient country will be complicated and costly. In some cases, it has become more so by virtue of government uncertainty regarding replenishment. This uncertainty was particularly severe in four cases, where project requests were evaluated and approved and project agreements were being negotiated, only to be cancelled by the governments involved. The Fund has estimated the cost in wasted resources at more than $200,000 (United Nations Revolving Fund for Natural Resources Exploration, 1979, p. 12).

The governments involved apparently hoped to obtain better terms elsewhere and therefore were unwilling to assume the 2 percent reimbursement obligation. The Fund since then has sought to protect itself to some extent against such occurrences by adopting a policy of stricter confidentiality with respect to the information it generates in the course of appraising project requests. It makes that information available to the government only after serious negotiations are in progress (United Nations Revolving Fund for Natural Resources Exploration, 1979, p. 10). However, there is little the Fund can do if the government chooses nonetheless to release the data to third parties.

Budget Allocation

As of 1983, UNRFNRE had obtained capital solely through donations and had received no replenishment contributions. The number of contributors had grown to fifteen and included developing as well as developed countries. Japan remained by far the largest donor, with contributions totaling $22 million. The cumulative resources of the Fund amounted to $46 million, and about $27 million of this sum had been spent or budgeted for specific projects. Deducting $1.8 million for preproject development and $6.1 million for administrative costs through 1983, roughly $11 million was available for additional programming (United Nations Revolving Fund for Natural Resources Exploration, 1983, annex III).[7]

---

[7]In establishing the amount of resources available for commitment to new projects, the Fund applies an actuarial formula based on the assumption that about one-third of all projects expend their total allocations, another third go only through the preliminary phases, and the remainder spend between these two extremes.

The Revolving Fund has committed slightly less than half of its resources to African countries. Latin American projects account for most of the remaining expenditures.

As noted earlier in the chapter, the replenishment concept was based upon the assumption that UNRFNRE would take on twenty new projects a year. Between its inception (in 1973) and 1982 the Fund had undertaken only sixteen projects. With one project agreement ratified in 1982 and a budget that would allow acceptance of five projects in 1983,[8] it was far from reaching this objective. The achievement of revolving status is rendered further unlikely if, as critics contend, the replenishment contribution principle has discouraged governments from submitting potentially viable projects.

Revolving Fund Exploration Projects

The operations of the Revolving Fund focus narrowly upon discovery and consequently do not cover major training or institution-building programs of the kind performed by UNDP. The Fund uses local staff and equipment on its projects, but only to the extent that they contribute to efficiency (United Nations Development Programme, 1983, p. 3). In addition, the title to all equipment, whether imported or domestically purchased, is retained by the Fund, which sells or reexports the equipment at the end of a project (United Nations Development Programme, 1983, p. 11). UNRFNRE project teams are composed of geologists, geochemists, and geophysicists drawn from an international hiring pool; much of the work is subcontracted to national and international mining services.

Work on a project may begin with detailed regional reconnaissance, using photogeological, geochemical, geophysical, and other techniques to define targets for detailed prospecting. The Fund's original mandate envisaged participation only up to the point of detailed evaluation, which could include large-scale mapping, geochemical and geophysical investigating, trenching, pitting, and limited drilling (United Nations Secretary General, 1974, p. 5). Although the definition of reported minerals qualifying for the replenishment contribution implies that the Fund will continue exploration until tonnage and grade are estimated, that requirement has been interpreted with a certain degree of flexibility. Moreover, the operational principles of the Fund provide that it may choose to terminate assistance when qualified public or private interests are willing to undertake exploration. However, it may do so only at the initiative of the government, upon which the obligation to pay replenishment contributions remains binding (United Nations Development Programme, 1983, p. 12).

The Revolving Fund's projects have involved target minerals of all types, ranging from precious metals to industrial raw materials such as kaolin

---

[8]The estimate that five new projects could be financed from the $11 million available for 1983 was based upon figures contained in the Revolving Fund's annual report for 1981 (United Nations Revolving Fund for Natural Resources Exploration, 1982, p. 8) and upon the fact that anticipated expenses for most projects are in the range of $2 million.

and phosphate. Precious metals tend to predominate, because of both their potential market value and the replenishment contribution. Weak prices for many other metals limit their commercial prospects, while the low unit value and end-use specificity of certain industrial minerals diminish their economic viability in the many developing countries with only limited domestic demand.

By the early 1980s, the Revolving Fund had begun to stress follow-up on promising findings. Some reports show particular interest in carrying activities through to the feasibility stage, which previously had been considered unsuitable for the Fund because of its high cost and close link to production (United Nations Secretary General, 1974, p. 5; United Nations Revolving Fund for Natural Resources Exploration, 1979, pp. 20–22, and 1983, p. 14). The budgetary constraint continues to hold, but the Fund has raised the possibility of financing such studies in conjunction with, for example, the recipient government or other international financial institutions.

The Revolving Fund produces annual technical assessments and final reports on all projects. These studies of both successful and unsuccessful projects have included detailed geologic, geophysical, and geochemical data as well as, when applicable, estimates of the mineralogy, grade, and tonnage of the minerals. On occasion, the UNRFNRE also has made recommendations about investment and development in its final reports.

The Fund stresses the need to transmit the information it has generated to potential investors. Its data and reports, like those of the UNDP, are considered confidential, and all information is submitted to the government and remains at its sole disposal, except as the two parties otherwise may agree (United Nations Development Programme, 1983, p. 13). At the government's request, the Fund may seek out particular mining enterprises or financial institutions with potential as investors (United Nations Revolving Fund for Natural Resources Exploration, 1983, pp. 3–4, 9). If the project appears to have reached a suitable point, the Revolving Fund may suggest that contact with potential investors be made. Conversely, when the Fund asserts that investment promotion would be premature and possibly detrimental to ultimate exploitation, it may exercise its right to veto a request from the government that it cease ongoing operations in favor of another party (United Nations Development Programme, 1983, p. 12). In general, discussions with firms or other organizations interested in developing a deposit do not begin until the final stages of a project.

Case Histories of Revolving Fund Projects

The following case histories illustrate the nature of UNRFNRE projects and the degree to which they, like UNDP projects, have encountered occasional difficulties in promoting follow-up.

Argentina

At Huemules, Argentina, UNRFNRE spent more than $3 million to carry out four years of exploration for copper, lead, zinc, silver, and gold. The project undertook geologic mapping, geochemical and integrated geophysical surveying, diamond and percussion drilling, and exploration tunneling on a high-grade gold prospect. Mineralogical and metallurgical tests were performed on composite samples, and consultants reviewed all results by computer.

The Revolving Fund evaluated the results as very promising and listed gold and silver as reported minerals. At Argentina's request, a summary report on the findings was written and distributed to potential investors and lending institutions. The Fund recognized, however, that the data on ore grade and reserves were insufficient for investors. After failing to find third-party financing, the Fund itself supplied an additional $100,000 to carry out the preliminary feasibility work, which included geologic and topographic mapping, soil and rock sampling, and limited preparation of access facilities. Subsequently, the government held an international tender for exploration and development rights. The Revolving Fund financed two consulting missions that assisted the government in drafting bidding documents, a work program, and a model contract. However, doubts about the investment climate appear to have kept a number of companies from responding to the tender (United Nations Revolving Fund for Natural Resources Exploration, 1983, pp. 3–4; Fozzard, Gurman, and Huhta, 1983, pp. 4–5).

Cyprus

In Cyprus, the Revolving Fund spent $1 million in three and a half years of exploration for copper, gold, and zinc. The targeted areas were investigated through a combination of mapping, drilling, and assaying, which produced evidence of three separate copper deposits. Upon agreement with the government, the Revolving Fund disclosed preliminary findings to several private mining enterprises. It continued to explore thereafter, but the data collected were not sufficient to demonstrate economic viability. The Fund decided that additional assistance would not improve those data to a point justifying continued funding. Its final report described its findings in detail but did not claim reported minerals. Nonetheless, the Fund assisted Cyprus in promoting additional exploration by the investors contacted (United Nations Revolving Fund for Natural Resources Exploration, 1983, p. 4).

Ecuador

Efforts to design an effective follow-up program have been less successful in the case of a $2.5 million project in Ecuador than in Argentina, despite a favorable investigation of a silver deposit at San Bartolome. In an effort to promote rapid investment follow-up, UNRFNRE presented the government with a special summary report during the course of project operations. Both it and the final report, which was submitted in 1979, contained detailed geologic, economic, and engineering data that demonstrated the deposit could generate profits sufficient to support a small mining operation after paying all capital and operating costs, as well as the expense of exploration for development of further reserves.

Although the Fund offered to help locate financing for follow-up, little was done until 1982 when Ecuador authorized UNRFNRE to discuss its findings with potential investors, including a regional development bank and a mining company. This delay may have been costly, as tunneling work could have been lost over time due to the instability of the ground and the possible collapse of supporting timbers (United Nations Revolving Fund for Natural Resources Exploration, 1979, annex 4, p. 2).

The Fund's experience in Ecuador was central to its decision to press for the authority to perform feasibility studies. However, it is doubtful that this would have gone much further than the final summary reports in alleviating governmental ambivalence. Given the deposit's small site and remote location, it could not have attracted investment even at the feasibility stage without strong government support (United Nations Revolving Fund for Natural Resources Exploration, 1979, annex 4, p. 2).

Results of Revolving Fund Exploration

Given the long lead times that characterize the mining industry, it is still far too early to evaluate the results of the Revolving Fund program. Nevertheless, some tentative assessments can be offered. As of 1983, a total of $9 million had been expended on eight completed projects, of which four were considered successful. While the Cyprus, Argentina, and Ecuador projects located precious metals and copper, the fourth project, in Benin, uncovered an allegedly economic deposit of kaolin. Although these projects appeared promising, relatively high levels of risk continued to characterize all four.

Evaluation of Multilateral Exploration Assistance Programs

In evaluating multilateral exploration assistance programs, it is useful to examine whether and how those programs have met the objectives put forth earlier in the chapter: enhancement of world mineral supply and promotion of economic development.

Increasing Mineral Supply

The supply-side rationale is premised upon the existence of a geographic distortion in the pattern of worldwide exploration investment. Fears that such a gap could result in supply imbalances and rapidly increasing prices came to the fore during the middle to late 1970s, following the first oil shock.

Today, the exploration gap theory is in disrepute. Private firms do appear to have cut back on investment in Third World exploration, but this decrease has been offset at least to some extent by the rise of new investors such as the state-owned enterprises of some developing countries (see chapter 2). The predicted price increases and shortages have not materialized; indeed, a number of mineral markets have excess supplies. Some observers argue that multilateral exploration assistance has contributed to this overabundance (Zorn, 1982, p. 127), but the combined resources available in 1983 to UNDP and the Revolving Fund—around $30 million—were too slim to have influenced world exploration investment to any appreciable extent. Moreover, the effect of UN exploration expenditures upon world mineral supply is contingent upon follow-up: deposits identified under UNDP auspices cannot add to mineral flows when, as is often the case, they lie neglected. As noted above, UNDP is responsible for just eight producing mines. These have not added to the flow of internationally traded minerals to such a degree as to exert any significant effect upon price or quantity (Carman, 1979, p. 45).

In sum, there is little evidence to support the existence of an exploration gap or the need for multilateral assistance to counter it. Developed countries also have come to doubt this rationale for exploration assistance programs: they cut back their contributions to the Revolving Fund from $7.5 million in 1977 to one third that amount in 1981 (United Nations Revolving Fund for Natural Resources Exploration, 1983, annex II).

Promoting Development

The development-promotion rationale also overstates the benefits of multilateral exploration assistance; nevertheless, it provides greater justification for such assistance than do the supply-side considerations. Here, too, many of the benefits in question are contingent ones that cannot be realized unless follow-up investment occurs. Even when follow-up does occur, the resulting mine probably makes a smaller contribution to development than asserted by proponents of resource-based growth. Research shows that, of the various benefits mining was thought to promise, only fiscal and foreign exchange receipts have materialized to any appreciable extent. Contributions to local industry and labor have been minimal due to the relatively high degree of capital intensity and the reliance upon imported personnel and machinery characterizing many projects (Radetzki and Zorn,

1979, p. 109; World Bank, 1979b, pp. i–iii, 1–5, 15; Giraud, 1983, p. 668). Of the producing mines discovered under the auspices of multilateral programs, a number are large operations producing for export, rendering them subject to these arguments. Others are smaller ventures, focusing upon industrial minerals deposits, and they may have more positive effects upon local employment and industry than their larger-scale counterparts (Rivington, 1976, p. 206; Thoburn, 1981, pp. 95, 160–161); however, the overall economic impact of such mining tends to be marginal (World Bank, 1979b, p. 5).

But multilateral exploration assistance also generates benefits that are not contingent upon follow-up and may be of considerable importance for the recipient's development. Training and institution-building carried out by UNDP directly and by UNRFNRE indirectly, through employment of local staff, impart knowledge that is fundamental to economic growth. Such gains are difficult to quantify; they will vary directly with the duration, range, and suitability of the skills and technology transferred.

It is impossible to generalize about the effectiveness of multilateral exploration assistance in promoting development. The gains from training, institutional reform, and potential fiscal and foreign exchange receipts from assistance in the mineral sector must be weighed against returns in other segments of the economy, which in some cases will be higher or less uncertain. The actions of UNDP recipients following adoption of the country programming system reflect their perception of the opportunity costs involved. Given a choice as to how to distribute their respective IPF funds, many countries emphasized projects with benefits more visible than those from mining, and so resources allocated to mineral exploration dropped as a percentage of the total UNDP budget (Bosson and Varon, 1977, p. 48).

■

The authors would like to express their gratitude to the United Nations Department of Technical Cooperation for Development, the United Nations Revolving Fund for Natural Resources Exploration, the governments of Burkina Faso and Burundi, as well as to officials of the World Bank, the American and German mining industries, the Federal Republic of Germany's Ministry for Economics, and our colleagues in the Institute for International and Foreign Trade Law. Although we have benefited from their assistance, the ideas expressed here are our own.

References

Arne, K. 1982. "Basic Concepts of Mine Financing," *Mining Magazine* (March) pp. 230–234.

Bergsten, C. 1974. "The New Era in World Commodity Markets," *Challenge* (September/October) pp. 34–42.

Bosson, R., and B. Varon. 1977. *The Mining Industry and Developing Countries* (New York, Oxford University Press).

Brandt Commission. 1980. *North-South: A Program for Survival* (Cambridge, Mass., MIT Press).

British-North American Committee. 1976. *Mineral Development in the Eighties: Prospects and Problems* (London/Washington, British-North American Committee).

Carman, J. 1977. "United Nations Mineral Activities," *Natural Resources Forum* vol. I (July) pp. 317–337.

————. 1979. *Obstacles to Mineral Development: A Pragmatic View* (New York, Pergamon Press).

Cheysson, C. 1980. "La Coopération Minière dans Lomé II," *Revue du Marché Commun* pp. 59–62.

Crowson, P. 1977. *Non-Fuel Minerals and Foreign Policy* (London, Royal Institute of International Affairs).

Diercke. 1981. *Weltwirtschaftsatlas* (World Commercial Atlas) (Munich, Deutscher Taschenbuch Verlag).

Europäische Investitionsbank. 1983. *EIB 1958–1983* (Luxembourg, European Investment Bank).

Fozzard, P., B. Gurman, and J. Huhta. 1983. "The United Nations Revolving Fund: Its Importance to Natural Resource Development." Paper presented at the Institution of Mining and Metallurgy Conference, Bangkok, Thailand.

Giraud, P. 1983. *Geopolitique des Ressources Minières* (*The Geopolitics of Mineral Resources*) (Paris, Economica).

Gocht, W. 1974. *Handbuch der Metallmärkte* (*Handbook of Metal Markets*) (Berlin, Springer Verlag).

Harkin, D. 1976. "Systematic Mineral Exploration," *Natural Resources Forum* vol. I (October) pp. 29–39.

Kirchner, C., E. Schanze, F. v. Schlabrendorff, A. Stockmayer, T. Waelde, M. Fritzsche, and R. Patzina. 1979. *Mining Ventures in Developing Countries, Part I: Interests, Bargaining Process, Legal Concepts* (Frankfurt am Main, Alfred Metzner Verlag).

Lewis, A. 1983. "The United Nations in Mineral Development," *Engineering and Mining Journal* (January) pp. 68–70.

Mikesell, R. 1974. "More Third World Cartels Ahead?" *Challenge* (September/October) pp. 24–31.

————. 1975. *Foreign Investment in Copper Mining* (Washington, D.C., Resources for the Future).

————. 1979. *New Patterns of World Mineral Development* (London, Washington, British-North American Committee).

Mining Annual Review. 1983. (London, Mining Journal).

Nooten, G. 1979. "Mineral Resource Development in Developing Countries with Reference to the Activities of the United Nations Revolving Fund" (M.S. thesis, Department of Geological Sciences, Queen's University, Ontario, Canada).

Prain, R. 1975. *Copper: The Anatomy of an Industry* (London, Mining Journal).

Radetzki, M., and S. Zorn. 1979. *Financing Mining Projects in Developing Countries* (London, Mining Journal).

Rivington, J. 1976. "Exploring for Industrial Minerals in Developing Countries." Paper presented at the Second International Congress on Industrial Minerals, Munich, Federal Republic of Germany.

Smith, D. 1977. "Information Sharing and Bargaining: Institutional Problems and Implications," in Garvey and Garvey, eds., *International Resource Flows* (Lexington, Mass., Lexington Books).

———, and L. Wells, Jr. 1975. *Negotiating Third World Mineral Agreements. Promises as Prologue* (Cambridge, Mass., Ballinger Publishing).

Stockmayer, A. 1982. *Projektfinanzierung und Kreditsicherung* (Frankfurt am Main, Alfred Metzner Verlag).

Thoburn, J. 1981. *Multinationals, Mining and Development: A Study of the Tin Industry* (Hampshire, England, Gower).

United Nations Department of Technical Cooperation for Development. 1983. "Natural Resources and Energy Division Minerals Branch: Operational Projects" (New York, United Nations).

United Nations Development Programme. 1983. "United Nations Revolving Fund for Natural Resources Exploration: Operational Procedures and Administrative Arrangements." UNDP Governing Council, February 10, 1983, U.N. Doc. DP/142/Rev. 1.

United Nations Revolving Fund for Natural Resources Exploration. 1979. "Report of the Administrator." UNDP Governing Council, April 3, 1979, U.N. Doc. DP/368.

———. 1982a. "Annual Report of the Administrator for 1981." UNDP Governing Council, April 1, 1982, U.N. Doc. DP/1982/40.

———. 1982b. "Standard Project Agreement (Natural Resources Exploration Project)" (New York, United Nations).

———. 1983. "Annual Report of the Administrator for 1982." UNDP Governing Council, April 15, 1983, U.N. Doc. DP/1983/34.

United Nations Secretariat. 1980a. "Two Decades in Mineral Resources Development." Paper presented at the Institution of Mining and Metallurgy Conference, London (New York, United Nations).

———. 1980b. "Mining Engineering in the Developing Countries: The Role of the United Nations." Paper presented at the Institution of Mining and Metallurgy Conference, London (New York, United Nations).

———. 1980c. "Mineral Engineering in the Developing World: The Role of the United Nations." Paper presented at the Institution of Mining and Metallurgy Conference, London (New York, United Nations).

———. 1981. "Two Decades of Mineral Resources Development: The Role of the United Nations," *Natural Resources Forum* vol. V (January) pp. 15–31.

United Nations Secretary General. 1974. "Report: The United Nations Revolving Fund for Natural Resources Exploration." UNDP Governing Council, April 8, 1974, U.N. Doc. DP/53.

Ventura, D. 1980. "Financial Structures for Mineral Exploration: The Problem of the Developing Countries" ("Structures de financement de la prospection miniere: le probleme specifique des pays en voie de developpement") (Thesis, l'Ecole Nationale Superieure des Mines de Paris, Paris, France).

Vernon, R. 1967. "Long-Run Trends in Concession Contracts," *Proceedings of the American Society of International Law* (Washington, D.C., American Society of International Law).

World Bank. 1977a. *Minerals and Energy in the Developing Countries* (Washington, D.C., World Bank).

_____. 1977b. *Annual Report* (Washington, D.C., World Bank).

_____. 1979a. *Annual Report* (Washington, D.C., World Bank).

_____. 1979b. *Development Problems of Mineral-Exporting Countries* (Washington, D.C., World Bank).

_____. 1980. *Annual Report* (Washington, D.C., World Bank).

_____. 1981. *Annual Report* (Washington, D.C., World Bank).

Zorn, S. 1982. "The Security of Mineral Supplies: Impacts on Developing Countries," in *Legal and Institutional Arrangements in Minerals Development* (London, Mining Journal) pp. 123–131.

8

French Mineral Exploration, 1973–82

WILLY CHAZAN

Both private and government-controlled organizations play significant roles in French mineral exploration. This chapter reviews French exploration from 1973 to 1982, a period in which exploration expenditures increased significantly. It begins by examining aggregate exploration expenditures for the period, as well as the allocation of funds according to geographic area and mineral type, and then identifies several influences upon these expenditure trends. The chapter concludes by looking at the productivity of French exploration and making tentative international comparisons.

Introduction

In this chapter, *mineral exploration* refers to all activities designed to identify new ore reserves or to augment previously known reserves. Exploration activities are thus clearly distinguished from mine development and actual mining activities, in which the goal is, on the contrary, to reduce and, ultimately, exhaust the reserves. Furthermore, *French mineral exploration* refers to all exploration on French territory, regardless of the nationality of the explorer, and to all exploration abroad conducted by French organizations, including minority interests. The data on exploration expenditures, therefore, are broken into three categories: those for (1) French territory, comprising metropolitan France and the French overseas departments and territories[1]; (2) the Franc Zone[2] and the FAC (Fund for Aid and Cooperation) Zone[3]; and (3) the rest of the world. All financial data in this study are in terms of constant 1982 French francs, unless noted otherwise.[4]

Of the some forty French organizations engaged in exploration during the decade, the eight most important were these: CEA–Cogema group; Bureau de recherches géologiques et minières (BRGM); Société nationale ELF-Aquitaine (SNEA); Imetal; Péchiney-Ugine-Kuhlmann (PUK); Minatome, created in 1975 to explore for uranium; Compagnie industrielle et minière (CIM), owned by Rhone-Poulenc; and the Dong-Trieu company, controlled until 1983 by Empain-Schneider.

These enterprises differ in terms of ownership and organization. CEA (the Atomic Energy Commission) is a public agency engaged in industrial and commercial activities, while Cogema (the General Company of Nuclear Materials) is a corporation entirely controlled by CEA. BRGM (the Geological Research and Mining Office) is also a public agency with industrial and commercial interests. It holds the majority of the shares of the French Mining Company (Coframines). Other companies also belong to the BRGM group, which in turn hold shares in numerous subsidiaries; among the latter is the Mining and Chemical Products Company of Salsigne, which operates a gold mine.

SNEA (the National Company ELF-Aquitaine) is a corporation, a majority of whose shares belong to the state. Imetal is a holding company whose numerous subsidiaries include the Mining and Metallurgical Company of Penarroya (SMMP), the French Company of Mokta (CFM), and Société le nickel (SLN), renamed Société métallurgique le nickel (SMLN) after a majority interest was acquired by the French public oil agency Entreprise de recherches et d'activités pétrolières (ERAP) and later on (in 1985) included in a new organization, Eramet-SLN.

Through ERAP, the government's interests are represented in the main French oil exploration and production companies. ERAP controls SNEA

---

[1]The French overseas departments are Guadeloupe, Guyana, Martinique, Mayotte, La Réunion, Saint Pierre et Miquelon; the French overseas territories are New Caledonia, French Polynesia, Wallis and Futuna, and the Southern Antarctic Territories.

[2]The Franc Zone includes those countries whose currency is tightly linked to the French franc: Benin, Burkina Faso, Cameroon, Central African Republic, Chad, the Comoros, Congo, Ivory Coast, Gabon, Equatorial Guinea, Mali, Niger, Senegal, and Togo.

[3]The FAC Zone includes those countries in the Franc Zone and, in addition, Burundi, Cape Verde, The Gambia, Guinea, Haiti, Madagascar, Mauritius, Mauritania, Rwanda, Seychelles, São Tomé and Principe, Zaire, the Lesser Antilles, Angola, Mozambique, and Guinea-Bissau.

[4]The interested reader can approximate the equivalent values in constant 1982 U.S. dollars by dividing the figures given in francs by 6.6, the average exchange rate of francs per dollar over 1982.

and its subsidiary Société nationale ELF-Aquitaine (production) (SNEA(P)), which developed mineral exploration activities after 1970. ERAP is an important shareholder of Eramet-SLN, whose subsidiary SMLN-SLN is the main producer of nickel ore in New Caledonia.

Empain-Schneider is an incorporated enterprise, as were (during the period under study) Rhone-Poulenc and PUK; the latter two companies were nationalized in recent years, but Rhone-Poulenc was returned to the private sector in 1986.

The French Petroleum Company (CFP), a state-controlled corporation, has complete control of Total compagnie minière (TCM), which recently absorbed both Minatome, a former subsidiary of CFP and PUK, and the Mining Company Dong-Trieu, to create (in 1986) Total compagnie minière France (TCMF).

The exploration expenditure data on which this study is based were collected annually by the Department of Mines of the French Ministry of Industry and Foreign Trade. They represent expenditures for all primary mineral deposits except liquid and gaseous hydrocarbons and water. Specifically included are expenditures for ferrous and nonferrous metals, coal, uranium, evaporites such as potash and phosphates, and other minerals such as borates, fluorine, sulfur, asbestos, and diamonds. However, in order to avoid unnecessary detail and to permit meaningful conclusions to be drawn, the chapter focuses on the principal interests of French explorers: uranium, base metals, alloy and specialty metals, and precious metals. Data for other minerals have been included under the headings "other minerals" or "general exploration." The latter category has very broad aims and serves as a stepping-stone to subsequent, more detailed activity.

Although exploration expenditures for coal, potash, seabed minerals, and general exploration are not discussed in detail, they represent sizable financial undertakings: for example, in 1980, 47 million francs (MF) were spent on general exploration; in 1981, 13 MF on potash and 73 MF on seabed minerals; and in 1982, 78 MF were spent on coal.

Exploration Expenditures

Expenditures for French mineral exploration between 1973 and 1982 totaled 8,573 MF, of which 4,260 MF were spent in French territory, 900 MF abroad in the Franc Zone, and 3,413 MF in the rest of the world (see figure 8-1). Annual expenditures peaked in 1981 with a record 1,153 MF and then declined to 1,131 MF in 1982, despite a record 606 MF in French territory. State-controlled organizations contributed 60 percent of total expenditures by the major concerns during the period; the major share came from the CEA-Cogema group, well ahead of BRGM, SNEA, and

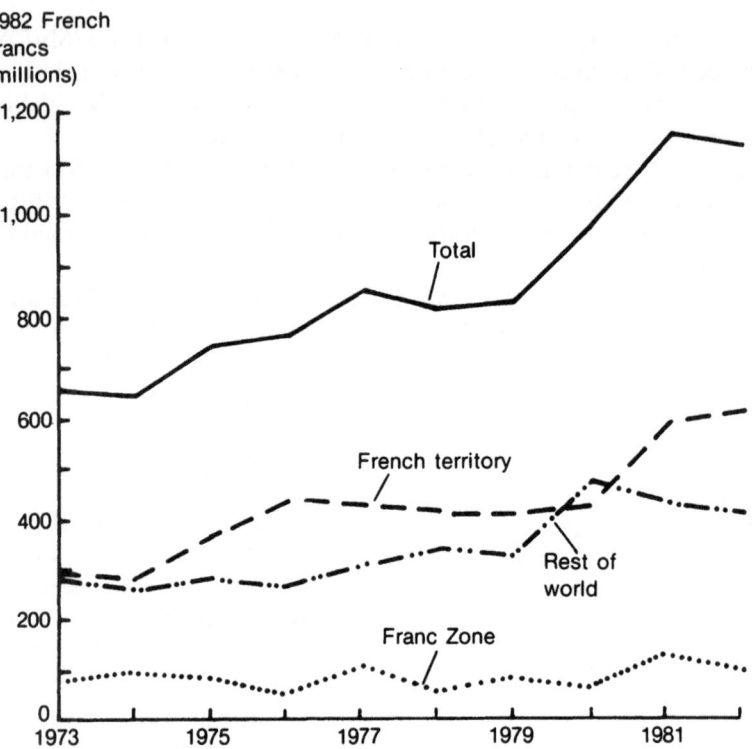

Figure 8-1.   Total French mineral exploration expenditures, 1973–82. *Source:* Appendix table 8-A-1 in this chapter, French Ministry of Industry and Foreign Trade.

Imetal.[5] In comparison, total French expenditures on oil and gas exploration between 1973 and 1982 were 47,700 MF.

Concerning exploration in French territory, expenditures by the eight majors constituted 82 percent of the total. Independent prospectors, small mining companies, and subsidiaries of foreign companies (including partners of French companies) were responsible for the remaining 18 percent. This consisted primarily of exploration for base metals in France, nickel in New Caledonia, and gold in Guyana. Expenditures by public enterprises made up 52 percent of the total; Cogema had the highest, followed by BRGM (including expenditures for the National Inventory of Mineral Resources), Imetal, and SNEA.

In the Franc Zone, the eight majors accounted for 97 percent of the expenditures. BRGM took the lead among French explorers and had the largest share of the 900-MF expenditures. BRGM benefits considerably from government subsidies, particularly from funds for development aid

[5]No account is taken of the contribution of Texasgulf, a subsidiary of SNEA since 1982.

and economic cooperation, which express France's support for the development of mineral resources in the mainly French-speaking developing countries. Cogema had the second-largest share; it explores worldwide for uranium in association with the relevant national agencies and other organizations. Imetal's third-largest share resulted primarily from uranium exploration by its subsidiaries in Niger and Gabon, as well as copper exploration with BRGM in Haiti. Other investments made up less than 15 percent of the franc total, including exploration for uranium, bauxite, rare earths, and other, mostly nonmetallic, ores.

As for exploration in the rest of the world, the eight majors represented 96 percent of total expenditures. The CEA-Cogema group had the largest expenditures, principally in Canada, the United States, and Australia. Next came SNEA in North America, Australia, and Europe, primarily in search of base metals and uranium. Minatome had the third-largest share of expenditures, with uranium exploration in the United States, Canada, and Australia (for the years 1978 and 1979, Minatome had the largest expenditure share). Imetal was fourth with its numerous activities abroad: in Europe, it explores extensively through its subsidiaries, and recently, in association with BRGM and a Portuguese company, it made a major discovery in Portugal. Imetal also has explored in Canada for uranium, and in the United States and Australia for copper. BRGM had the fifth-largest share of expenditures, primarily in the United States, Canada, Peru, Australia, and Portugal, where, in cooperation with Penarroya and its Portuguese associates, it took part in discovering the Neves-Corvo copper deposit.

Exploration Targets

French exploration has been directed toward some forty minerals. Only the principal targets, divided into five groups, have been included for detailed discussion in this chapter: the groups are (1) uranium, (2) base metals, (3) alloy and specialty metals, (4) precious metals, and (5) other minerals, mostly nonmetallics. Exploration activities for coal, lignite, potash, salt, and seabed minerals are not discussed in detail.

Between 1973 and 1982, uranium exploration expenditures amounted to 4,057 MF, or 47 percent of total expenditures (see figure 8-2). Of this, 1,581 MF (39 percent) were spent in French territory, 459 MF (11 percent) abroad in the Franc Zone, and 2,016 MF (50 percent) in the rest of the world. Overall expenditures for uranium increased dramatically from 229 MF in 1973 to a peak of 596 MF in 1980, before declining to 455 MF in 1982. Between 1978 and 1981, uranium exploration expenditures exceeded expenditures for all other minerals combined. The rest of the world (excluding the Franc Zone and French territory) had the largest geographic

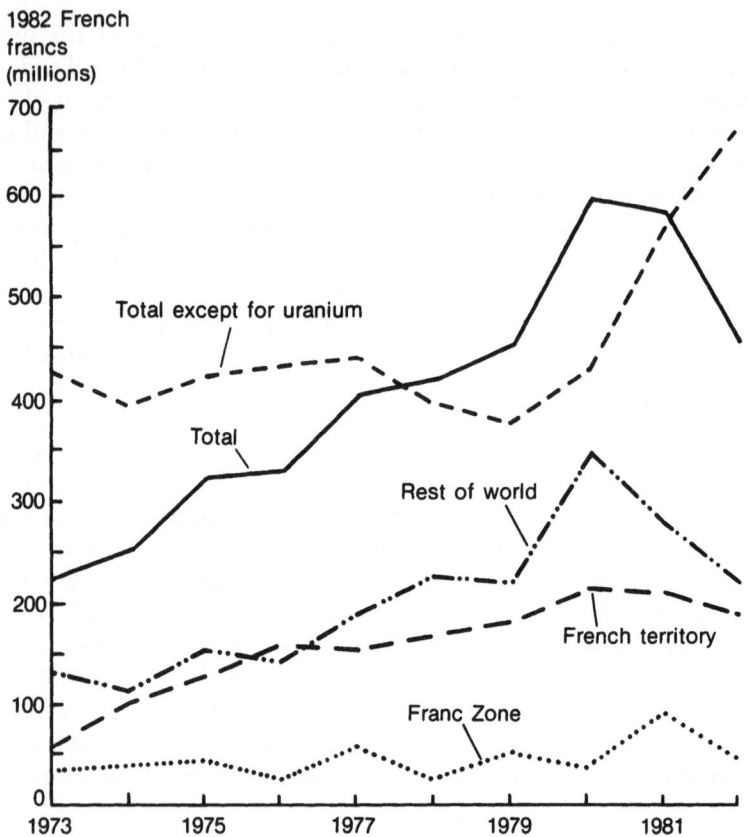

1982 French
francs
(millions)

Figure 8-2.   French uranium exploration expenditures, 1973–82. *Source:* Appendix table 8-A-2 in this chapter, French Ministry of Industry and Foreign Trade.

share of total expenditures throughout the period, except in 1976 when French territory had the largest share.

The principal organizations involved in uranium exploration were (in order of importance): the Cogema group, Minatome, the Imetal group, SNEA, Dong-Trieu, and CIM. These companies explore for uranium deposits associated with granitic and metamorphic rocks. Common objectives are also detrital formations and concentrations along unconformities. Some exploration has been done in shallow formations of the calcrete type.

Base metals exploration includes expenditures for lead, zinc, tin, and copper (including molybdenum, because the two metals often are associated with one another). Bauxite, which could have been included in this group, was excluded deliberately because it is only a minor target; bauxite exploration was conducted almost exclusively by the PUK group, and

1982 expenditures represented only 0.3 percent of 1973 expenditures for bauxite.

Total expenditures for base metals exploration from 1973 to 1982 were 1,954 MF, 846 MF (43 percent) in French territory, 172 MF (9 percent) in the Franc Zone, and 936 MF (48 percent) in the rest of the world (see figure 8-3). Although annual expenditures hovered near 200 MF for the entire period, they increased substantially in French territory from 51 MF in 1973 to 108 MF in 1981, before sliding back to 93 MF in 1982. In the Franc Zone, expenditures never exceeded the high mark of 28 MF in 1974 and in 1982 were a mere 9 MF. For the rest of the world, expenditures from 1977 to 1981 remained lower than those in French territory, while in 1982 they exceeded expenditures in French territory for the first time since 1976.

The three organizations principally involved in base metals exploration were (in order of importance): BRGM, SNEA, and the Imetal group, which together accounted for 93 percent of total base metals exploration.

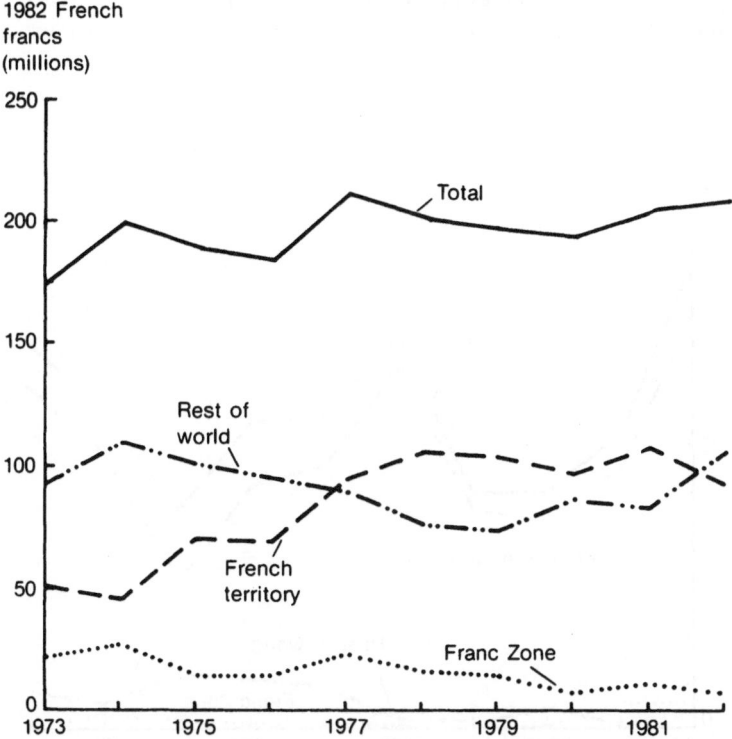

Figure 8-3. French base metals exploration expenditures, 1973–82. *Source:* Appendix table 8-A-3 in this chapter, French Ministry of Industry and Foreign Trade.

The balance was conducted by the PUK group and CIM. Their exploration efforts were divided mainly between volcanogenic- and sedimentary-hosted sulfide deposits, and stratified or vein-type hydrothermal flows. Tin ores (frequently associated with tungsten, included in the alloy metal group) also were sought near the edges of acidic granites. Secondary concentrations of cassiterite also have become potentially economic targets.

Alloy and specialty metals include nickel, cobalt, chromium, tungsten, antimony, and rare earths. Total exploration expenditures for these metals were 664 MF; 579 MF (87 percent) were in French territory, largely as a result of nickel-cobalt exploration in New Caledonia (see figure 8-4). The Franc Zone accounted for 23 MF (4 percent), and the rest of the world for 62 MF (9 percent). The ten-year expenditure trend, determined largely by exploration in New Caledonia, is cyclical: expenditures fell from 98 MF in 1973 to 49 MF in 1976, rose to 82 MF in 1977, fell again to 46 MF in 1980, and finally increased to 80 MF in 1982. The Imetal group, in French territory, and the BRGM, abroad, were the principal explorers for this group of metals.

The precious metals category includes only exploration in which precious metals were the main target. Therefore, expenditures for precious

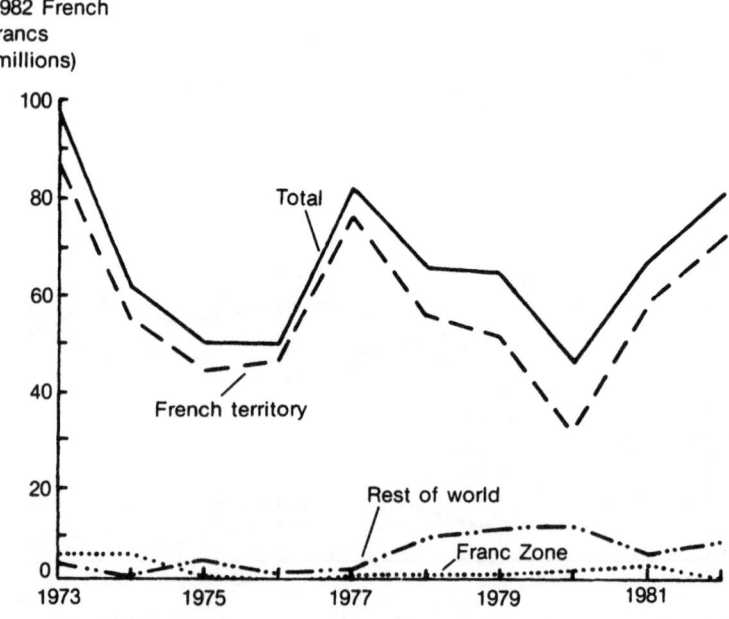

Figure 8-4. French exploration expenditures for alloy and specialty metals, 1973–82. *Source:* Appendix table 8-A-4 in this chapter, French Ministry of Industry and Foreign Trade.

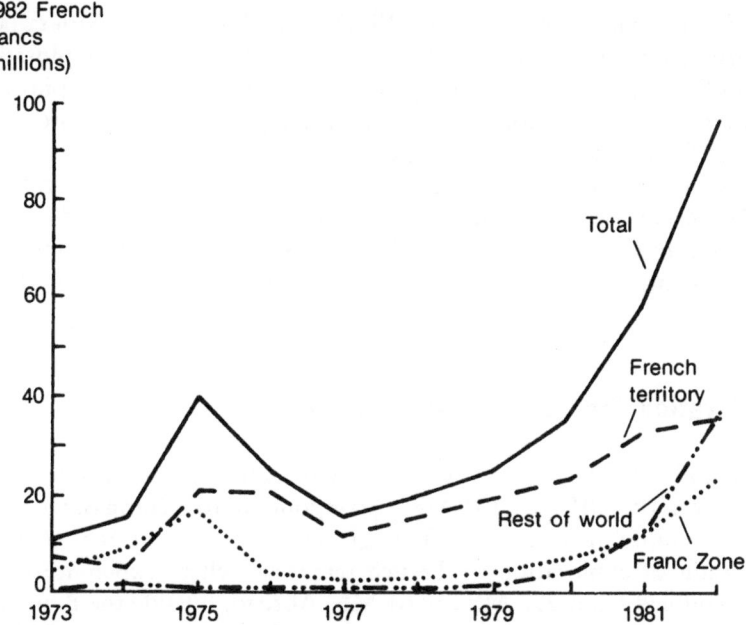

1982 French
francs
(millions)

Figure 8-5.   French precious metals exploration expenditures, 1973–82. *Source:* Appendix table 8-A-5 in this chapter, French Ministry of Industry and Foreign Trade.

metals (see figure 8-5) for the most part are limited to gold. Because silver usually occurs in association with lead and zinc sulfides, most silver exploration expenditures are included in the base metals category.

Of the cumulative total precious metals expenditures of 338 MF, the majority (191 MF, or 56 percent) occurred in French territory, primarily in France and Guyana where many gold deposits exist. Expenditures in the Franc Zone, which were largely in French-speaking Africa, amounted to 84 MF (25 percent), while in the rest of the world they totaled 63 MF (19 percent). Since 1978, activities in the rest of the world increased considerably, with expenditures rising from 1 MF in 1978 to 37 MF in 1982. Despite a marked increase, the total for French territory reached only 36 MF.

The BRGM has the largest share of French precious metals exploration, well ahead of Salsigne and the Imetal group. In addition, small-scale gold prospectors are important in Guyana, and the Dong-Trieu company commenced gold exploration in France in 1982.

Among other minerals of interest to French explorers, fluorspar exploration accounted for 104 MF from 1973 to 1982, of which 98 MF were devoted to French territory since many deposits are known in metropolitan

France; the main explorers were BRGM, the PUK group, Mokta (a sub-sidiary of Imetal), CIM, and Dong-Trieu. As for diamond exploration, a total of 75 MF was spent during the period. Exploration for phosphates was also important, totaling 38 MF, principally for phosphate rock in West Africa and submerged deposits in Polynesia; the most important companies were CIM, BRGM, and the PUK group, which led exploration on the Mataiva atoll in French Polynesia (subsequently transferred to BRGM control). Finally, exploration expenditures aimed at developing seabed minerals, primarily deep-sea polymetallic modules, totaled 417 MF.

Exploration Trends and Determinants

From 1973 to 1982, French mineral exploration expenditures nearly dou-bled, from 655 MF to 1,131 MF, corresponding to a comparable increase in the volume of operations. This growth in expenditures is principally the result of exploration on French territory, where expenditures more than doubled from 292 MF to 604 MF. Abroad, outside the Franc Zone, expenditures grew by 44 percent from 283 MF to 420 MF. Within the Franc Zone, in contrast, expenditures varied around a nearly horizontal trend of about 100 MF, with a low of 57 MF in 1976 and a high of 134 MF in 1981.

Uranium exploration expenditures, which grew from 229 MF in 1973 to 455 MF in 1982, account for much of the increase in total expenditures. By 1982, uranium expenditures on French territory had grown to 190 MF from 66 MF in 1973, but were still less than expenditures abroad (265 MF). During the same period, expenditures for base metals increased by 23 percent, while precious metals expenditures increased almost ninefold from 11 MF to 97 MF. For alloy and specialty metals, the trend is less clear—expenditures fluctuated between a high of 98 MF and a low of 46 MF.

It is important to note that the growth in overall expenditures for French mineral exploration is not limited to the decade 1973–82; by 1973, total French exploration expenditures had grown to 655 MF from 444 MF in 1968, an increase of 48 percent (175 MF in French territory, 165 MF in the Franc Zone, and 104 MF in the rest of the world).

Several reasons explain and justify this widespread growth in French mineral exploration. First it should be recalled that France long has been a mining country. Beginning in the nineteenth century it provided for the growth of its heavy industry through a secure supply of raw materials mostly from mines within France. Thereafter, the mining industry began to develop in territories controlled by France, especially in Africa. After the intense reconstruction efforts following the world wars, however, most of these traditional sources of mineral supply entered a period of declining production. It therefore was necessary to restore the health of

France's own mining firms, so that they could maintain their contribution to the supply of raw materials for French heavy industry.

For various reasons, a similar situation of dependence developed in the area of petroleum products, provoking the world energy crisis. French leaders, realizing that their country was at the mercy of oil-exporting countries, decided to equip France with nuclear power centers. Virtually overnight, uranium exploration became a top priority, encouraged by favorable geologic environments in several parts of France. Beginning in 1973, uranium became the most important metal target for French exploration.

The postwar transition to independence of many Third World countries, notably in Africa, profoundly changed relations between the new states, which wanted more control over the development of their mineral resources, and mining companies operating within their borders. In order to replace minerals from mines they no longer controlled, as well as to locate new mines, French companies resolutely turned toward other countries that offered them new opportunities for success in mineral exploration. Along with geologic potential, these countries had to have a judicial and administrative system that was sufficiently stable to assure investors of the security of their operations, total or partial ownership of production, and the possibility of repatriating profits. In such countries, French companies faced competitors who had access to sizable technical and financial resources, requiring the French to rise to the level of this competition.

In 1968 the French government instituted an annual inquiry into French mineral exploration, which covered technical and financial measures of annual activity, data on newly discovered reserves, and plans for future exploration. The statistics dramatically revealed that insufficient exploration was taking place to ensure French industry an adequate long-term supply of mineral raw materials. This disquieting finding prompted several reactions by government authorities and private mining organizations. First, the government provided money to support mineral exploration by private French companies as well as by other companies exploring on French territory. These funds were to be repaid out of revenues gained through successful exploration. Furthermore, prospecting on all French territory was strengthened by the creation of a French inventory of mineral resources. This inventory became the responsibility of the BRGM, which conducts grass-roots prospecting and exploration aimed at identifying promising targets, on which interested French mining companies then undertake detailed exploration at their own expense. Finally, scientific and technologic research in a broad range of disciplines—from geosciences to mineral processing—aimed at enhancing mineral supply has been encouraged.

In addition to these determinants of French mineral exploration, discoveries made by French explorers in France and abroad served to stimulate further exploration.

Exploration Results

Before presenting what is a sort of profit-and-loss account for French mineral exploration between 1973 and 1982, the following constraints and limitations need to be highlighted.

The only way to assess the productivity of exploration expenditures for the entire range of minerals is in terms of value, and therefore constant 1982 French francs have been used as the basis for evaluation.

The following restrictions apply to the reserves included in this study. Only mineral reserves that are surely exploitable and were discovered between January 1, 1973, and December 31, 1982, are included; reserves acquired by purchase are thus excluded. These reserves, regardless of their nature and grade, have been valued using the average international 1982 prices of their relevant metal and mineral content. Such values are of necessity gross rather than net, because they do *not* account for losses in extraction and processing or for the costs of these processes. Where exploration outside French territory was carried out jointly by mining groups that include French companies, only the relevant share of reserve value is assigned to each French participating organization.

Data on the metric tons (tonne) of raw reserves established for the minerals of this study between 1973 and 1982 are displayed in table 8-1. Prices for the minerals are shown in table 8-2. The data in both tables are used to evaluate the effectiveness of French mineral exploration during this period, as shown in table 8-3.

The seven principal organizations exploring for uranium spent 3,979 MF in discovering reserves worth 91,204 MF, yielding a ratio of gross discovery value (GDV) to expenditures of 22.9. In French territory, in which expenditures totaled 1,522 MF and discoveries are valued at 40,495 MF, the ratio is 26.6. In the Franc Zone, exploration expenditures of 453 MF yielded discoveries valued at 41,591 MF, resulting in a very satisfactory ratio of 91.8. For the rest of the world the ratio is 4.5, based on expenditures of 2,000 MF and a discovery value of 9,000 MF.

For base metals, the five largest organizations together account for 2,005 MF of the total expenditure of 2,084 MF, and together they made discoveries valued at 60,481 MF. The worldwide ratio of discovery value to expenditures for the five majors is 30.2, attributable largely to discoveries made outside of French territory and the Franc Zone by Imetal and BRGM. The ratio for the rest of the world, therefore, works out to a satisfactory 55.8.

As for alloy and specialty metals, this category encompasses a variety of metals, particularly nickel-cobalt, chromium, tungsten, antimony, and rare earths. The aggregate benefit−cost ratio of 22.3 obscures the range of values for particular metals. New Caledonian nickel, the main target of exploration in this metal category, has a benefit−cost ratio of 27.4.

Table 8-1.   French Mineral Exploration: Reserves Discovered, 1973–82

Mineral	Location of discovery	Quantity (metric tons, as annotated)
Uranium	French territory (all in France) The majority were discoveries by Cogema; the rest (in order of importance) by Dong-Trieu, Minatome, CIM, and the Imetal group.	80,988[a]
	Franc Zone (discoveries by CEA-Cogema, Imetal, PUK)	83,180[a]
	Rest of world (discoveries by Minatome, CEA-Cogema, Imetal, PUK)	18,235[a]
Total uranium		182,403[a]
Base metals		
Zinc	French territory	309,000[b]
Lead	French territory	49,000[b]
Copper	French territory	7,500[b]
Bauxite	French territory	1,000,000[c]
Copper	Franc Zone	83,000[b]
Silver	Franc Zone	490[b]
Tin	Franc Zone	11,700[b]
Copper	Rest of world	2,229,000[b]
Lead	Rest of world	3,094,000[b]
Zinc	Rest of world	3,118,000[b]
Associated silver	Rest of world	4,507[b]
Bauxite	Rest of world	7,000,000[c]
Alloy and specialty metals		
Nickel (in New Caledonia)	French territory	1,300,000[b]
Precious metals		
Gold	French territory	7[b]
Silver	French territory	47[b]
Gold	Franc Zone	7[b]
Gold	Rest of world	3[b]
Other minerals		
Phosphates (34 percent P_2O_5)	French territory	3,600,000[c]
Acid fluorspar contained	French territory	1,170,000[d]
Metallurgic fluorspar contained	French territory	925,000[d]
Acid fluorspar	Rest of world (excluding Franc Zone)	1,000,000[d]

Note: Tonnages represent, where appropriate, French shares of reserves.
Source: The reserve figures were obtained from individual companies, adjusted to ensure consistency.
[a]Metric tons of contained uranium.
[b]Metric tons of contained metal.
[c]Metric tons of ore.
[d]Metric tons of concentrates.

Table 8-2. Mineral Prices, 1982

Mineral	Price
Uranium	500,000 francs/metric ton[a]
Copper	9,600 francs/metric ton[a]
Lead	4,000 francs/metric ton[a]
Zinc	4,600 francs/metric ton[a]
Tin	83,200 francs/metric ton[a]
Bauxite	
France	100 francs/metric ton[b]
Greece	154 francs/metric ton[b]
Nickel (value in an ore with 2.5 percent	
nickel content)	7,270 francs/metric ton[a]
Gold	100,000 francs/kilogram [c]
Silver	1,845 francs/kilogram [c]
Phosphates (34 percent P_2O_5)	162 francs/metric ton[b]
Fluorspar	
Metallurgic	460 francs/metric ton[d]
Acid	700 francs/metric ton[d]

Note: The values shown in this table only indicate orders of magnitude as a basis for comparison. The author calculated the prices by estimating average quotations for each commodity.

Source: Based on information provided by various departments of the French Ministry of Industry and Foreign Trade, primarily the Bureau de documentation minière and its successor, the Observatoire des matières premières, editors of the *Annales des Mines*'s special yearly issue devoted to statistical and other information on mineral industries activities.

[a]Value refers to metric tons of metal contained in ore.
[b]Value refers to metric tons of ore.
[c]Value refers to kilograms of metal content.
[d]Value refers to metric tons of concentrate.

Precious metal discoveries by the major organizations in France, Africa, and Canada are valued at 1,794 MF, yielding a benefit–cost ratio of 6.4, when compared with expenditures of 281 MF (out of total precious metal expenditures of 338 MF). The benefit–cost ratio for the Franc Zone is 8.5, and for the rest of the world is 7.14.

To summarize, if only the eight major organizations are considered, total exploration expenditures for all minerals represent 4.6 percent of the gross value attributed to discoveries between 1973 and 1982: the corresponding benefit–cost ratio is 21.7. In the Franc Zone, only seven of the majors were active in exploration, and their expenditures totaled only 1.94 percent of the reserves discovered. For the rest of the world the ratio is 4.97 percent. The corresponding benefit–cost ratios are, respectively, 51.6 and 20.12.

International Comparisons

It is tempting to try to make an objective comparison of French mineral exploration and corresponding activities elsewhere, particularly in countries with significant and active mining industries. But given the differing

Table 8-3. French Mineral Exploration Effectiveness, 1973–82 (Principal Companies Only)

Mineral targets	French territory			Foreign						Total		
				Franc Zone			Rest of world					
	Expenditure (1982 MF)	GDV[a] (1982 MF)	Benefit-cost ratio	Expenditure (1982 MF)	GDV[a] (1982 MF)	Benefit-cost ratio	Expenditure (1982 MF)	GDV[a] (1982 MF)	Benefit-cost ratio	Expenditure (1982 MF)	GDV[a] (1982 MF)	Benefit-cost ratio
Uranium (7 companies)	1,522	40,495	26.60	453	41,591	91.81	2,004	9,118	4.55	3,979	91,204	22.92
Base metals (5 companies)	824	1,789	2.17	177	2,671	15.09	1,004	56,021	55.80	2,084	60,481	30.17
Alloy and specialty metals (7 companies)	345	9,450	27.39	22	—	—	56	—	—	423	9,450	22.34
Precious metals (5 companies)	157	794	5.06	82	700	8.54	42	300	7.14	281	1,794	6.38
TOTAL (8 companies)	3,443	54,152	15.73	871	44,962	51.62	3,287	66,139	20.12	7,601	165,253	21.74

[a]GDV = gross discovery value.

Table 8-4. Breakdown of Exploration Expenditures by Commodity, United States and France, 1977

(percent)

Mineral group	United States	France
Uranium	23	48
Base metals	14	32
Alloy and specialty metals	12	12
Precious metals	2	4
Coal, lignite	43	1
Other	6	3
Total	100	100

Source: U.S. percentages: Crowson (1983); French percentages: French Ministry of Industry and Foreign Trade.

ways in which relevant data are collected, recorded, and published, such comparisons are, for all practical purposes, impossible. Indeed, any comparison of the available data must be approached with the utmost caution.

France is one of the few countries with a market economy in which, for both the private sector and state-owned organizations, data are systematically collected on an annual basis for financial and technical investments in mineral exploration, the minerals sought, and the reserves discovered. For various other countries and the Organisation for Economic Cooperation and Development (OECD), statistical surveys of exploration spending are made, as Crowson (1983)[6] points out. The European Community's liaison committee for the nonferrous metals industry publishes statistics on mineral exploration expenditures by its member organizations, but the only mineral for which the figures are listed separately is uranium. Occasionally, information is published in the English-language trade press, but usually only in summary form, upon which a comparative study cannot be based. This is particularly true of information from the United States, probably the source of the bulk of worldwide exploration finance, where the relevant data are not necessarily passed on to the government. It also is difficult to obtain detailed information about mineral exploration in countries with centrally planned economies and in the developing countries.

Despite data that are not always comparable and are usually incomplete, certain general comparisons are possible. For example, the French share of mineral exploration by the European Community rose from 20 to 38 percent between 1977 and 1982; and the French share of uranium exploration increased from 19 to 49 percent between 1976 and 1982, a year in which, if the data are correct, the European Community's total expenditures were only half of what they were in 1980.

A relatively recent study by Cranstone (1980) contains annual exploration expenditures in Canada from 1946 to 1977, permitting comparison

[6]The final version of this study is chapter 2 of this volume.

Table 8-5.   Uranium Exploration Effectiveness, 1972–80

Country	Exploration expenditure (1982 MF)	Discovery value (1982 MF)	Ratio of discovery value to exploration expenditure
Australia	1,971	284,000	144.0
Brazil	1,379	99,000	71.8
Canada	4,106	216,000	52.6
French territory	1,228	36,000	29.3
India	394	27,000	68.5
Mexico	197	3,500	17.8
United States	17,936	391,000	21.8

Note: Worldwide, French organizations spent 2,856 MF and discovered reserves valued at 82,000 MF, yielding a discovery value to exploration expenditure ratio of 28.7.

Source: The French expenditure figures are derived from annual statistics compiled since 1968 by the Department of Mines of the Ministry of Industry; the expenditures for the years 1973 through 1982 are presented in appendix table 8-A-2 in this chapter. (*See* Chazan, 1970, 1975, 1981, and 1982.) These statistics, both detailed and comprehensive, are included in annual reports of the Department of Mines, which are classified matter. The figures for the other countries are from Crowson (1983).

with French expenditures between 1973 and 1977. The data indicate that 5,026 MF, or 504 francs per square kilometer, were spent in Canada. Over the same five-year period, 1,826 MF, or 2,775 francs per square kilometer, were spent in French territory.

Data from Crowson (1983) indicate that exploration expenditures in the United States totaled 3,285 MF in 1970 and 5,256 MF in 1977 (that is, about five or six times as much as French expenditures). Table 8-4 compares the shares of total expenditures devoted to particular mineral groups in the United States and France in 1977. Coal and lignite represented a much larger share in the United States than in France, whereas uranium and base metals accounted for a much larger share in France. For Australia between 1973 and 1982, Crowson's figures show total exploration expenditures of 24,933 MF, or 3,238 francs per square kilometer, while in French territory the comparable figures are 4,260 MF and 6,474 francs per square kilometer.

Crowson (1983) also uses data collected by the OECD on uranium exploration expenditures to calculate the relative effectiveness of expenditures in various countries:

**Constant 1982 U.S. Dollar Expenditures per
Metric Ton of Uranium Discovered Between
1972 and 1980**

Australia	530
Brazil	1,060
India	1,150
Canada	1,440
United States	3,490
Mexico	4,000

Crowson also calculates a figure of $6,960 (constant 1982 U.S.) for France, but this is apparently based on incomplete data on expenditures and discovery values. Complete data yield the following costs per metric ton of uranium discovered as a result of French exploration: U.S.$2,465 for French territory and U.S.$2,652 overall. Finally, Crowson's data can be used to calculate the ratio of discovery value to expenditure for uranium exploration between 1972 and 1980 for various countries (see table 8-5).

Some interesting, if tentative, figures have been published for overall mineral exploration effectiveness, and they can be used, not without reservation, in the following comparisons. According to Cranstone (1980), the value of deposits discovered per Canadian exploration dollar was C$16.6 between 1972 and 1974, and C$22.2 between 1975 and 1977. The respective ratios are 6.0 and 4.5 percent. For the United States, Rose (1982) calculated ratios of between 1.0 and 2.5 percent between 1955 and 1974.

Bearing in mind the limitations imposed by incomplete and disparate data and by the different geographic, economic, and legal backgrounds, it may be concluded that French mineral exploration, with an average ratio over ten years of 4.60 percent, has been more or less as effective as similar activities in North America.

■

The author wishes to thank the many organizations that provided essential data for this chapter, as well as those colleagues who kindly read and commented on the original French version. He is also grateful to Roderick G. Eggert for his many suggestions and efforts to improve the English version.

References

Chazan, W. 1970. "Motivations et coùt de la recherche minière." *Annales des Mines* (February).

———. 1975. "La recherche minière en France." *La Recherche* no. 60 (October).

———. 1981. "La recherche minière française de 1968 à 1979: analyse des investissements." *Chronique de la Recherche Minière* no. 463 (November-December).

———. 1982. "La recherche minière française en 1980." *Geologues* no. 62-63.

Cranstone, D. A. 1980. "Canadian Ore Discoveries 1946–1978: A continuing record of success," *CIM Bulletin* vol. 73, no. 817 (May) pp. 30–40.

Crowson, P. C. F. 1983. "A Perspective on Worldwide Exploration for Nonfuel Minerals." Paper presented at the International Institute for Applied Systems Analysis, Laxenburg, Austria, December 1983.

Rose, A. W. 1982. "Mineral Adequacy, Exploration Success and Mineral Policy in the United States," *Journal of Geochemical Exploration* vol. 16.

Appendix 8-A
French Exploration Expenditures, 1973–82

Table 8-A-1. Total French Exploration Expenditures, 1973–82
(millions of 1982 French francs)

| Year | French territory | Foreign | | Total |
		Franc Zone	Rest of world	
1973	292	80	283	655
1974	288	98	261	647
1975	369	92	282	743
1976	441	57	269	767
1977	436	106	310	852
1978	407	62	346	815
1979	406	91	337	834
1980	427	75	474	976
1981	588	134	431	1,153
1982	606	105	420	1,131
Total	4,260	900	3,413	8,573

Note: Includes ferrous and nonferrous metals, coal, uranium, evaporites, other minerals (such as borates, fluorine, sulfur, asbestos, diamonds). Excludes liquid and gaseous hydrocarbons and water.
Source: French Ministry of Industry and Foreign Trade.

Table 8-A-2. French Uranium Exploration Expenditures, 1973–82
(millions of 1982 French francs)

| Year | French territory | Foreign | | Total |
		Franc Zone	Rest of world	
1973	66	34	129	229
1974	100	37	114	252
1975	127	43	155	325
1976	162	27	143	332
1977	156	61	190	407
1978	171	26	225	422
1979	182	53	221	456
1980	215	37	344	596
1981	212	93	278	583
1982	190	48	217	455
Total	1,581	459	2,016	4,057

Source: French Ministry of Industry and Foreign Trade.

Table 8-A-3. French Base Metal Exploration Expenditures, 1973–82
(millions of 1982 French francs)

| Year | French territory | Foreign | | Total |
		Franc Zone	Rest of world	
1973	51	22	94	167
1974	46	28	125	199
1975	71	16	102	189
1976	71	17	96	184
1977	96	25	90	211
1978	107	18	77	202
1979	105	16	75	196
1980	98	9	87	194
1981	108	12	84	204
1982	93	9	106	208
Total	846	172	936	1,954

Note: Includes lead, zinc, copper, molybdenum, and tin.
Source: French Ministry of Industry and Foreign Trade.

Table 8-A-4. French Exploration Expenditures for Alloy and Specialty Metals, 1973–82
(millions of 1982 French francs)

| Year | French territory | Foreign | | Total |
		Franc Zone	Rest of world	
1973	88	6	4	98
1974	55	6	1	62
1975	44	2	4	50
1976	47	—	2	49
1977	76	1	5	82
1978	56	1	9	66
1979	52	1	11	64
1980	32	2	12	46
1981	58	3	6	67
1982	71	1	8	80
Total	579	23	62	664

Note: Includes nickel, cobalt, chromium, tungsten, antimony, and rare earths.
Source: French Ministry of Industry and Foreign Trade.

Table 8-A-5. French Precious Metals Exploration Expenditures, 1973–82
(millions of 1982 French francs)

Year	French territory	Foreign		Total
		Franc Zone	Rest of world	
1973	7	4	—	11
1974	5	8	2	15
1975	21	17	1	39
1976	20	3	1	24
1977	12	2	1	15
1978	15	3	1	19
1979	19	4	2	25
1980	23	7	5	35
1981	33	12	13	58
1982	36	24	37	97
Total	191	84	63	338

Source: French Ministry of Industry and Foreign Trade.

9

The Canadian Mineral Discovery Experience Since World War II

DONALD A. CRANSTONE

This chapter analyzes costs and rates of mineral discovery in Canada during the thirty-seven-year period following World War II. The analysis involves the roughly 900 Canadian discoveries that were made between January 1, 1946, and January 1, 1983, as a result of various mineral exploration activities. In conjunction with the findings presented, the chapter reappraises two questions of major significance in assessing the Canadian mineral experience since World War II: whether Canadian mineral deposits are being depleted faster than new ones are being discovered and whether Canadian ore has become more expensive to discover in constant dollar terms.

Subsequent sections present the record of costs and discoveries for the 1946–82 period and discuss the significance of the discovery of major deposits, the rates of discovery for thirteen individual metals, and the rates of discovery for six major metals by geologic deposit type. After considering the influence of developments in exploration technology and geologic concepts on rates of ore discovery, the chapter closes with an assessment of the relevance of the Canadian mineral discovery experience for other nations.

Before delineating its scope and methodology, one particular feature of this study needs to be mentioned—namely, that it was designed to eliminate a bias in favor of the past which other studies of rates of ore discovery have generally incorporated. This built-in bias made the rate of discovery in the past appear higher than that of more recent years, and it tended to lead to the conclusion that the cost of the discovery of metals in mineral deposits was increasing in constant dollar terms.

The reasons for this bias, relating to the lack of available information on newly or recently discovered deposits, are these:

1. Since some recent discoveries are not yet publicly known, even well-informed students of this field are not likely to be aware of every ore body discovered in Canada in recent years.

The existence of this bias in favor of the past was proved by a detailed review of earlier studies (Cranstone, 1982; Cranstone and Martin, 1973; Derry, 1968, 1970, 1972; Derry and Booth, 1978). For example, in the case of Cranstone and Martin (1973), a thorough search in 1973 of all information about Canadian ore discoveries in the five-year period 1966 through 1970 resulted in a total count of thirty-seven. In 1981, using exactly the same criteria to define a discovery but with the benefit of an additional eight years of information, it was found that another twenty-nine deposits could be added to this 1966–70 discovery list. For 1971 alone, Cranstone and Martin (1973) listed four ore discoveries, but by 1981 another seven were apparent, four of which were by then in production or committed for production. These examples are not exceptional; many others could be cited.

Although one can quibble about exact discovery years or about which deposits should in retrospect have been counted as new discoveries, the fact that these studies did not account for large numbers of recent discoveries inevitably led to the conclusion that it was becoming more difficult to find new ore bodies.

2. Some recent discoveries cannot yet be recognized as the economically exploitable ore bodies that they may later prove to be.

3. Some authors have counted each discovery as one, even though not all ore bodies are of identical size. The amount of metal in each discovery should be taken into account, but, of course, this is very difficult when dealing with recent discoveries, because many years will have elapsed before the precise dimensions of most newly discovered deposits are known. Furthermore, the amount of ore that is ultimately recovered in many cases considerably exceeds the estimate at the time of discovery or even at the start of production. As explained below, the analysis presented in this chapter makes use of multipliers to offset this tendency toward initial underestimation.

Scope and Methodology of the Study

This chapter presents the findings of an analysis of some 900 mineral discoveries in Canada made as a result of various exploration activities. Taken into account as discoveries are deposits of all minerals—except those of lithium, iron ore, industrial minerals, and coal—discovered between

January 1, 1946, and January 1, 1983, for which tonnage[1] and grade were reported.

Deposits of the minerals noted above were excluded for various reasons. For example, there is little market for lithium minerals and no clearly defined market price for them. Thus, the fifteen or more significant lithium deposits discovered in Canada since 1946 were not counted in the total value of discoveries.

Iron ore deposits were excluded because development of new iron mines is not generally related to new discoveries but to marketing opportunities and because most known Canadian iron deposits were discovered many years ago. Also, inclusion of the vast tonnages of iron in known iron deposits, if valued at nominal prices for iron ore, would have seriously distorted the results of this study.

Asbestos deposits were not included because the estimation of deposit grades and of gross "value" is complicated by a lack of adequate data on asbestos content, grade of fiber, recovery factor, and the market value of the asbestos fiber contained in the deposits.

In the case of industrial minerals other than asbestos—e.g., limestone, clays, salt—production is usually related more to the level of demand and to transportation costs than to local availability of raw materials and the timing of discoveries. Moreover, these industrial commodities are not usually sought by typical mineral exploration programs. For these reasons, deposits of limestone, clays, coal, barite, fluorspar, gypsum, apatite, potash, silica, sodium sulfate, salt, talc, and the like were excluded.

In this chapter the term *discovery* refers to a mineral deposit sufficiently attractive to have warranted the expenditure necessary to establish its tonnage and grade. Discoveries are subdivided into two groups, as follows:

Group 1 consists of

- deposits that were in production at the end of 1983,
- those being prepared for production at the end of 1983,
- other deposits included by producing companies in their ore reserves at the end of 1982, and
- deposits that had been but were no longer in production at the end of 1983 (less minable tonnages that were left behind, which are included in group 2).

Group 2 consists of

- deposits not brought into production in spite of being within reach of profitable production at or near the metal prices prevailing in late 1983,

[1]Although Canada is now in the process of conversion to the metric system, nearly all of the available historical data concerning mineral deposit tonnages and grades and many current data are still in short tons and troy ounces per ton. Therefore, the figures in this chapter are in short tons and troy ounces. Conversion factors to the metric system are as follows: 1 short ton = 0.90718474 metric ton; 1 troy ounce = 31.10348 grams.

- those judged by this author to be unprofitable to mine at or near metal prices prevailing in late 1983, and
- minable tonnages left behind in deposits formerly in production (see group 1, category 4).

The term *date of discovery* as used in this chapter usually indicates the year in which the first drill hole intersected a mineral zone that was recognized relatively soon thereafter to be part of a mineral deposit of economic interest. Thus, a "discovery" is not necessarily a newly found showing.

Rather than applying a rigid definition of this term, an attempt was made to consider all relevant information about each deposit so as to arrive at the most logical discovery date. The following hypothetical examples illustrate how discovery dates are assigned:

- A surface showing was staked and trenched in 1905 but remained idle until 1965 when diamond drilling established an ore body. This would be counted as a 1965 discovery.
- A new ore body was discovered at an old mine previously considered to be mined out. This would qualify as a new discovery.
- A deposit discovered in 1900 was explored underground at that time. Tonnage and grade were established, but the discovery was not brought into production. Assume that a production decision was made in 1965 on the basis of the former exploration work; even if some additional geologic mapping and new drilling were carried out to check the old work, the deposit would not be considered a new discovery. On the other hand, if in 1965 new ore was discovered which led to production from a much larger ore body than was known from the earlier exploration, this deposit would be counted as a 1965 discovery.

To add up tonnages of various metals in various mineral deposits, one must convert metal tonnages into the common denominator of dollars. For this study the "values" of individual deposits were summed for all discoveries within successive three-year periods and were plotted as the *gross metal value of discoveries.* Hence, *metal value* (content of metals in deposits discovered, multiplied by appropriate metal prices) as used here is representative of an aggregate tonnage of various metals. The analysis was made using two different sets of Canadian dollar metal prices:[2] (1) prices that prevailed in late January 1979 and (2) average prices that prevailed for each metal in each three-year period, using an appropriate price series for each metal, converted into constant 1979 dollars.[3]

---

[2]Over the thirty-seven-year period (1946 through 1982) covered in this chapter, the average annual noon exchange rate, quoted by the Bank of Canada, was $1.00 (in Canadian dollars) = $0.9574 (in U.S. dollars).

[3]Other chapters in this volume use 1982 dollars. The factor for conversion of 1979 to 1982 dollars is $1.00 (in 1979 Canadian dollars) = $1.353 (in 1982 Canadian dollars).

Three-year periods are used for this analysis because discovery totals for shorter periods are so erratic that trends are not clearly discernible. Also, when only one or two major deposits were discovered in any particular year, proprietary information as to deposit sizes would be revealed in a one-year discovery total, whereas three-year discovery totals preserved the confidentiality of such information.

Since it is not possible to predict the size of newly discovered deposits, an effort was made to develop a systematic approach toward estimating their size. An analysis was made of about 100 Canadian mines, of which some have been mined out and others are still in production. For each mine, account was taken of the initial ore reserves when production began; of historical, year-by-year ore production; of the current ore reserves; and of the potential for additional ore. A multiplier was derived for each mine; applied to the metal value of the initial ore reserves, this multiplier gives the metal value of the total amount of ore that was ultimately produced (in the case of mined-out deposits) or that is likely to be ultimately produced (in the case of deposits still in production).

From this analysis, multipliers were derived for the following four generalized deposit types:

Deposit type	Multiplier
1. Open-pit porphyry copper, copper-molybdenum, and molybdenum deposits	2
2. Nonporphyry, nonvein base metal sulfide deposits (excluding nickel-copper deposits)	2
3. Vein deposits of gold, silver, and other metals	3
4. Nickel-copper deposits (other than those at Sudbury, Ontario, and the Thompson Nickel Belt in Manitoba, which were dealt with separately)	1.3

While these deposit types do not cover all mineral deposits discovered in Canada since January 1, 1946, the choice of appropriate multipliers for other deposit types with few Canadian examples was influenced by the types and multipliers listed above. For uranium deposits a different approach was taken. Resource estimates made by the Uranium Resource Appraisal Group (URAG) of the Canadian government department Energy, Mines and Resources Canada were used as a basis for estimating deposit sizes. Multipliers ranging from 1.0 to 1.3 were chosen for these deposits.

The multipliers derived for a generalized deposit type were applied to those discoveries of deposits yet to be mined (mostly of group 2, some of group 1) for which not enough information was available to allow a better judgment. For many producing mines, enough is known by Energy, Mines and Resources Canada to permit reasonable estimates of how much ore beyond current reserves is likely to be produced; such information was taken into account in figuring deposit sizes.

In general, the multipliers chosen are deliberately on the conservative side, and thus recent ore discoveries are still likely to have been underestimated. In attempting to eliminate bias favoring the past through the use of multipliers, care was taken not to create a bias favoring recent discoveries.

Record of Discoveries

Figures 9-1 and 9-2 show the gross metal value of Canadian mineral discoveries from 1946 through 1982. Tables 9-1 and 9-2 list the metal prices used to obtain these values.

In spite of the multipliers used, discovery values for the most recent periods are undoubtedly still too low, first, because it is likely that not all recent discoveries are known to the author and, second, because the multipliers used are on the conservative side. In figures 9-1 and 9-2 the multiplier additions are shown separately only for group 2; for group 1, more direct ways of estimating deposit sizes were used in most cases.

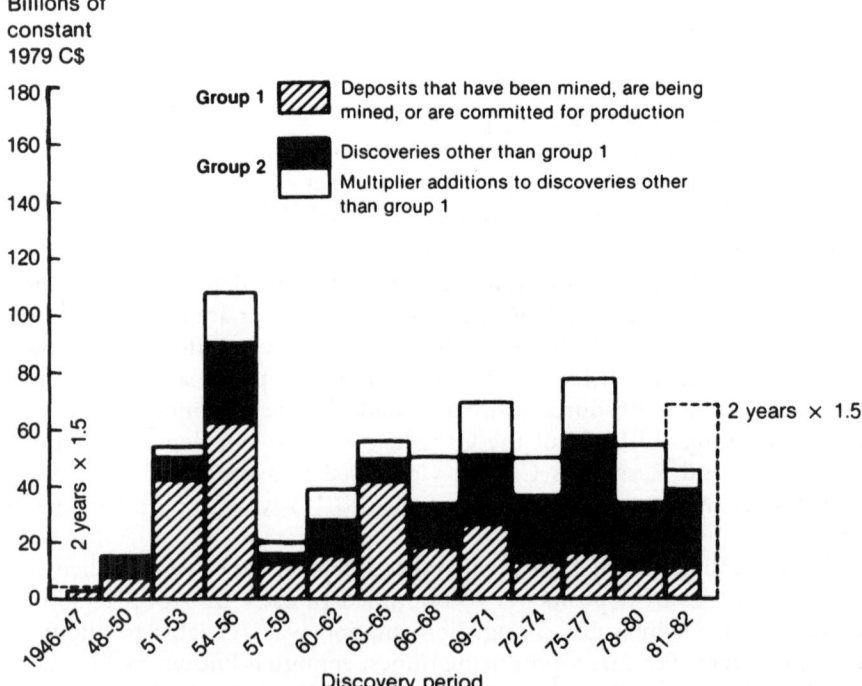

Figure 9-1.   Gross metal value of Canadian mineral discoveries at average prices per three-year period, 1946–82 (in constant 1979 Canadian dollars).

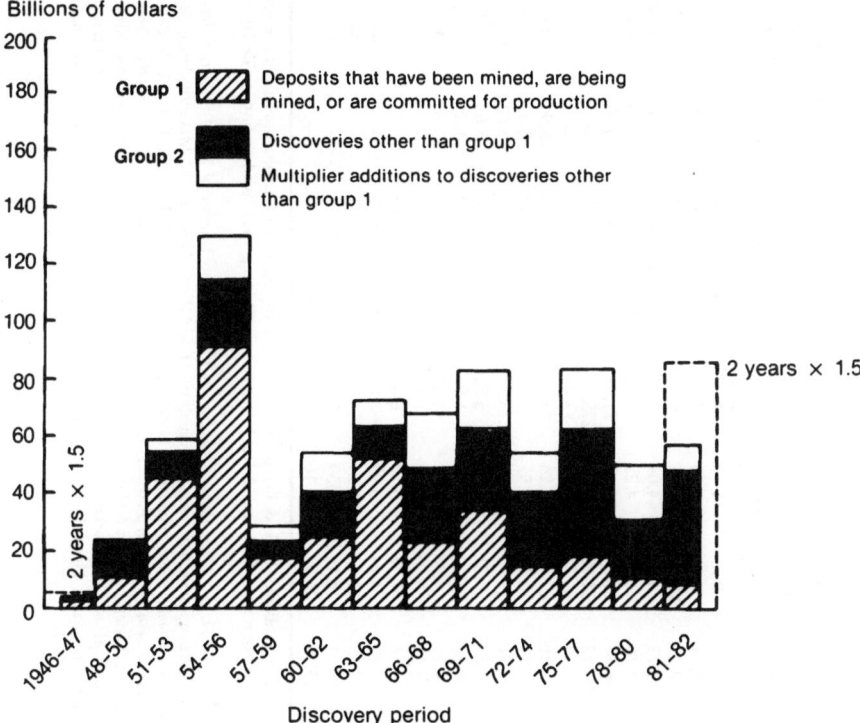

Figure 9-2.   Gross metal value of Canadian mineral discoveries per three-year period, 1946–82 (at January 1979 Canadian dollar metal prices).

A striking feature of these two figures is the outstanding success of the 1954–56 period. Major discoveries took place in that period—for example, most of the uranium deposits at Elliot Lake, Ontario; the Thompson nickel deposit in Manitoba; the Heath Steele, Anaconda Caribou, and several other zinc-lead-copper-silver deposits in the Bathurst area of New Brunswick; seven large niobium deposits in Quebec and Ontario; and several copper-zinc deposits in the Flin Flon–Snow Lake area of Manitoba.

A likely explanation for the outstanding success of the 1954–56 period is provided by Derry (1968, table II; 1970, table II), who lists the discovery methods for many ore deposits. As he shows, there was a sudden switch, around 1951, from discoveries made mostly by conventional prospecting to those made mostly by geologic intuition or geophysical prospecting or a combination of the two. The 1954–56 successes, then, represented the first major results of the much greater application of geologic intuition to the search for new mineral deposits, combined with the first application of airborne geophysical methods. To a certain extent just plain luck may also have been a contributing factor, as these newly developed methods

Table 9-1. Canadian Metal Prices: Averages for Three-Year Periods, 1946–82 (constant 1979 Canadian dollars)

Metal	Weight unit	1946-47 (2 years)	1948-50	1951-53	1954-56	1957-59	1960-62	1963-65	1966-68	1969-71	1972-74	1975-77	1978-80	1981-82 (2 years)
Nickel	lb	1.66	1.59	1.83	1.92	2.05	2.18	2.23	2.20	2.65	2.64	2.88	3.07	2.75
Copper	lb	0.69	0.83	0.92	1.10	0.81	0.83	0.89	1.10	1.12	1.09	0.84	0.85	0.74
Zinc	lb	0.38	0.52	0.49	0.37	0.32	0.34	0.35	0.33	0.32	0.43	0.46	0.40	0.39
Lead	lb	0.39	0.60	0.51	0.44	0.35	0.29	0.35	0.33	0.31	0.30	0.31	0.48	0.30
Molybdenum	lb	2.13	2.05	3.22	3.29	3.34	3.75	4.05	4.04	3.72	3.23	4.31	8.44	4.07
Silver	tr. oz.	5.52	2.90	2.80	2.67	2.51	2.76	3.58	4.22	3.72	4.89	5.77	18.79	8.77
Gold	tr. oz.	165.81	137.69	112.83	105.21	96.92	98.56	98.84	90.18	81.86	174.86	183.78	410.49	396.15
Platinum	tr. oz.	280.30	314.53	298.63	274.62	214.02	234.87	251.28	275.04	275.78	261.20	218.40	—	—
Palladium	tr. oz.	113.69	94.42	77.40	67.08	55.42	70.39	83.57	96.73	83.96	137.87	89.06	—	—
Uranium	lb U_3O_8	33.19	33.19	33.19	32.17	28.85	23.81	22.41	20.24	13.70	12.99	44.96	45.90	20.91
Niobium	lb Nb_2O_5	2.41	3.24	7.94	6.03	3.07	2.95	2.68	2.51	2.28	2.41	2.52	3.09	3.07
Tantalum	lb Ta_2O_5	15.48	12.13	20.59	24.49	12.34	19.74	19.60	25.07	15.83	13.77	22.59	73.62	58.00
Tungsten	lb WO_3	4.09	3.87	9.32	4.56	2.11	2.35	2.31	5.48	5.52	5.25	6.76	7.60	5.25
Antimony	lb	1.11	1.30	1.28	.91	0.85	0.88	1.10	1.11	1.93	1.64	2.22	1.81	1.15
Cobalt	lb	7.44	6.83	7.54	7.80	5.36	4.26	4.36	4.51	4.57	5.03	6.05	23.38	14.34
Bismuth	lb	7.62	8.20	7.26	6.76	6.23	6.34	8.00	10.12	12.57	9.39	9.19	—	—
Cadmium	lb	6.55	7.87	7.33	5.14	4.22	4.44	7.38	6.50	6.43	5.77	3.89	—	—
Tin	lb	3.06	3.86	3.71	2.88	2.71	3.10	4.24	3.93	3.67	4.43	5.10	—	—
Mercury	lb	5.70	4.09	8.53	10.75	8.55	7.40	13.18	16.24	11.66	5.86	2.37	—	—
Selenium	lb	8.57	8.87	10.97	29.38	19.85	17.57	14.14	11.39	17.33	23.72	22.94	—	—
Tellurium	lb	8.22	6.88	5.64	5.26	5.32	13.36	16.94	15.18	13.07	11.47	15.72	—	—

Notes: Dash = prices not required for analysis. Canadian producer price series used where available. U.S. dollar price series most applicable to Canadian producers selected and converted to Canadian dollars using the Bank of Canada quote for the average noon exchange rate. Constant dollar prices calculated using the Canadian implicit deflator of the gross national expenditure, as compiled by Statistics Canada.

Sources: Data are from the following periodical publications (various issues) and books: *Non Ferrous Metal Data* (Secaucus, N.J., American Bureau of Metal Statistics, issued annually); *Metallstatistik* (formerly *Metal Statistics*) (Frankfurt am Main, Metallgesellschaft Aktiengesellschaft, issued annually); *Metals Week* (New York, McGraw-Hill); *Metal Statistics* (New York, Fairchild Publications, issued annually); *Mineral Commodity Summaries* (*Commodity Data Summaries* in 1977 and earlier years) (Washington, D.C., U.S. Bureau of Mines, issued annually); Roskill Information Services Ltd., *Uranium, Plutonium and the Growth of Nuclear Power 1975–2000* (London, 1975); *Monthly Report on the Uranium Market* (Menlo Park, Calif., Nuclear Exchange Corporation); V. B. Schneider, *Molybdenum*, Mineral Report 6 (Ottawa, Department of Mines and Technical Surveys, Mineral Resources Division, 1963); records of Energy, Mines and Resources Canada; W. R. Barton, *Columbium and Tantalum: A Materials Survey*, Information Circular IC 8120 (Washington, D.C., U.S. Bureau of Mines, 1962); J. C. Burrows, *Tungsten, An Industry Analysis* (Lexington, Mass., Heath Lexington Books, 1971); A. M. Lansch, *Selenium and Tellurium: A Materials Survey*, Information Circular IC 8340 (Washington, D.C., U.S. Bureau of Mines, 1967).

Table 9-2. Canadian Metal Prices in January 1979
(in Canadian dollars)

Metal	Price	Metal	Price
Nickel	2.47/lb	Bismuth	2.50/lb
Copper	0.92/lb	Tungsten	7.00/lb WO_3
Zinc	0.39/lb	Cadmium	2.25/lb
Lead	0.48/lb	Platinum	357.00/oz
Molybdenum	6.97/lb	Tin	8.05/lb
Silver	7.50/oz	Barite[a]	0.00/lb
Gold	273.00/oz	Lithium[a]	0.00/lb
Uranium	51.40/lb U_3O_8	Palladium	95.09/lb
Columbium	3.03/lb Nb_2O_5	Thorium[a]	0.00/lb
Tantalum	45.17/lb Ta_2O_5	Selenium	17.83/lb
Antimony	1.49/lb	Tellurium	25.55/lb
Cobalt	23.77/lb	Mercury	3.17/lb

Note: U.S. dollar prices converted using the Bank of Canada average noon exchange rate.

Sources: The Northern Miner (January 20, 1979), and various other sources shown in table 9-1 of this chapter.

[a]No value was assigned to barite in base metal deposits because this mineral has not usually been recovered from such deposits; nor were values assigned to lithium and thorium because markets for minerals of these elements are limited.

were used in areas that produced remarkable results the first time serious exploration efforts were applied.

Certainly many of the major discoveries of the early and mid-1950s were found by the first thorough exploration program for the deposit type found in the respective regions. For example, the Thompson Nickel Belt had never been extensively explored for nickel until the late 1940s or early 1950s, nor the Blind River (Elliot Lake) area in Ontario for uranium until the early 1950s. Similarly, although base metal occurrences had been known in the Bathurst area for almost 100 years, the first major discovery was the outcome of ground follow-up of an anomaly revealed by an airborne geophysical survey in the early 1950s. The discovery of several major mining districts (Bathurst, Elliot Lake, and Thompson) along with several lesser ones (such as Snow Lake, Manitoba, and Manitouwadge, Ontario) within a period of only four or five years represented a combination of events that was both unusual and perhaps never to be repeated. Only the period before 1900 that includes the discovery of several of the larger Sudbury-area nickel-copper deposits might exceed the 1954–56 period in value of discoveries.

Thus, the 1954–56 period stands alone. There has been no steady decline since then. In fact, if the record of that period is viewed as anomalous, it can be stated on the basis of figures 9-1 and 9-2 that the ore discovery record in Canada since World War II shows, on the whole, a slightly rising trend.

Discoveries Versus Subsequent Development of Mines

In considering figures 9-1 and 9-2 (which portray separately the mineral deposit discoveries that have been mined, are being mined, or are committed for production in group 1 and other discoveries in group 2), it is apparent that over the past decade or so a significantly smaller proportion of the new discoveries has attained production compared to discoveries of earlier years. It has always taken time to turn a mineral discovery into a producing mine. Studies from the Mineral Policy Sector of Energy, Mines and Resources Canada (Martin, Cranstone, and Zwartendyk, 1976; Martin and Jen, in this volume) found that, over the period from 1946 through 1982, the average time between discovery of an ore body and initial production was six years. The six-year average included the one to three years required to prepare a deposit for production after a production decision has been made; the lag between deposit discovery and production decision averaged four years.

The reasons for the slow development of new mines in recent years relate either to the quality and location of the new discoveries or to the general economic situation of the period. That is, the size, grades, or location of recent discoveries may have been less attractive than those of earlier discoveries. Or, development may have been hindered by general economic factors such as low prices or other unfavorable market conditions, high development and production costs, higher levels of taxation, and more stringent environmental requirements.

It has been found (Cranstone, 1980a), however, that up to the end of the 1977 discovery period, the sizes and grades of recent discoveries were generally comparable or even superior to those of discoveries of earlier years that were amenable to similar mining methods. Therefore, the reasons for the delays in development of new deposits must be sought among economic conditions of the period: they included low world demand and consequently low prices for metals, changes and instability in Canadian tax laws during the 1970s, increases in production costs and interest rates with which prices did not keep pace, more stringent and costly environmental regulations, and the general uncertainty created by all these changes. The dominant problem was the depressed market conditions for base metals that existed throughout most of the 1970s and that seem likely to continue for the foreseeable future.

Cost of Discoveries

Figure 9-3 portrays expenditures in Canada on "outside or general" exploration for the same three-year periods shown in figures 9-1 and 9-2. (The figure is based on annual expenditures listed in table 9-3. To eliminate

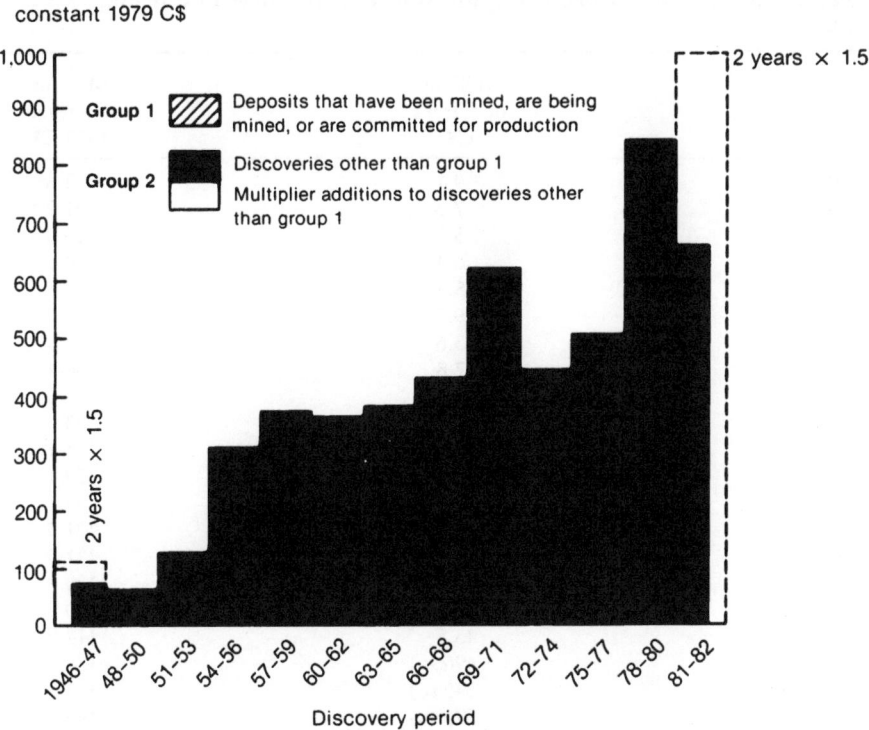

Figure 9-3. Canadian exploration expenditures per three-year period, 1946–82 (in constant 1979 Canadian dollars). See text for definition of expenditures included in this figure. *Source:* Based on table 9-3 in this chapter.

the effects of inflation, the exploration expenditures were converted to constant 1979 Canadian dollars.)

Outside or general exploration refers to the search for new mineral deposits. It encompasses both surface and underground exploration but does not include mine or on-property exploration costs incurred on producing properties or on those being prepared for production. Once a production decision has been made, further expenditure or more detailed follow-up exploratory work is counted among the many items constituting the costs of mining; such expenditures are not included in figure 9-3.

Figures 9-4 and 9-5 portray the metal value of discoveries per exploration dollar by three-year periods (that is, the metal value of discoveries shown in figures 9-1 and 9-2 divided by the exploration expenditures of figure 9-3). The anomalous 1954–56 period conspicuous in figures 9-1 and 9-2 stands out in these figures as well. The metal value of discoveries per exploration dollar in the 1948–50 and 1951–53 periods are relatively high

Table 9-3.   Annual Mineral Exploration Expenditures in Canada, 1946–82

Year	Millions of current C$	Millions of constant 1979 C$
1946	9.1	43.6
1947	7.3	31.9
1948	5.1	20.0
1949	6.4	24.1
1950	5.4	19.6
1951	9.2	30.3
1952	13.6	42.9
1953	17.8	56.4
1954	26.8	83.5
1955	26.9	83.3
1956	48.4	144.4
1957	54.4	159.1
1958	32.5	93.7
1959	43.0	121.5
1960	43.6	121.5
1961	43.5	120.8
1962	43.8	120.0
1963	43.5	117.0
1964	49(est.)	128.6
1965	54(est.)	137.3
1966	59(est.)	143.7
1967	52.8	123.6
1968	72.7	164.8
1969	96.6	209.8
1970	115.2	239.0
1971	86.4	173.8
1972	70.8	135.6
1973	84.0	147.4
1974	106.1	161.5
1975	119.7	164.6
1976	121.0	148.2
1977	164.2	192.6
1978	194.9	215.0
1979	248.9	248.9
1980	403.6	363.3
1981	484.7	394.4
1982	362.4P	267.9

Notes: P = preliminary; est. = estimated.
Sources: Data for 1946–63 and 1968–82 from Statistics Canada; for 1964–67 from D. A. Cranstone and H. L. Martin, "Are Ore Discovery Costs Increasing?" in *Canadian Mineral Exploration, Resources and Outlook*. Mineral Bulletin MR 137 (Ottawa, Energy, Mines and Resources Canada, 1973) pp. 5–13.

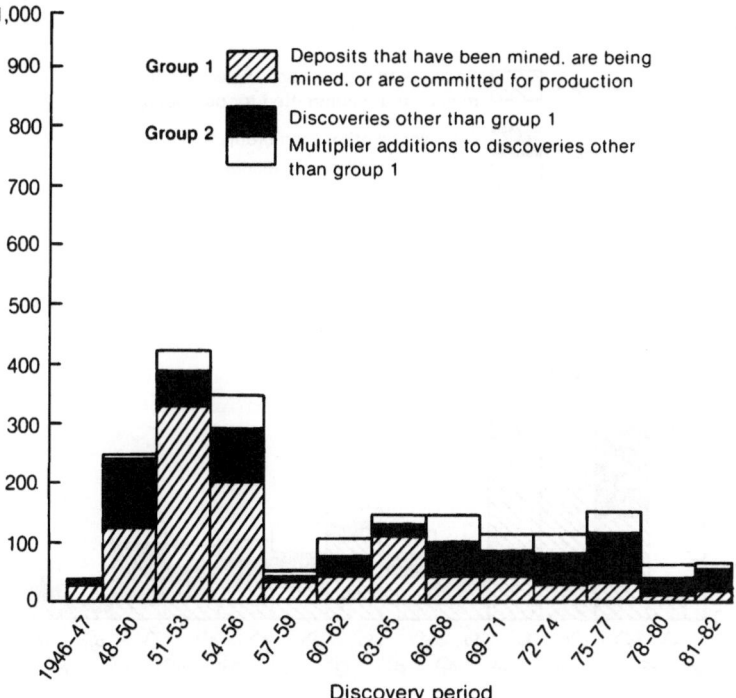

Gross $ value of metals
contained in mineral deposits
discovered in 3-year periods,
per exploration $

Figure 9-4. Canadian mineral discoveries: metal value discovered per exploration dollar at average prices per three-year period, 1946–82 (in constant 1979 Canadian dollars).

in figures 9-4 and 9-5 because of low exploration expenditures during these periods.

As the metal value of the deposits discovered over the last ten or fifteen years covered in this study will probably be larger than estimated here, it is reasonable to conclude from figures 9-4 and 9-5 that, after the remarkable exploration successes of the mid-1950s, the value of metals discovered per dollar of exploration expenditure was fairly steady from then on, certainly until the end of 1977. It would be premature to conclude that the lower values since 1978 are clear evidence of a permanent long-term increase in Canadian ore discovery costs.

The notably lower discovery value per exploration dollar in the 1978–80 and 1981–82 periods reflects a decline in the average size rather than

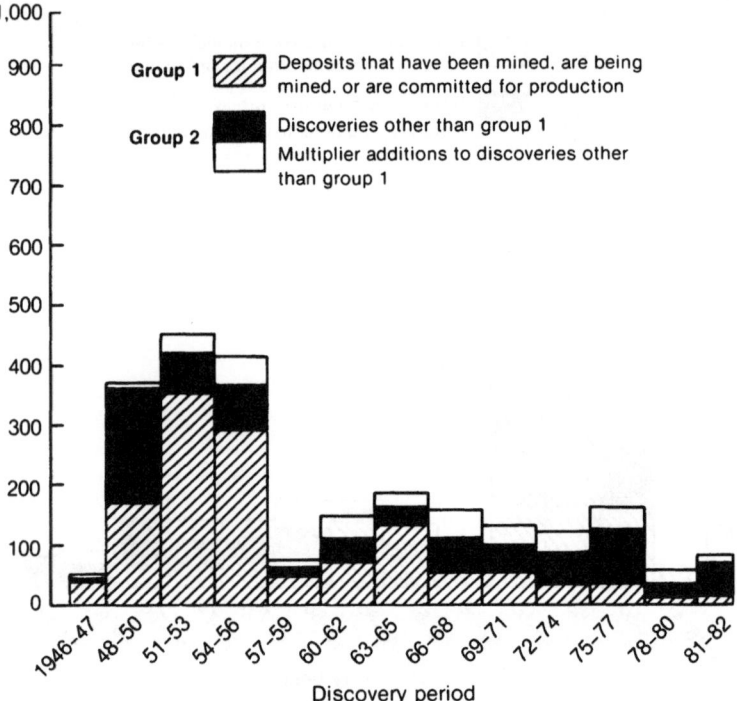

Figure 9-5. Canadian mineral discoveries: metal value discovered per exploration dollar per three-year period, 1946–82 (at January 1979 Canadian dollar metal prices).

in the number of deposits discovered. Some 140 new mineral deposits were discovered in Canada in the five years 1978 through 1982, compared to just under 100 deposits in the preceding five years, 1973 through 1977. Over the 1946–82 period, there were some 900 deposits discovered in Canada—an average of about 24 discoveries per year. In comparison, an average of 28 (and probably more) discoveries were made per year from 1978 through 1982.

The Significance of the Discovery of Major Deposits

A large percentage of the world's output of most metals comes from a relatively small proportion of the world's mines. Similarly, only a few nations provide the major portion of total world output of most mineral commodities. Although Canada has a relatively higher percentage of small to medium-sized mines than do nations such as Chile, Australia, and South Africa, much of Canada's mineral output comes from a few relatively large mines.

Figure 9-6 shows that almost half the gross metal value of the 900 metal deposits discovered in Canada from 1946 through 1982 is accounted for by the 30 largest deposits and 80 percent by the 120 largest deposits. From figure 9-7 it can be seen that about 75 percent of the total gross value of all metal discoveries made in Canada over the same period occurs in only sixteen major mining districts or major individual deposits.

Rates of Discovery for Individual Metals

Analysis of the rates of discovery of individual metals presented in this section and that of the rates of discovery of major metals by geologic deposit type in the next section help provide better understanding of the overall record of Canadian mineral discoveries. (Available data do not permit analysis of the rates of discovery of minor metals such as selenium, tellurium, indium, and rhenium, because assays for such by-product metals are not normally available.)

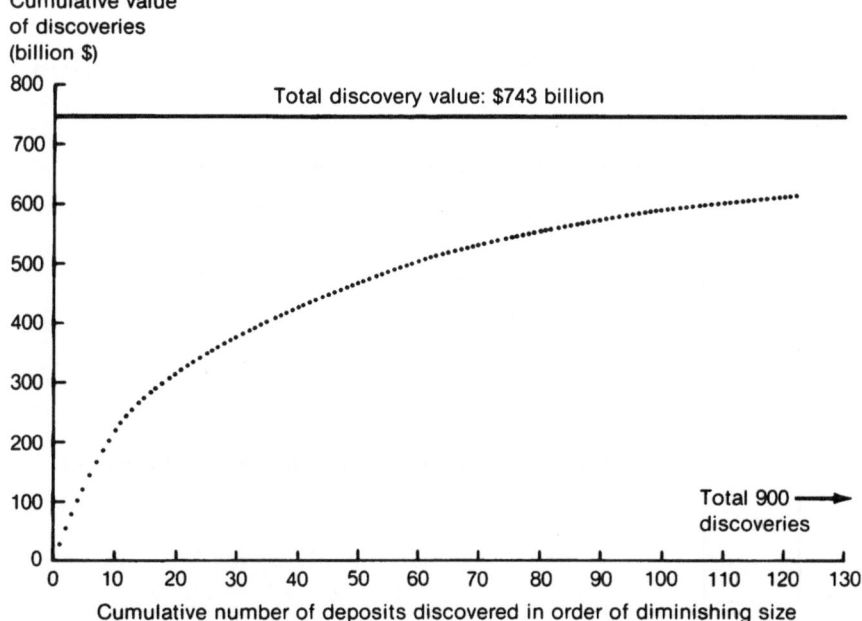

Figure 9-6. Cumulative value of Canadian mineral discoveries, 1946–82 (at January 1979 Canadian dollar metal prices).

298

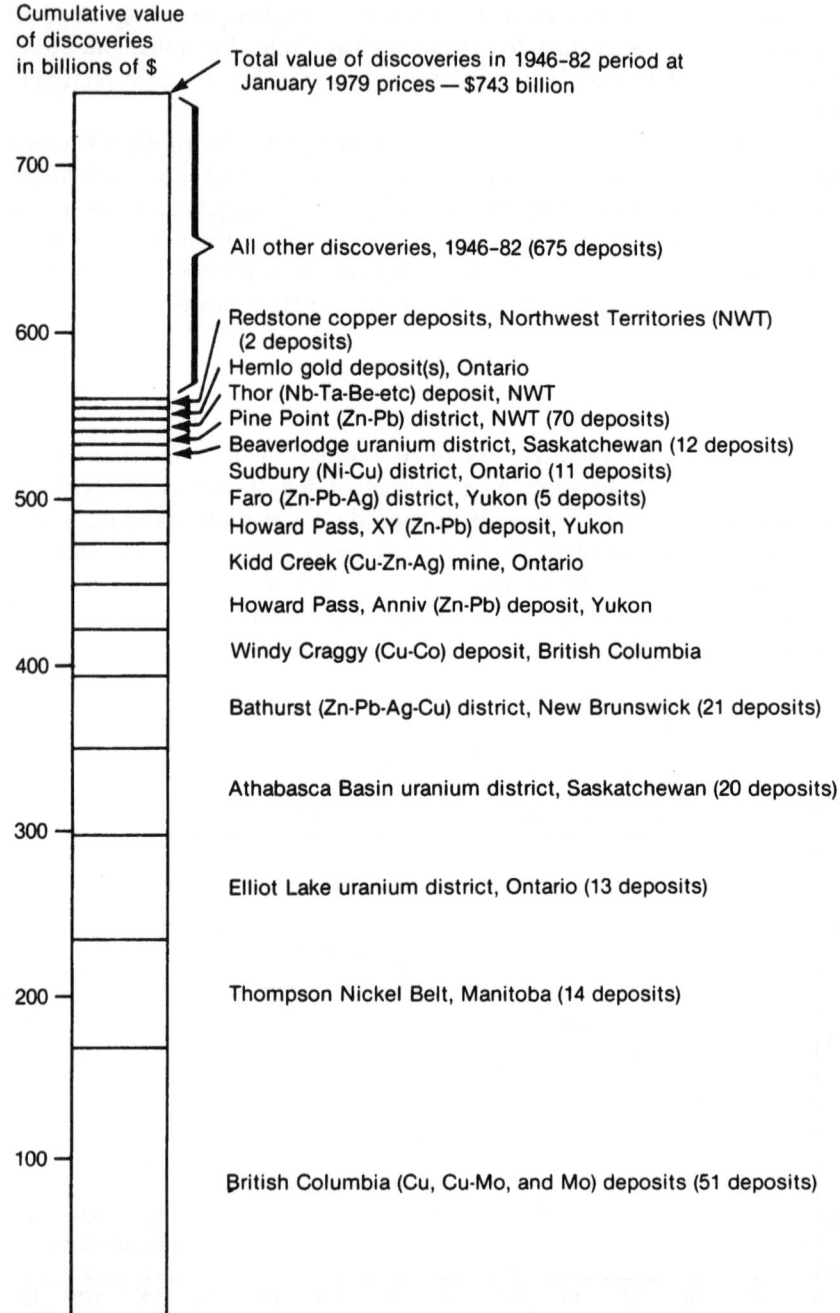

Figure 9-7. Proportion of total value of Canadian mineral discoveries consisting of deposits in major mining districts and of major individual deposits, 1946–82.

Nickel.   Most of the nickel discovered in Canada since 1945 has been found in the Thompson Nickel Belt of Manitoba and at Sudbury, Ontario. The large but low-grade deposit at Mystery Lake near Thompson contains most of the nickel discovered in the 1948–50 period (figure 9-8). The Thompson mine represents about 80 percent of the nickel discovered in 1954–56. The Pipe No. 2 mine, also near Thompson, accounts for more than 75 percent of the nickel discovered in 1957–59, and the Birchtree and Soab mines for about half of the nickel discovered in 1960–62. Two additional nickel deposits in the Thompson district—the Mel deposit discovered about 1968 and a deposit beneath Ospwagan Lake discovered about 1974—are not included in the totals because available information does not permit a reliable estimate of their metal content. The two deposits probably each contain between 200,000 and 1 million tons of nickel.

Copper.   The sudden increase in copper discoveries beginning in the 1960–62 period is the most striking feature of figure 9-9. The discovery peak from 1960 to 1974 represents the rapid discovery of most of the more than

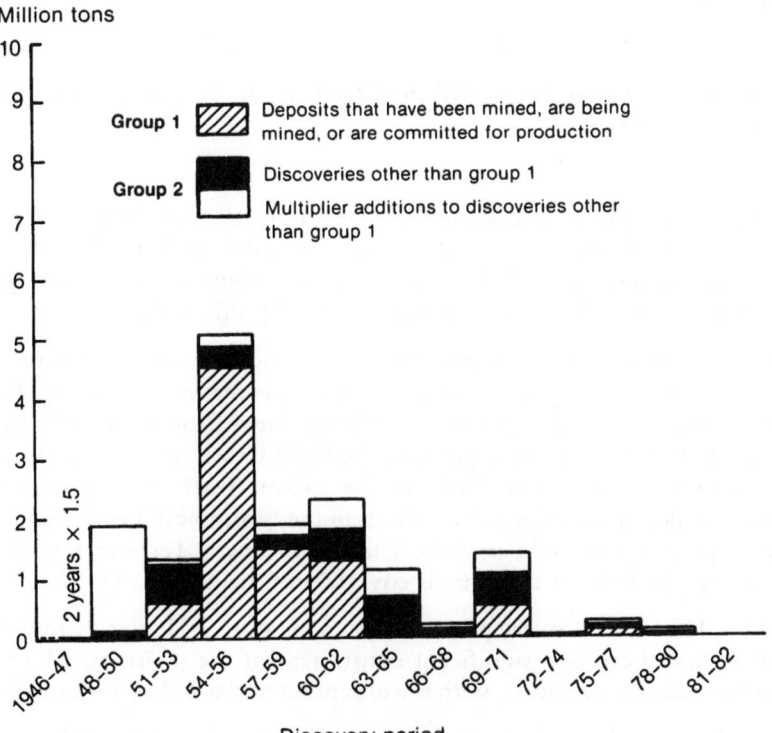

Figure 9-8.   Nickel discovered in Canada per three-year period, 1946–82 (in millions of tons).

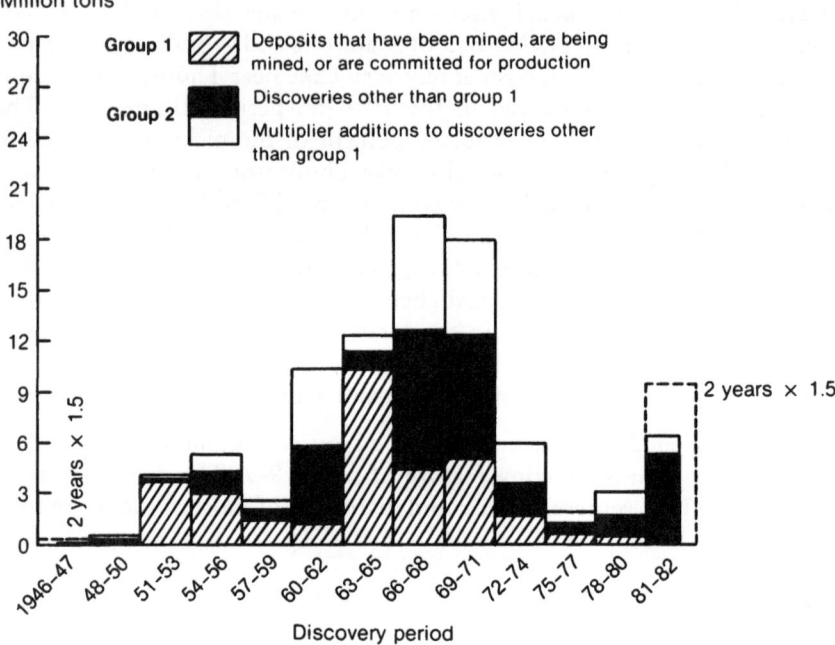

Figure 9-9.   Copper discovered in Canada per three-year period, 1946–82 (in millions of tons).

thirty-five porphyry copper and copper-molybdenum deposits that are now known in British Columbia. The ore bodies of Bethlehem Copper Corporation, found in 1955, were the first porphyry deposits of significance discovered during the period covered by this analysis.

Zinc.   Three major discovery periods for zinc stand out in figure 9-10. The six-year 1951–56 discovery peak represents major discoveries of zinc, lead, copper, silver, and gold near Bathurst, New Brunswick. The second peak, in 1963–65, chiefly represents the Kidd Creek ore body in Ontario and the Cyprus Anvil ore body in the Yukon Territory, together with much smaller discoveries at Pine Point in the Northwest Territories. Two large deposits at Howard Pass at the Yukon–Northwest Territories' boundary constitute the bulk of the zinc discovered in Canada in 1972–77.

Lead.   The lead discovery pattern, shown in figure 9-11, closely resembles that of zinc. Lead is a significant constituent of the major zinc-lead discoveries mentioned above, with the exception of the Kidd Creek ore body.

Molybdenum.   As with copper, a major period of discovery that began with the recognition of the porphyry deposits in British Columbia is evident in figure 9-12. Molybdenum and copper-molybdenum deposits, with most of the latter containing less than 0.02 percent molybdenum,

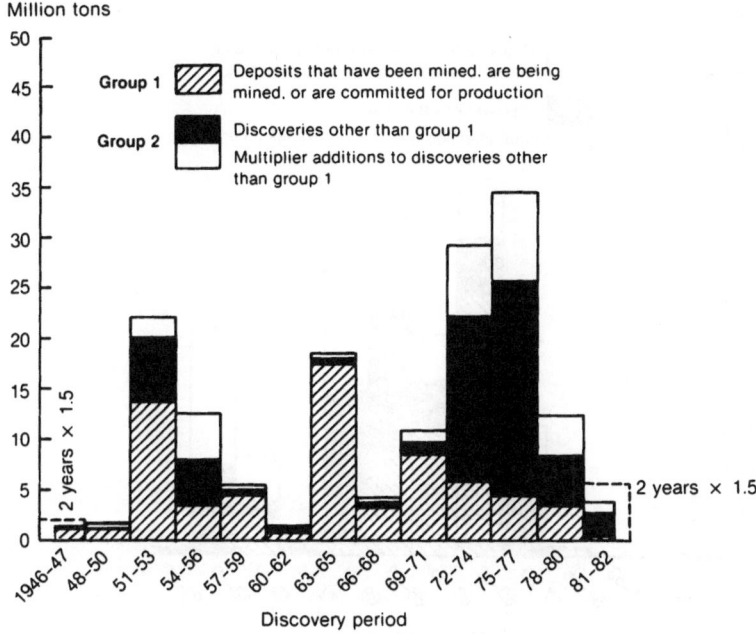

Figure 9-10. Zinc discovered in Canada per three-year period, 1946–82 (in millions of tons).

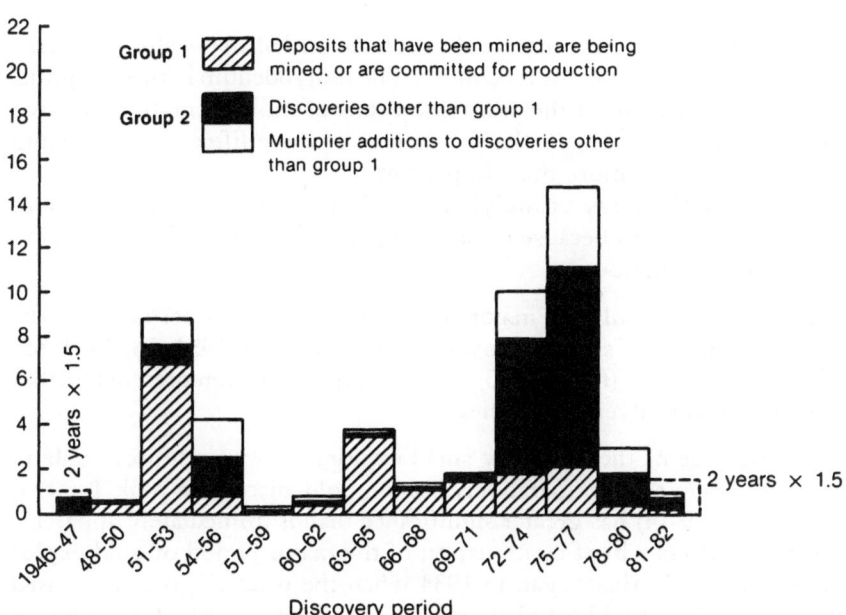

Figure 9-11. Lead discovered in Canada per three-year period, 1946–82 (in millions of tons).

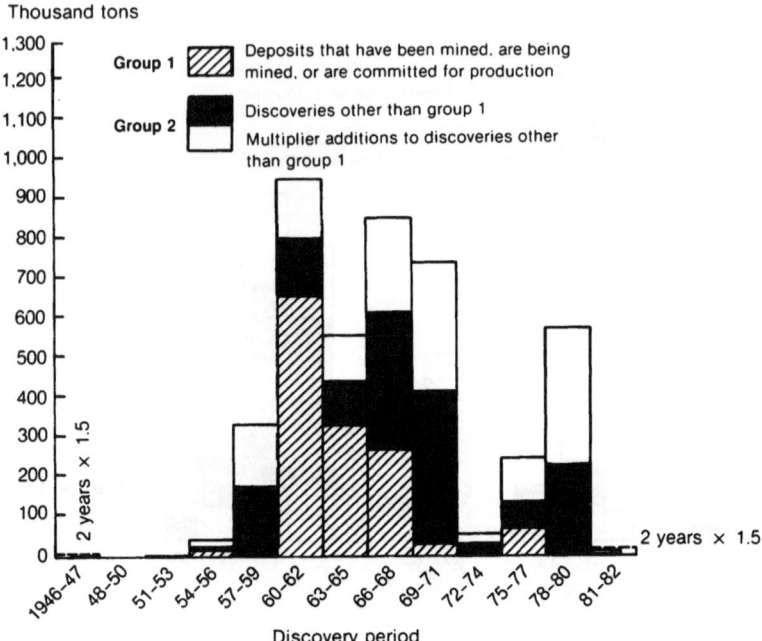

Figure 9-12. Molybdenum discovered in Canada per three-year period, 1946–82 (in thousands of tons).

arc represented. A total of fifteen, almost one-third of all porphyry deposits discovered, contain only molybdenum. The molybdenum in these deposits constitutes 55 percent of the 4.2 million tons of molybdenum discovered in Canada since 1946, with the five largest of these fifteen molybdenum deposits containing more than 40 percent.

The rate of discovery of molybdenum declined in the 1972–74 period, but it increased again because of one major discovery in each of the 1975–77 and 1978–80 periods.

Silver. A total of only ten major silver-bearing deposits accounts for the unusual quantities of silver discovered in the 1951–53, 1954–56, 1963–65, and 1969–71 periods (figure 9-13). The next major section presents further discussion of the silver discoveries.

Gold. Because of the relatively small exploration expenditures made in 1946 and 1947, the comparatively minor gold discovery peak for that period (figure 9-14) has greater significance than is immediately apparent. This peak reflects a brief continuation of the major period of gold exploration and mining that began in 1934 when the price of gold was raised from U.S.$20.67 to U.S.$35.00 per ounce. That period of exploration activity was interrupted by World War II and resumed briefly at the war's

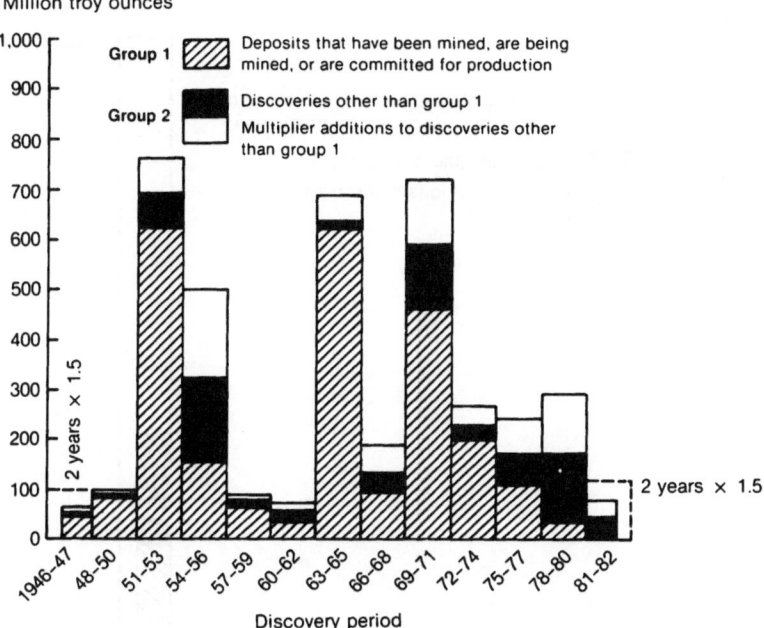

Figure 9-13. Silver discovered in Canada per three-year period, 1946–82 (in millions of troy ounces).

end before being slowed by inflation and by the fixed price of U.S.$35.00 per ounce.

Most of the gold discovered in Canada during the 1950s and 1960s consisted of minor amounts of gold per ton of ore in large tonnages of base metal ores. In the 1960–62 period five gold deposits of significance were discovered, despite the relative unattractiveness of gold mining at that time. As of 1968 the price of gold was allowed to seek its own level. It had risen significantly by 1971, and so had gold exploration in Canada and the rate of gold discovery. Most of the gold discovered in Canada since that time is in gold deposits rather than in base metal deposits.

The Hemlo gold deposit, discovered in 1981 in northwestern Ontario along the Trans-Canada Highway, has been the most significant gold discovery in Canada since the discovery of the Hollinger mine in Ontario in 1910. Hemlo probably contains 25 million or more troy ounces of gold.

Niobium. Until 1980 all significant deposits of niobium discovered in Canada had consisted of the mineral pyrochlore as a relatively minor constituent of large bodies of carbonatite rock. The major niobium discovery peaks (figure 9-15) represent the discovery of about ten pyrochlore-bearing deposits at Oka, Quebec, in the 1954–56 period and a major deposit in Ontario and another in Quebec in 1966–68. In 1978–80 a major

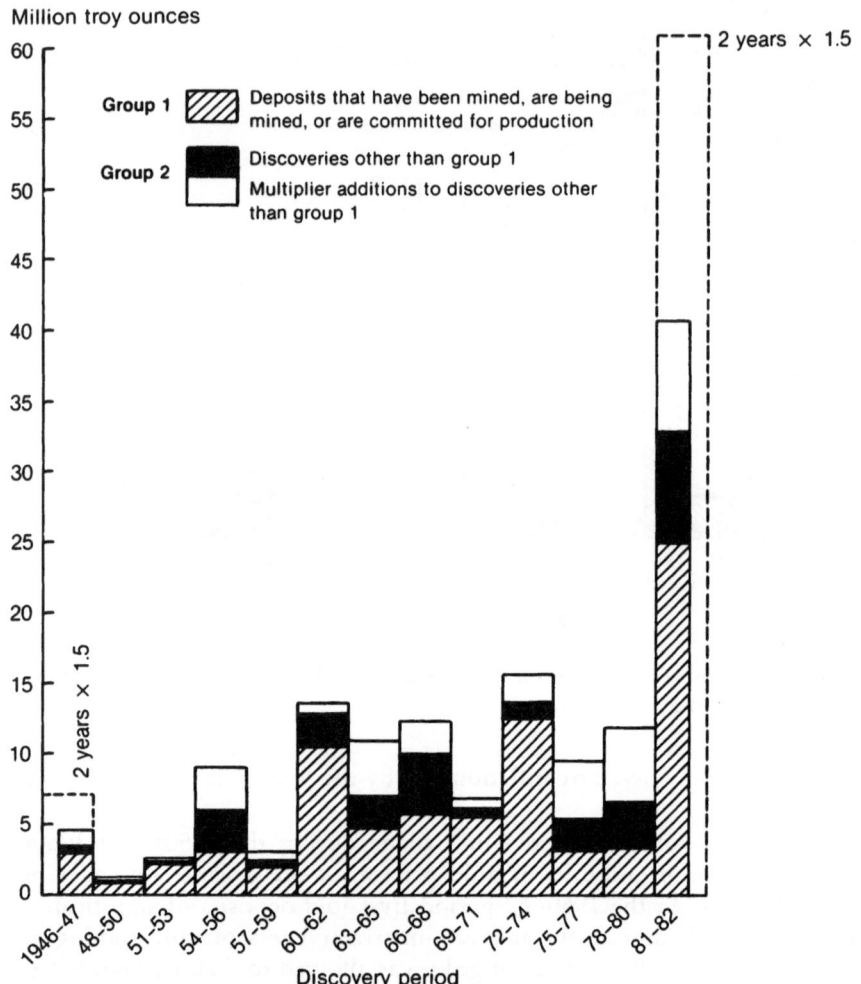

Figure 9-14.   Gold discovered in Canada per three-year period, 1946–82 (in millions of troy ounces).

columbite-bearing deposit was discovered in the Northwest Territories and another pyrochlore-type deposit was discovered on the Labrador–Quebec boundary; both of these occur in granitic rather than carbonatite rocks. The niobium discovered in 1981–82 is all contained in one carbonatite-type pyrochlore deposit in Ontario.

Uranium.   The Canadian uranium discovery pattern has been governed largely by perceived world requirements for uranium. Before World War II there were only minor uses for uranium; it was used mostly as a coloring agent for glass and ceramic glazes. The few uranium deposits then known

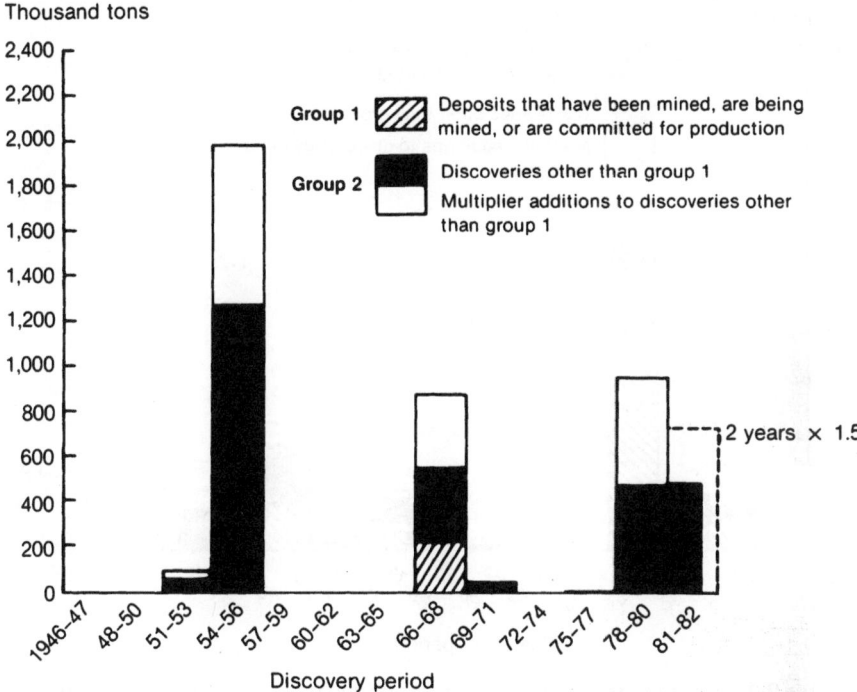

Figure 9-15.   Niobium (Nb$_2$O$_5$) discovered in Canada per three-year period, 1946–82 (in thousands of tons).

throughout the world were mined primarily for their radium content. With the development of nuclear weapons in the 1940s came sudden demand for uranium, and world requirements increased again when the first commercial nuclear power reactors were developed in the 1950s. From 1944 to 1947, uranium exploration in Canada was restricted exclusively to the Geological Survey of Canada and to the government-owned uranium company Eldorado Mining and Refining Limited. Exploration by these agencies led to the discovery in 1948 of Eldorado's second uranium mine at Beaverlodge Lake, Saskatchewan (see figure 9-16). The government-offered price, C$2.75 per pound of U$_3O_8$, was not attractive enough to bring any Canadian deposits into production, even after it was increased to C$7.25 per pound by the addition of development and processing allowances. When the price was raised to C$10.50 per pound, several Canadian uranium deposits were brought into production, and there was a marked increase in Canadian uranium exploration activities.

Major new discoveries of uranium were made in the 1950s near what is now Elliot Lake, Ontario. Ten mines were brought into production. Other new mines were developed at Bancroft, Ontario, and in northern Saskatchewan near Beaverlodge Lake. By the late 1950s world uranium

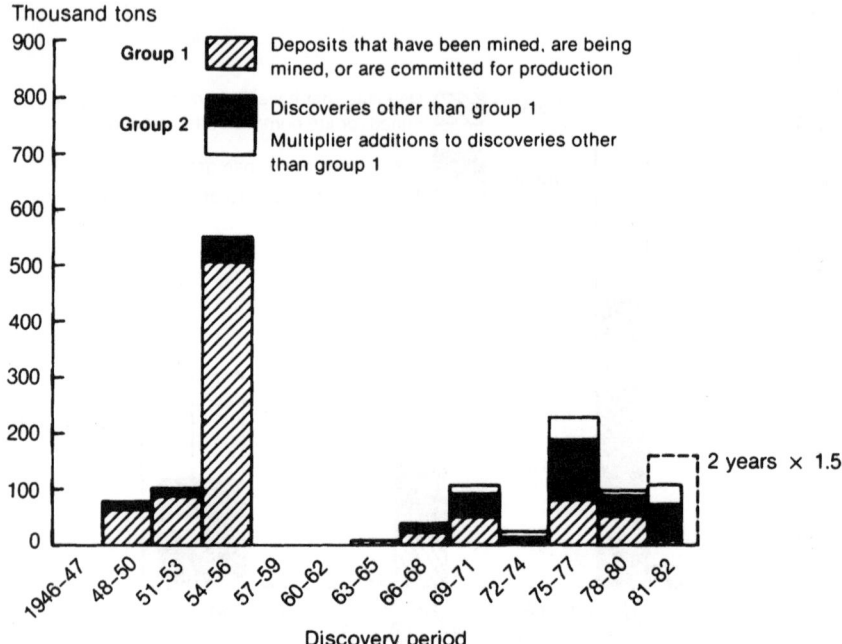

Figure 9-16.   Uranium (U₃O₈) discovered in Canada per three-year period, 1946–82 (in thousands of tons).

discoveries had proved more than adequate to meet foreseeable world requirements, and Canadian uranium exploration virtually ceased. Exploration for uranium was revived briefly in the mid-1960s in anticipation of renewed commercial demand for uranium, and some additional discoveries were made. But when the anticipated market demand did not materialize, little more exploration for uranium was carried out in Canada until the early 1970s. Significant new discoveries of uranium were made in the late 1970s and early 1980s.

The most important Canadian uranium discoveries since those made near Elliot Lake have been unconformity deposits in the Province of Saskatchewan near the base of the Athabasca sandstone basin, which is of late Precambrian age. More than twenty deposits of economic significance were discovered in that region between 1968 and 1982, and additional discoveries there seem almost certain.

The involvement of petroleum companies in much of Canadian uranium exploration began in the mid-1960s and expanded in the 1970s. The hundreds of millions of dollars of exploration expenditures by these companies led to many of the new discoveries. However, uranium is again in excess supply throughout the world, and many Canadian petroleum companies seem to be losing interest in mining exploration. In addition, because of low prices for uranium and many other metals, mining companies have recently lacked both the economic incentive and the funds for exploration. Thus, Canadian uranium exploration has dropped off sharply, and, despite the fact that the recent discoveries in Saskatchewan contain the highest-grade and most price-competitive uranium in the world, the uranium discovery rate in Canada is likely to decline.

Tungsten.   Seventeen tungsten deposits have been discovered in Canada since 1945. The major discovery peak for tungsten—1963 through 1965—(figure 9-17) is explained by the discovery of the Mactung deposit in the Yukon Territory and the Fire Tower zone near Mount Pleasant in New Brunswick. The second major discovery peak, 1975–77, reflects the discovery of four tungsten deposits in northern British Columbia and the Yukon.

Antimony.   There were three significant discoveries of antimony in Canada from 1946 through 1982 (figure 9-18): the Equity silver mine in British Columbia and the Lake George antimony mine in New Brunswick were discovered in the 1969–71 period, and the Minimiska Lake deposit in Ontario in 1981.

Cobalt.   Although figure 9-19 depicts most of the cobalt discovered in Canada, additional amounts were discovered in many Canadian nickel-copper deposits for which cobalt grades are not available. Most of the

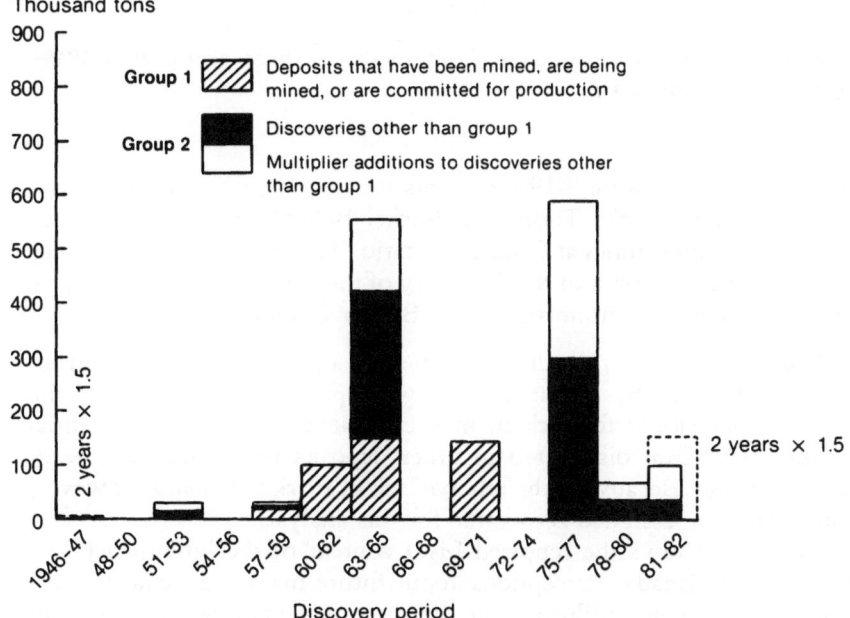

Figure 9-17.   Tungsten (WO₃) discovered in Canada per three-year period, 1946–82 (in thousands of tons).

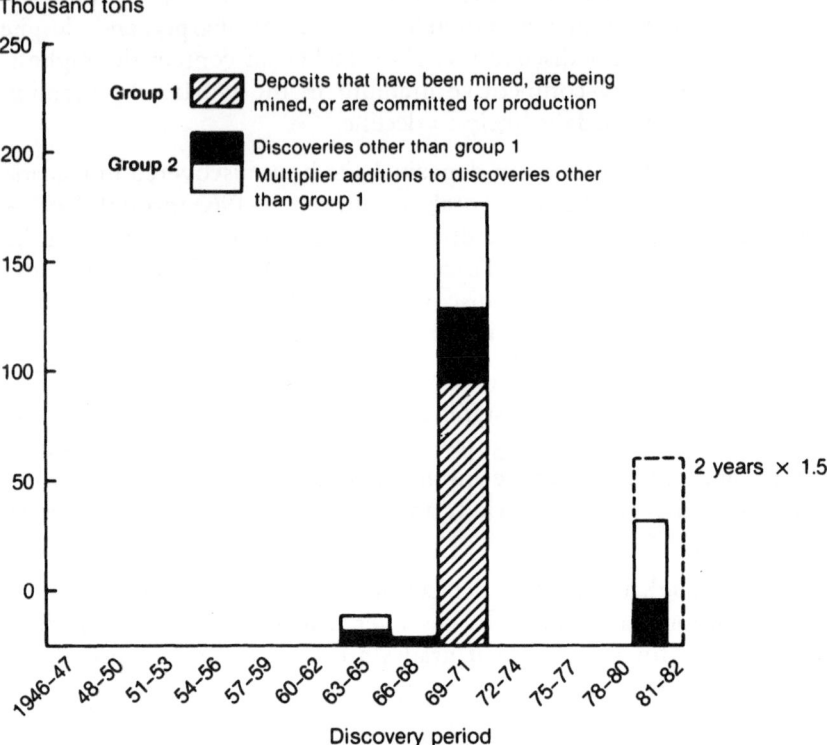

Figure 9-18.   Antimony discovered in Canada per three-year period, 1946–82 (in thousands of tons).

amount shown in figure 9-19 represents the cobalt content of nickel deposits discovered in the Thompson Nickel Belt of Manitoba. A minor amount is in silver mines at Cobalt, Ontario. The major cobalt discovery peak, 1981–82, is a result of the discovery of the very large Windy Craggy copper-cobalt deposit in northwestern British Columbia.

Lithium.   Only one producing lithium mine operated in Canada during the period covered by this analysis; it was open for a short time in the late 1950s but closed for lack of markets. Nevertheless, the substantial amounts of lithium discovered in other deposits may ultimately be of economic value. Because of the relative lack of market demand, however, lithium has been assigned zero value for this analysis.

Figure 9-20 shows the reported Li_2O content of the lithium deposits discovered in Canada. Perceptions about future market demand for this element had a major influence on the discovery pattern: soon after an interest in the element developed in the early 1950s, fifteen significant Canadian lithium deposits were discovered. Some of these had been noted previously by prospectors, but, as mere mineralogical curiosities, they had

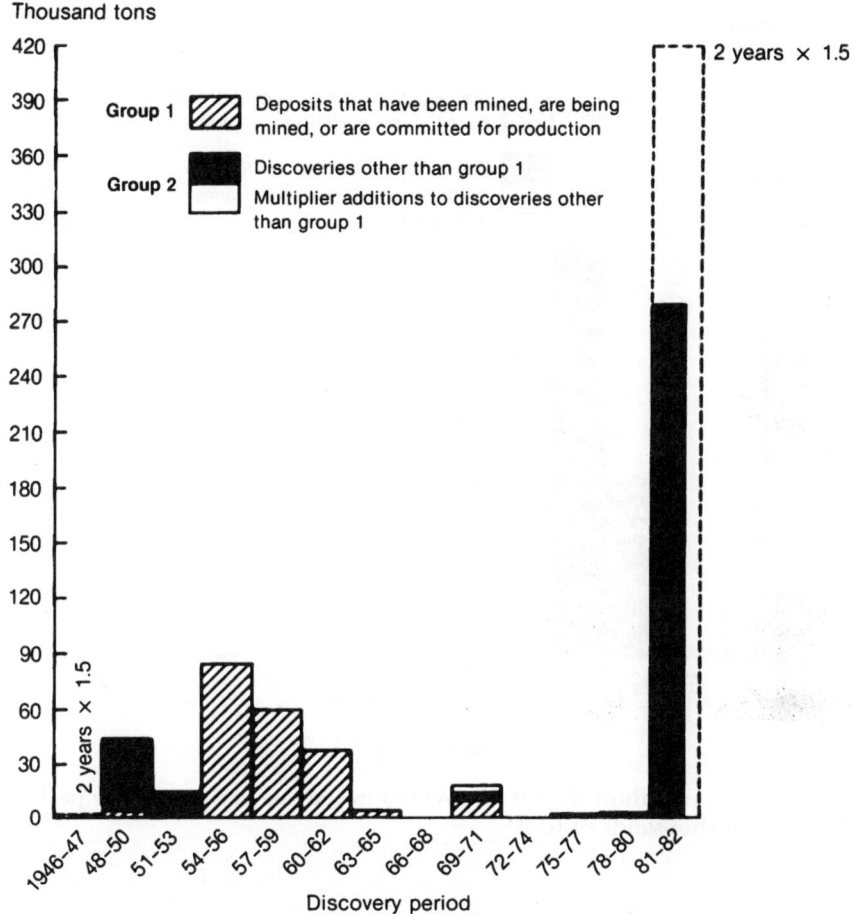

Figure 9-19.   Cobalt discovered in Canada per three-year period, 1946–82 (in thousands of tons).

never been drilled. No additional lithium discoveries have been reported in Canada since the 1950s.

Discovery Rates for Selected Major Metals, by Type of Geologic Deposit

The rates of discovery of copper, zinc, lead, silver, gold and uranium in Canada by geologic deposit type are depicted in figures 9-21 through 9-26 (see subsections on respective minerals, below). Some of these figures illustrate how the relative importance of discovery of specific types of deposits has changed through time. For example, most discoveries of

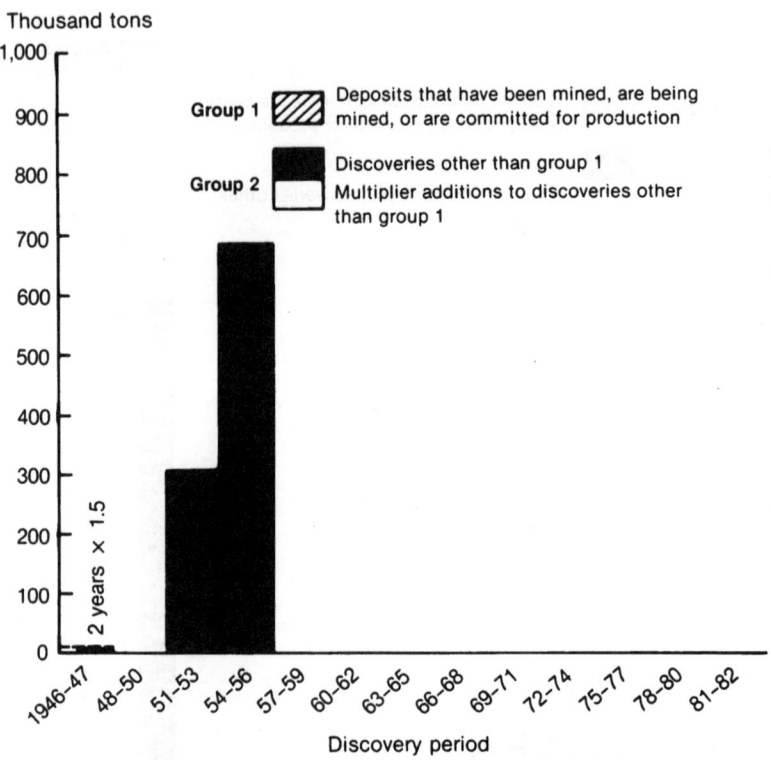

Figure 9-20. Lithium (Li$_2$O) discovered in Canada per three-year period, 1946–82 (in thousands of tons).

porphyry copper deposits in British Columbia were made in the period from 1960 through 1971, of red-bed copper deposits in the period from 1966 through 1977, of Mississippi Valley-type zinc-lead deposits between 1963 and 1982, of zinc and lead in sediment-hosted exhalative-sulfide deposits from 1972 through 1980, and of unconformity-type uranium deposits from 1968 through 1982.

Figures 9-21 through 9-26 also show discovery peaks in particular geologic deposit types. Some of these peaks are the result of a chance discovery that rather rapidly led to further discoveries of the same type nearby or elsewhere in Canada. Other peaks, such as the one resulting from the discovery of paleoplacer uranium deposits at Elliot Lake, Ontario, have been the result of the application of a particular geologic concept. For example, the recognition by geologist Franc Joubin of the similarity between the geologic environment at Elliot Lake and that of the Witwatersrand gold-uranium deposits in South Africa led to the rapid discovery of more than ten ore bodies of essentially the same type at Elliot Lake in 1953–55.

Economic factors have always affected rates of discovery in particular deposit types. Several Mississippi Valley-type zinc-lead deposits were discovered at Pine Point in the Northwest Territories in the late 1920s, and the potential for additional discoveries of this type was recognized. However, the remote location, the depressed economic conditions of the 1930s, and the adequacy of the world's known zinc-lead ore reserves led to suspension of exploration for these minerals at Pine Point until the late 1940s and early 1950s, when several more deposits were discovered. Exploration apparently then ceased until a decision was made in about 1963 to construct a railroad to Pine Point and to develop deposits there for production. Since then, further discoveries of the same deposit type have been made regularly at Pine Point.

For some deposit types the mechanism of ore deposition and the source of the metals are not yet understood by geologists. This is true for the majority of Canadian gold deposits, despite the fact that Canada has been a major producer of gold for more than seventy years. Also, it was not generally recognized in Canada until about 1965 that the far more than 200 volcanic-associated exhalative-sulfide deposits discovered in Canada over the past seventy or more years were deposited by hot water solutions emanating from volcanic vents in the ocean floor. Earlier hypotheses—for example, the theory of a "replacement" origin (replacement of older rock by younger sulfide mineralization) of such deposits—apparently did not hinder the discovery of more than 100 such deposits in Canada between 1946 and 1965 (an average of 5 deposits per year). Conversely, the new understanding of their origin did not result in a noticeable increase in the rate of discovery of such deposits in Canada: from 1966 through 1982, another 65 or so of these volcanic-associated deposits were discovered, an average of fewer than 4 deposits per year.

Copper.   Canadian copper deposits can be subdivided into the following simplified geologic deposit types: porphyry, volcanic-associated exhalative sulfide, red-bed, nickel-copper, vein, contact metasomatic (skarn), and sediment-hosted exhalative sulfide.

The first Canadian copper discoveries of significance were nickel-copper deposits at Sudbury, Ontario, many of which were discovered between 1884 and 1916; volcanic-associated exhalative-sulfide ore bodies in western Newfoundland; and a variety of copper deposits in British Columbia. Subsequently, major copper discoveries were made in 1915 at Flin Flon, Manitoba, and another in 1923 at Noranda, Quebec, that became the Horne mine. Both the Flin Flon and Noranda-Horne ore bodies were considered for many years to be of the sulfide "replacement" type, but they were recognized in the mid-1960s to be of the volcanic-associated exhalative-sulfide type. Discovery of significant quantities of copper in deposits of this type continued through the 1946–82 period (figure 9-21).

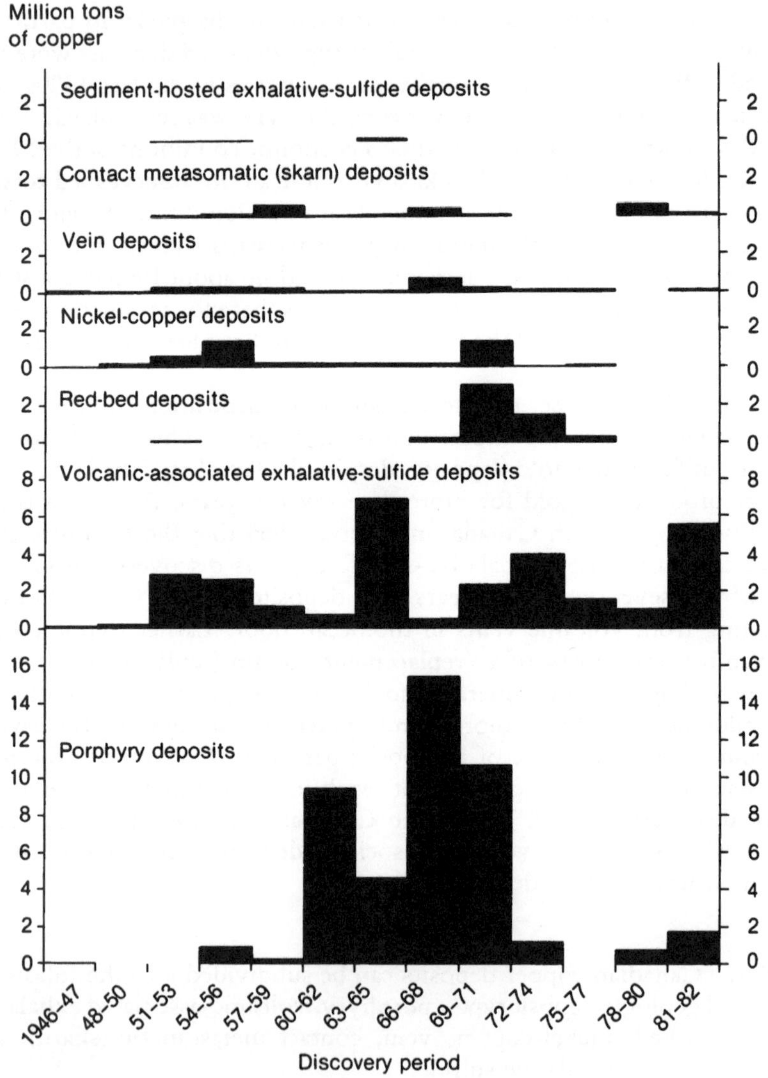

Figure 9-21. **Copper discovered in Canada per three-year period, 1946–82, by geologic deposit type (in millions of tons).**

In the 1950s the apparent lack of porphyry copper (and molybdenum) deposits in the Canadian Cordillera (the Canadian section of the major zone of mountains in the western part of North America and South America) was puzzling, as such deposits were major sources of these metals in the United States, Mexico, Peru, and Chile. Later it became apparent that Canada did have porphyry deposits, which had not generally been recognized as such. Part of one such deposit, Copper Mountain, British

Columbia, had been mined, but the great majority of the many Canadian porphyries now known either had not been discovered or had not been recognized as porphyries before the early 1960s. The rapid discovery of more than fifty such deposits in the Canadian Cordillera between 1955 and 1981 (mostly in the 1960s) was facilitated by the understanding of porphyry deposits gained in other parts of the world earlier in this century. In Canada these deposits contained some 40 million tons of copper and 4 million tons of molybdenum.

Why the long delay in recognizing the existence of porphyry deposits in Canada? Some of the Canadian deposits had been recognized as copper showings in the early part of the twentieth century but were not then considered worthy of further investigation because of their low grades. With few exceptions the four periods of continental glaciation that covered much of Canada during the past million years have removed any super-gene-enriched upper zones (near-surface areas enriched by redistribution of metals during weathering) that may have existed. As a result, only low-grade protore, generally averaging between 0.25 and 0.6 percent copper, remained at most Canadian porphyry copper deposits. Such low grades had little or no recognized economic significance in the early twentieth century. Porphyry-type discoveries appear to have dropped off sharply in the second half of the 1970s, but it is not yet clear whether this happened because the easy-to-discover deposits have already been found or because exploration for such deposits has essentially ceased as a result of the unfavorable price outlook for both copper and molybdenum.

As shown in figure 9-21, a discovery peak for copper in red-bed deposits occurred between 1968 and 1975.

More than 10 million tons of copper were discovered in nickel-copper deposits in the Sudbury district between 1883 and the end of World War II. Since 1945, the amount of copper discovered in such deposits in Canada has been relatively small (figure 9-21), and that discovered in vein, skarn, and sediment-hosted exhalative-sulfide deposits since 1945 has been even smaller.

Zinc. Canadian zinc deposits belong, with minor exceptions, to the following geologic deposit types: sediment-hosted exhalative sulfide, volcanic-associated exhalative sulfide, Mississippi Valley-type, vein, contact metasomatic (skarn), and sandstone.

The rate of zinc discovery by geologic deposit type over time is shown in figure 9-22. Before 1946 the major Canadian sediment-hosted exhalative-sulfide discovery was the Sullivan mine of Cominco Limited in British Columbia. To date, all significant discoveries of this type have been in British Columbia and the Yukon. The following major deposits are represented in the sediment-hosted exhalative-sulfide portion of figure 9-22: 1951–53: Tom and Vangorda Creek, Yukon; 1963–65: Anvil and Swim Lake, Yukon; 1972–74: X-Y and Grum, Yukon; 1975–77: Anniv, DY,

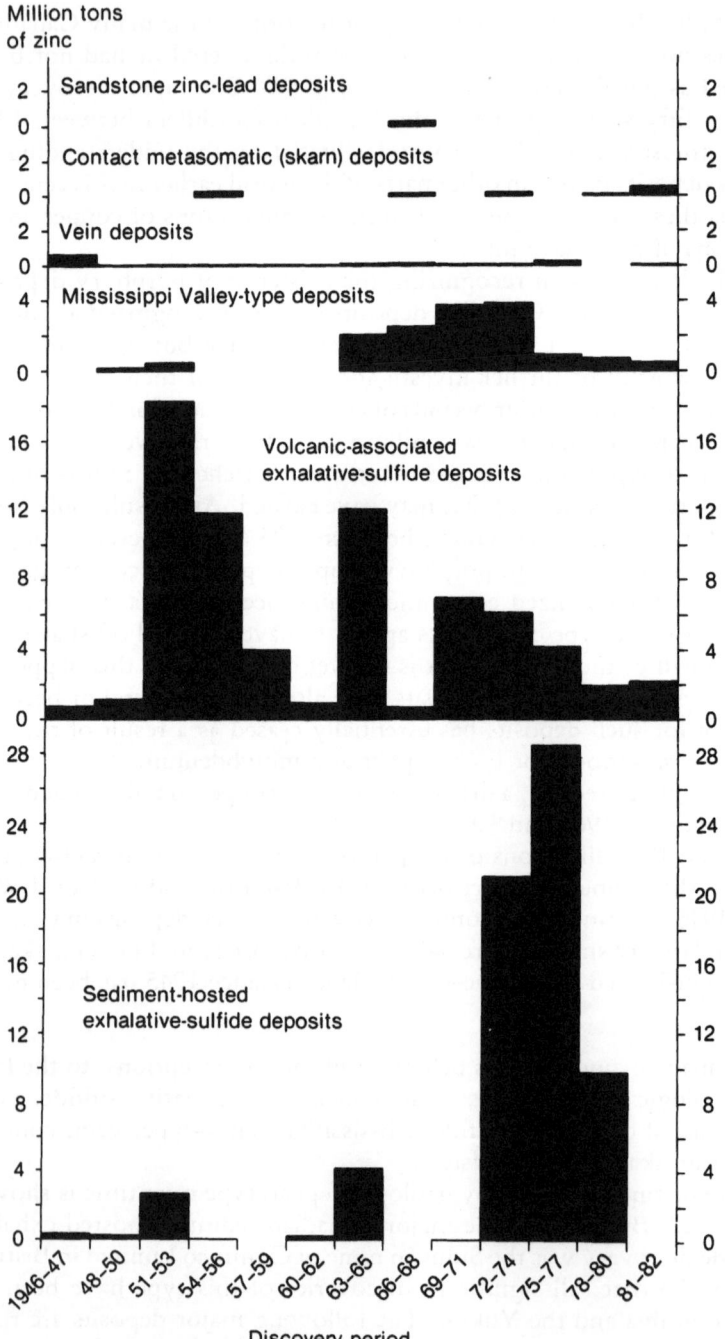

Figure 9-22. **Zinc discovered in Canada per three-year period, 1946–82, by geologic deposit type (in millions of tons).**

and Jason, Yukon; 1978–80: Cirque and several nearby deposits, British Columbia.

The Flin Flon mine in Manitoba, discovered in 1915, was the major volcanic-associated exhalative-sulfide zinc discovery in Canada before 1946. The major discoveries of this type in the 1951–53 and 1954–56 periods were the Brunswick No. 12, Heath Steele, Caribou, and other deposits near Bathurst, New Brunswick, plus other less significant discoveries, including the Geco mine in Ontario and several deposits at Snow Lake, Manitoba. The major volcanic-associated zinc discovery in the 1963–65 period was the Kidd Creek mine in Ontario. The moderate rate of discovery from 1969 through 1977 resulted from the discovery of seven moderate-sized deposits in various parts of Canada over those years.

Most of Canada's Mississippi Valley-type zinc-lead deposits were found near Pine Point, Northwest Territories. Such deposits tend to be relatively small, with 1 million to 2 million tons of ore. The Arvik mine of Cominco Limited in the Arctic Islands, a very large deposit of Mississippi Valley type, accounts for the major proportion of the zinc discovered in 1969–71. Four deposits in various Canadian localities made up the 1972–74 discovery peak.

As figure 9-22 indicates, the quantities of zinc discovered in Canadian vein, skarn, and sandstone deposits are relatively insignificant.

Lead. With the exception of the comments on volcanic-associated deposits, the preceding discussion of discovery patterns for zinc also applies to lead (see figure 9-23), although smaller quantities of lead are involved. The only volcanic-associated exhalative-sulfide deposits with significant lead content are the Paleozoic deposits in the Bathurst district of New Brunswick, mostly discovered from 1951 through 1956.

Silver. Silver-bearing ore in Canada occurs in the following geologic deposit types: volcanic-associated exhalative sulfide, vein-type base metal-silver-(gold) and vein-type silver, sediment-hosted exhalative sulfide, porphyry copper, red-bed copper, contact metasomatic (skarn), nickel-copper, and gold.

The silver content of gold and nickel-copper deposits discovered in Canada since 1945 is generally not publicly known. Silver in Canadian gold deposits generally amounts to 10 percent, or at most 20 percent, of the gold content, so that Canadian gold deposits discovered since 1945 probably contained about 10 million, or at most 15 million, troy ounces of silver. Canadian nickel-copper deposits discovered since 1945 likely contained less than 25 million troy ounces of silver. These amounts are small in comparison to the much larger quantities of silver contained in most other deposit types.

Most of the silver discovered in Canada since 1945 has been in volcanic-associated exhalative-sulfide base metal deposits (figure 9-24). The major volcanic-associated discoveries of this type made in the six years from

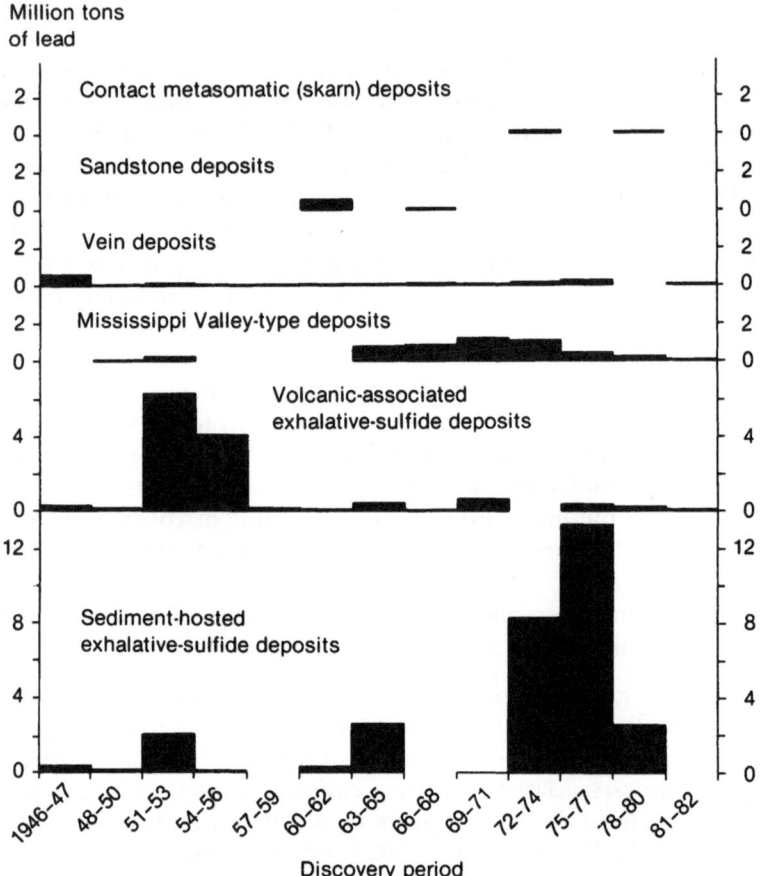

Figure 9-23.   Lead discovered in Canada per three-year period, 1946–82, by geologic deposit type (in millions of tons).

1951 through 1956 are zinc-lead-copper-silver deposits near Bathurst, New Brunswick, and the Geco mine in Ontario. Nearly all the volcanic-associated exhalative silver discovered in 1963–65 was in ores of the Kidd Creek mine in Ontario. The 1969–71 discovery peak chiefly represents the discovery of the Bathurst Norsemines deposit in the Northwest Territories and the Mattabi and Lyon Lake mines in Ontario.

Most of the remainder of the silver discovered in Canada over the 1946–82 period was contained in vein deposits and in sediment-hosted exhalative-sulfide deposits. The most notable vein-type discovery peak is the result of the discovery of the Equity silver-copper deposit, a replacement deposit discovered in British Columbia in 1969. The various discovery peaks for silver in sediment-hosted exhalative-sulfide deposits are accounted for by the discovery of the same ten or so deposits listed earlier

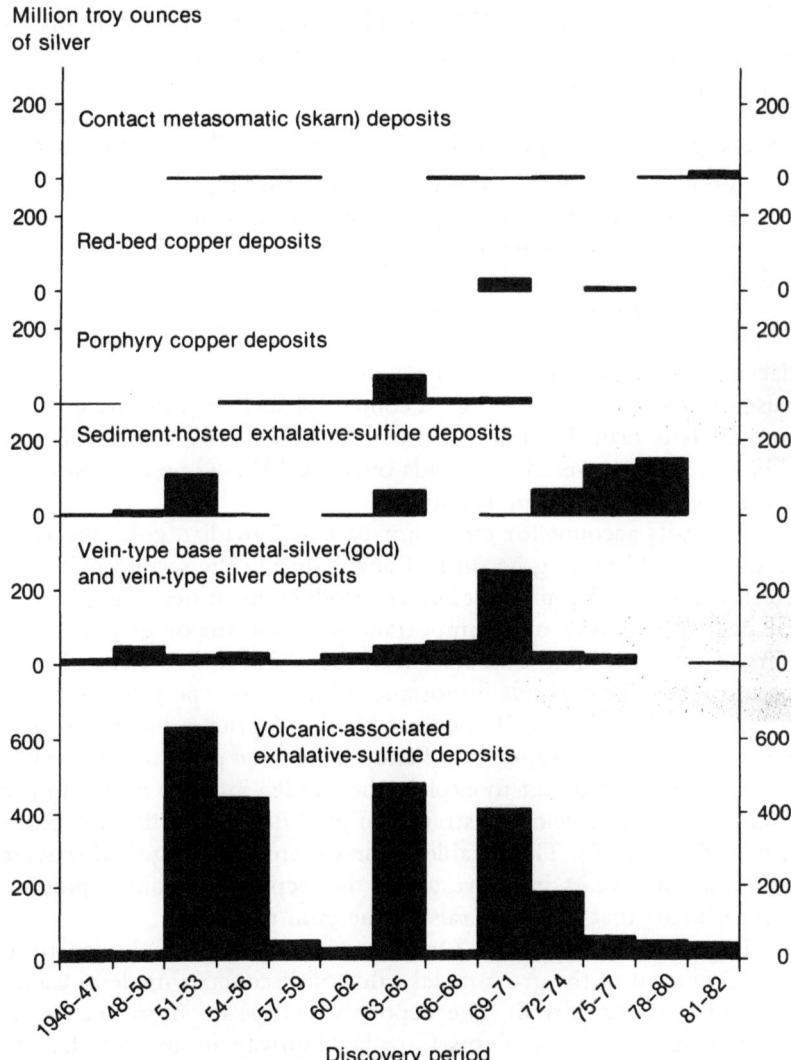

Million troy ounces
of silver

Figure 9-24.   Silver discovered in Canada per three-year period, 1946–82, by geologic deposit type (in millions of troy ounces).

for zinc, discovered in the Yukon and British Columbia from 1951 through 1979.

Gold.   Canadian gold deposits can be classified as follows:

1. Geologic deposit types with gold as the dominant element of value:
    a. Vein and other structurally controlled deposits
    b. Stratiform gold deposits

c. Hemlo-type deposits (The Hemlo deposit, a major recent discovery, is of as-yet-uncertain type. It may be stratiform.)

d. Placer deposits

2. Geologic deposit types in which gold is generally a by-product:

a. Volcanic-associated exhalative-sulfide base metal deposits
b. Porphyry copper (and copper-molybdenum) deposits
c. Vein-type base metal deposits
d. Contact metasomatic (skarn) deposits
e. Sediment-hosted exhalative-sulfide deposits

Placer gold deposit discoveries have not been considered in this chapter because the discovery date and gold content of such deposits are generally unknown. It is probable that, at most, 2 million troy ounces of placer gold have been discovered in Canada over the 1946–82 period. Nearly all of it has been in the Yukon Territory.

Vein deposits account for the major part of Canadian gold discovered since serious gold mining began in Canada during the second half of the nineteenth century. By-product and co-product quantities of gold in base metal deposits became more important with the major growth of the nonferrous metal mining that started in the 1920s and accelerated in the 1950s and 1960s. The relative importance of the vein-type gold discoveries declined after World War II, because the gold price remained fixed at U.S.$35.00 per ounce in spite of inflation. The fixed price was maintained until 1968, so it is difficult to explain the sudden increase in the amount of gold discovered in vein and stratiform gold deposits in the first part of the 1960s (figure 9-25). The notable rise in the amounts of gold discovered in 1972 and later years is the result of the stepped-up gold exploration effort in Canada that followed raising the gold price.

The large Hemlo deposit in Ontario, discovered in 1981, constitutes almost 25 percent of the gross metal value of discoveries made in Canada during the 1981–82 period. The deposit outcrops no more than a few hundred meters from the Trans-Canada Highway in an area that was relatively thoroughly prospected for gold in earlier years. The gold at Hemlo seems to occur in sedimentary rocks; the quartz veins that cut them are essentially barren.

As is the case with copper and zinc, the first gold-bearing volcanic-associated exhalative-sulfide base metal deposits of major significance discovered in Canada were the Flin Flon and Horne ore bodies. Other volcanogenic copper-zinc-silver-gold deposits were discovered elsewhere in Canada in the 1920s and 1930s, but the great majority were found after World War II (figure 9-25). These deposits contain important quantities of gold as recoverable by-product. While the gold content per ton of such ores is generally quite low, hundreds of millions of tons of these ores have been discovered.

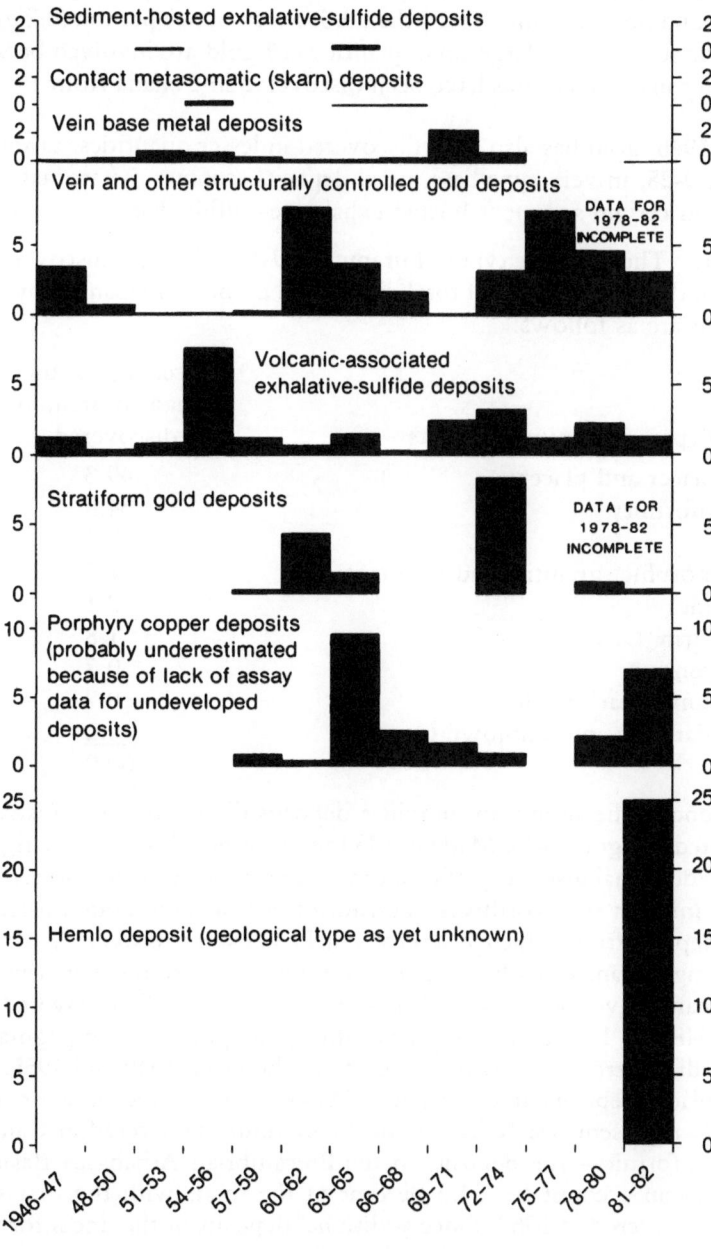

Figure 9-25. Gold discovered in Canada per three-year period, 1946–82, by geologic deposit type (in millions of troy ounces).

Significant quantities of gold also occur in porphyry copper and copper-molybdenum deposits. Although Canadian copper porphyry deposits contain less, and normally much less, than 0.006 troy ounce per ton (0.2 gram per metric ton) of ore, large total quantities of gold are involved because 11 billion tons of such ores have been discovered in Canada from 1955 to 1981.

Since 1946, gold has also been discovered in lesser quantities, as shown in figure 9-25, in vein-type base metal deposits, contact metasomatic deposits, and certain sediment-hosted exhalative-sulfide deposits.

Uranium. The geologic types of uranium (U_3O_8) deposits discovered in Canada and the percentage of total discovered uranium in Canada by type of deposit are as follows:

Types of uranium deposits	Percentage of total Canadian uranium discovered
Paleoplacer and placer	49.3
Unconformity	34.5
Vein	7.7
Metamorphic, granitic, and pegmatitic	2.6
Volcanic	2.1
Stratiform	1.8
Sandstone	0.7
Uranium in carbonatites	0.7
Miscellaneous and unknown types	0.2
	100.0

All but one of the significant uranium deposits discovered in Canada are represented in figure 9-26. Made in 1930 and thus not shown in the figure, this one additional discovery of economic significance was the former Port Radium mine in the Northwest Territories, a vein-type uranium-silver-copper deposit from which 7,000 tons of U_3O_8 were recovered.

Early significant Canadian uranium discoveries were those in vein deposits near Beaverlodge Lake, Saskatchewan, nearly all discovered between 1948 and 1954 and those in metamorphic, granitic, and pegmatitic deposits discovered near Bancroft, Ontario, between 1949 and 1955.

Paleoplacer deposits at Elliot Lake, Ontario, discovered between 1953 and 1955, represent nearly half of all the uranium discovered in Canada, and unconformity-type deposits in the Precambrian Athabasca Basin of Saskatchewan account for 34.5 percent of the total, with the first such deposit discovered in 1968. Since additional deposits of the unconformity type are likely to be discovered in the same area, such deposits may soon account for about half of all the uranium discovered in Canada.

Most volcanic-type uranium has been discovered in Labrador, but other deposits are known in British Columbia and the Northwest Territories.

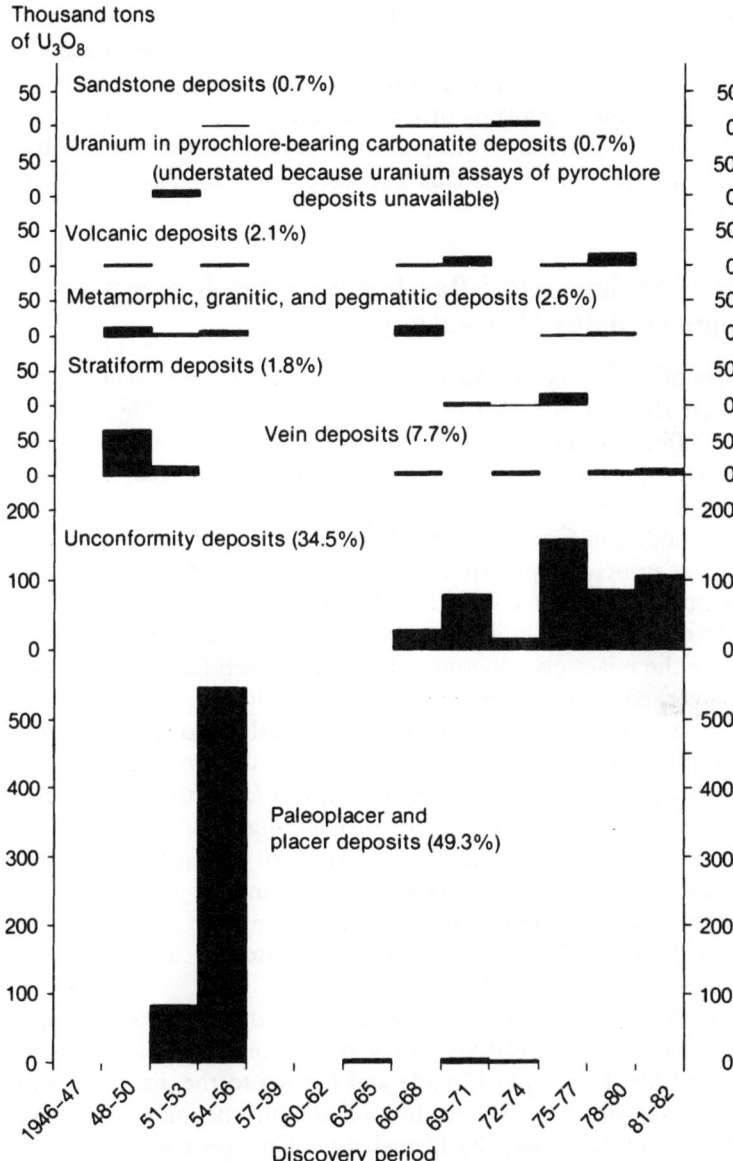

Figure 9-26. Uranium (U_3O_8) discovered in Canada per three-year period, 1946–82, by geologic deposit type (in thousands of tons). Percentages of the total amount of U_3O_8 discovered in Canada, 1946–82, by deposit type are shown in parentheses.

Sandstone uranium deposits, which are of major importance in the western United States, do not make up a significant percentage of the uranium discovered in Canada. Such deposits could be present at depth in the western interior plains of Canada and would therefore be difficult to find. Sandstone deposits may also be present in the Arctic Lowlands and Innuitian regions of Canada but have not been discovered in those regions to date.

Effects of Technological Developments and Geologic Concepts on Rates of Discovery

The rates of discovery of Canadian mineral deposits and types of deposits have been affected in recent decades by factors such as the development of more effective exploration technology, the growth in geologic information through ongoing exploration, and changes in geologic concepts.

Conventional prospecting was almost the only effective way of discovering ore bodies in Canada until the development of effective geophysical methods of mineral exploration, which were adapted from military technology immediately after World War II. Before that time, not much more could be done, especially in the exploration of areas covered by overburden, once the relatively obvious deposits had been found by prospectors.

However, conventional prospecting has not disappeared with the advent of airborne and ground-based geophysical methods of exploration. Even now it is difficult to carry out geophysical surveys in mountainous regions, so that prospecting (much of it more sophisticated than in earlier years) remains the most effective way of discovering new ore bodies in such terrain. Some of the major deposits discovered in Canada during the last fifteen years covered by this study were found by prospecting. These include the huge X-Y and Anniv zinc-lead deposits at Howard Pass straddling the Yukon–Northwest Territories' boundary and the very large Hemlo gold deposit in Ontario.

The development of effective airborne and ground electromagnetic methods has made it relatively easy to explore for volcanic-associated massive sulfide deposits in Canada and has led to the discovery of many such deposits. Similarly, the finding of uranium deposits has been facilitated by the development, by Canadians, of the portable Geiger counter in the 1940s and of the portable scintillometer in the early 1950s, followed by that of increasingly sensitive and sophisticated gamma ray spectrometers capable of airborne and ground radiometric surveys.

The application in Canada of accumulated geologic knowledge of ore deposits and concepts of ore deposition developed in other countries have also affected the Canadian ore discovery record. For example, the relatively rapid discovery of more than fifty porphyry copper and molybdenum deposits in Canada during the 1960s and 1970s was greatly facilitated by the application of the extensive geologic understanding of such deposits that had developed in other countries over the preceding sixty years.

The Gaspé copper mine in Quebec, discovered by conventional prospecting in the late 1930s, is a contact metasomatic (skarn) deposit. It occurs in a porphyry copper type of environment, but none of the copper occurs in the granitic porphyry. Exploration in the vicinity of the mine resulted in the discovery of another ore body of this type deep beneath the mine townsite some thirty years after a production decision was originally made. This discovery came about through the application of a geologic deposit model developed both at the mine and elsewhere for the search for contact metasomatic (skarn) copper deposits. Use of such models in various parts of the world could result in the discovery of similar contact metasomatic copper ore bodies in limestone formations cut by copper-bearing granitic intrusions.

Many opportunities still exist for the development of more advanced or completely new mineral exploration technology that could be applied not only in Canada but in many other parts of the world. For example, there is a need for more discriminating geochemical and geophysical exploration techniques, for new geochemical and geophysical technology, and for geophysical equipment or techniques able to detect mineral deposits at greater depths than is currently feasible. There has, as yet, been little practical exploration technology developed for some types of mineral deposits in Canada. For example, over the years, seven red-bed copper deposits of significant tonnage and grade have been discovered. Thus, such deposits are known to exist in Canada, but they have not generally been explored for because they are difficult to recognize in outcrop and because there is no effective geophysical technology for locating them. Similar problems affect exploration for other types of mineral deposits— for example, vein-type gold deposits covered by overburden.

Significance of the Canadian Mineral Discovery Experience for Other Nations

What is the significance of the Canadian mineral discovery experience for other nations? Being the world's second-largest nation in area (almost 10 million square kilometers), Canada covers widely varying geologic environments. Some areas have been found to contain a relative abundance of mineral deposits, while others appear to be essentially barren. It is possible that an apparent lack of mineral deposits is related to a lack of exploration, which in turn is related to geologically unfavorable perceptions: a self-fulfilling prophecy.

Since Canada has such a variety of geologic settings, its mineral endowment is probably comparable in total to that of other areas of equal size. Therefore, the Canadian mineral discovery experience (rates and costs of discovery, quantities of metals likely to be found, size distribution of deposits, rates of discovery by geologic deposit type, and the like) is probably roughly indicative of the discovery potential for comparable exploration effort in some parts of the world, although comparability with nations having less varied geology is less reliable.

Total mineral endowments may be similar elsewhere, but the costs of discovering deposits may differ considerably in countries where the type of geologic terrane, on the average, is much different. Canadian discovery costs may be comparable to those in the Soviet Union, Norway, Sweden, or Finland, for example, but are likely to differ from those in deeply weathered areas such as in Australia and Brazil.

Most of Canada has been glaciated several times in the past million years (most recently, only about 10,000 years ago). This continental glaciation has carried away most of the deeply weathered rocks at the surface, including weathered (and sometimes supergene-enriched) upper portions of ore bodies. As a result, most Canadian ore deposits that reach the surface are relatively unweathered and therefore tend to be relatively easy to discover by prospecting.

Deposits that are covered by overburden can be located by airborne magnetic and electromagnetic exploration techniques followed up by investigations with ground-based geophysical techniques and diamond drilling. But in Canada most areas contain graphitic conductors and magnetic rocks that are devoid of valuable minerals, although these rocks show up as anomalies similar to mineral deposits. It is therefore difficult to decide which anomalies to drill, and anomalies resulting from mineral deposits can easily be rejected without ever being drilled.

Surface prospecting and airborne or ground-based electromagnetic geophysics are normally effective only where rocks are not deeply weathered. For example, as is the case in Australia, conductivity measurements may be masked by highly saline groundwaters, which are conductive. Therefore, electromagnetic techniques in particular, which have revealed so many conductive sulfide ore bodies in Canada, would not work equally well in nonglaciated countries.

Airborne magnetic exploration techniques, which are used in Canada, measure the vertical variation in the earth's magnetic field. In equatorial regions the magnetic field is essentially horizontal, so ore bodies with certain orientations cannot be detected by magnetic methods, and interpretation of magnetic surveys is therefore much more difficult.

For the reasons stated above, some of the mineral exploration techniques that are most effective in Canada would be less effective elsewhere. Mineral exploration programs in such regions need to rely largely on other techniques, which in turn may not be especially applicable to exploration in Canada. Such differences obviously affect discovery costs. And beyond the geological and technological considerations there are, of course, differences from country to country in exploration expertise, wages, and history of exploration and mining.

Analysis of the costs of mineral discovery similar to that presented in this chapter could be carried out for any region or country for which there are adequate data on exploration expenditures and discoveries. In such cases this presentation of the Canadian experience would be particularly relevant.

Summary and Conclusions

This historical analysis of the costs of finding ore deposits in Canada is based on the roughly 900 deposits of all minerals—except lithium, iron ore, industrial minerals, and coal—discovered in Canada between January 1, 1946, and January 1, 1983. A *discovery* is defined as a deposit sufficiently attractive to have warranted the expenditure necessary to establish its tonnage and grade.

In many studies of the record of Canadian ore discovery, a built-in bias makes the rate of discovery in the past appear higher than that of more recent years. The chief reasons for this common bias are that (1) some recent discoveries are not yet publicly known; (2) some recent discoveries cannot yet be recognized as the economically exploitable ore bodies that they may later prove to be; and (3) the amount of ore that will ultimately be recovered in many cases considerably exceeds the estimate at the time of discovery or even at the start of production. This bias leads to unduly alarming conclusions about the most recent results of exploration activities. The analysis presented here makes use of multipliers to offset this tendency toward underestimation. Individual discovery "values"—the content of metals in deposits discovered, multiplied by appropriate metal prices—were summed over three-year periods from 1946 through 1982. To make up for the likely underestimation of the size of new discoveries, adjustments were made by means of specially derived multipliers.

The multipliers are based on previous experience. For each of some 100 Canadian mines, a multiplier was established to express the difference between the initial estimate of the size of a reserve and what the producible tonnage actually turned out to be or is likely to be. Multipliers were deduced for four generalized types of deposit and were applied unless better information was available for individual deposits.

More than 140 of the 900 deposits discovered in the 1946–82 period were found in the last five years of that period, and additional discoveries unknown to this writer have also probably been made—perhaps as many as 20 more, experience would suggest—but the bulk of them will likely be small and have little effect on the gross metal value of discoveries or on the discovery value per exploration dollar. They may, however, have a notable effect on the discovery pattern of some of the metals involved, especially when considered by type of geologic deposit.

As discussed in this chapter, there were remarkable exploration successes in Canada in the mid-1950s. From then certainly until the end of 1977 the gross value of metals discovered per dollar of exploration expenditure was fairly steady. Although discovery value per exploration dollar was relatively low for the five years 1978 through 1982, the average number of discoveries per year was at least twenty-eight, compared to an average of twenty-four per year over the thirty-seven-year period 1946 through 1982, and about twenty per year over the immediately preceding five years (1973 through 1977).

About 75 percent of the gross metal value of the 900 discoveries made in Canada over the 1946–82 period occurs in only sixteen major mining districts or individual deposits. More than half of the gross metal value of the 900 deposits is accounted for by the 30 largest deposits.

Besides the aggregate gross metal value of discoveries and the discovery value per exploration dollar as measures of exploration success, the variations in the rates of discovery of the various individual metals through time are of interest. For most metals the rate has been far from uniform through time and tends to exhibit one or more discovery peaks. These may represent the discovery of one major deposit or that of an entire mineral district. The largest such district discovered during the period under study encompasses the porphyry copper and molybdenum deposits scattered throughout a major part of British Columbia.

The discovery record of the individual metals copper, zinc, lead, silver, gold, and uranium also tends to exhibit discovery peaks when analyzed by geologic deposit types.

The development of airborne and ground-based electromagnetic, magnetic, and radiometric geophysical methods of exploration has greatly increased the rate of Canadian discovery for a variety of geologic deposit types, especially deposits of nickel, uranium, and the volcanic-associated exhalative-sulfide deposits containing copper, zinc, and lead. Without such methods, discovery costs would have been much higher. Nonetheless, significant deposits are still being discovered by conventional prospecting methods, and by geologic inference, in various parts of Canada. New geologic concepts have led to the discovery of many ore bodies, but many deposit types known elsewhere in the world have not been thoroughly searched for in Canada. These include red-bed copper deposits and contact metasomatic (skarn) copper deposits.

Certain exploration techniques—for example, conventional prospecting and airborne and ground-based geophysical surveys—have been very successful in Canada because deeply weathered rocks have been removed by continental glaciation. Such techniques would probably be effective in other glaciated regions of the world, but less so elsewhere. Not only for this reason, but also in view of differences in local exploration expertise, wages, and history of exploration and mining, caution should be used in attempting to project the Canadian mineral discovery experience to other areas of the world. However, if other nations or regions of the world have enough data on mineral exploration expenditures, discovery records (dates, tonnages, and grades), and geologic classification of the deposits discovered, then analytic techniques similar to those used in this chapter can readily be applied in measuring their exploration success and costs of ore discovery.

■

My colleague W. H. Laughlin provided most of the information on Canadian discoveries for the 1978–82 period presented in this chapter. Various members of the Geological Survey of Canada, especially Drs. R. T. Bell, Sunil S. Gandhi, R. W. Kirkham, Donald F. Sangster, and Ralph I. Thorpe assisted in the determination of geologic deposit types for the 900 deposits discovered in Canada since 1945 and also provided me with other ideas. As author, I accept full responsibility for the somewhat simplified deposit-type classifications presented in this chapter and for any errors or omissions.

Many individuals in the Canadian mineral exploration industry provided information on recent discoveries. Although they cannot be named individually, the writing of this chapter would not have been possible without their help, which is sincerely appreciated, as is the considerable editorial assistance provided by Dr. Jan Zwartendyk of the Mineral Policy Sector, Energy, Mines and Resources Canada.

References and Bibliography

Albers, J. P. 1977. "Discovery Rate and Exploration Methods for Metallic Mineral Deposits in the U.S., 1940–76," *Engineering and Mining Journal* vol. 178, no. 1 (January) pp. 71–73.

Annis, R. C., D. A. Cranstone, and M. Vallée. 1978. *A Survey of Known Mineral Deposits in Canada That Are Not Being Mined*. Mineral Bulletin MR-181 (Ottawa, Energy, Mines and Resources Canada).

Anonymous. 1954. "Where Do the New Mines Come From?" *Northern Miner* Annual Review Number (November 25) p. 3.

Bailly, Paul A. 1977. "Changing Rates of Success in Metallic Exploration." Unpublished paper presented at the Joint Annual Meeting of the Geological Association of Canada/Mineralogical Association of Canada/Society of Economic Geologists/Canadian Geophysical Union, Vancouver, B.C., April 25–28.

———. 1978. "Exploration and Future Demand for Minerals." Unpublished paper presented at the Joint Annual Meeting of the Geological Association of Canada/Mineralogical Association of Canada/Geological Society of America, Toronto, October 23–26.

Barton, W. R. 1962. *Columbium and Tantalum: A Materials Survey.* Information Circular IC 8120 (Washington, D.C., U.S. Bureau of Mines).

Burrows, J. C. 1971. *Tungsten, An Industry Analysis* (Lexington, Mass., Heath Lexington Books).

Cranstone, D. A. 1980a. *Canadian Ore Discoveries 1946–1978: A Continuing Record of Success.* Mineral Policy Sector Internal Report MRI 80/5 (Ottawa, Energy, Mines and Resources Canada).

———. 1980b. "Canadian Ore Discoveries 1946–1978: A Continuing Record of Success," *C.I.M. Bulletin* vol. 73, no. 817 (May) pp. 30–40.

———. 1980c. "Mineral Exploration and Ore Discovery in Canada 1946–1977," *CRS Perspectives* no. 7 (July) (Kingston, Ontario, Centre for Resource Studies, Queen's University).

———. 1982. "An Analysis of Ore Discovery Cost and Rates of Ore Discovery in Canada Over the Period 1946 to 1977" (Ph.D. thesis, Harvard University, Cambridge, April) (Available from University Microfilms, Ann Arbor, MI 48106).

———, and H. L. Martin. 1973. "Are Ore Discovery Costs Increasing?" in *Canadian Mineral Exploration, Resources and Outlook.* Mineral Bulletin MR 137 (Ottawa, Energy, Mines and Resources Canada) pp. 5–13.

Derry, D. R. 1968. "Exploration," *The Canadian Mining and Metallurgical Bulletin* vol. 61, no. 2 (February) pp. 200–205.

———. 1970. "Exploration Expenditure, Discovery Rate and Methods," *The Canadian Mining and Metallurgical Bulletin* vol. 63, no. 3 (March) pp. 362–366.

———. 1972. "Exploration Statistics Show Despite Funds Spent Barely Enough New Ore Being Found for Future Needs," *The Northern Miner* (April 6) p. 14.

———, and Booth, J. K. B. 1978. "Mineral Discoveries and Exploration Expenditure—A Revised Review, 1966–1976," *Mining Magazine* vol. 138, no. 5 (May) pp. 430–433.

Energy, Mines and Resources Canada. 1977. *A Summary View of Canadian Reserves and Additional Resources of Nickel, Copper, Zinc, Lead, Molybdenum.* Mineral Bulletin MR-169 (Ottawa).

Lang, A. H. 1967. "Discovery Methods of Post-1955 New Producers," *Canadian Mining Journal* vol. 88, no. 1 (January) pp. 47–50.

Lansch, A. M. 1967. *Selenium and Tellurium: A Materials Survey.* Information Circular IC 8340 (Washington, D.C., U.S. Bureau of Mines).

Martin, H. L., D. A. Cranstone, and J. Zwartendyk. 1976. *Metal Mining in Canada 1976–2000.* Mineral Bulletin MR-167 (Ottawa, Energy, Mines and Resources Canada).

Roscoe, W. E. 1971. "Probability of an Exploration Discovery in Canada," *The Canadian Mining and Metallurgical Bulletin* vol. 64, no. 3, (March) pp. 134–137.

Roskill Information Services Ltd. 1975. *Uranium, Plutonium and the Growth of Nuclear Power 1975–2000* (London).

Schneider, V. B. 1963. *Molybdenum*. Mineral Report 6 (Ottawa, Department of Mines and Technical Surveys, Mineral Resources Division).

10

Exploration in the United States

ARTHUR W. ROSE
RODERICK G. EGGERT

How successful has metallic mineral exploration been in the United States? This chapter examines that question and related issues using data on U.S. exploration expenditures and discoveries available in 1984 when the chapter was completed. Reviewing the limitations of these data as part of the analysis, the following section presents two measures of exploration success: (1) gross value of discoveries from 1940 to 1982 and (2) success ratios from 1955 to 1982. It also analyzes the distribution of success among various types of companies and geologic deposit types.

The next section focuses on two factors that have importantly influenced recent exploration success: (1) geologic models of idealized deposit characteristics and (2) the nature of exploration planning and management. Although not necessarily the most important determinants of success trends, these factors are often neglected in discussions of exploration. Taking a broader, historical look at mineral exploration in the United States, the next section suggests that exploration and discovery of particular metals are inherently episodic. The final section summarizes the major conclusions of the chapter.

Exploration Success

Economists might measure exploration success as the discounted net value of exploration activities by comparing total exploration costs to the net financial returns from mineral deposits discovered and brought into production. Mackenzie and Woodall (in this volume) use this type of calculation for appraising mineral exploration activities in Australia and Canada. Unfortunately, the data for such an analysis are completely lacking for the United States. This study, therefore, presents two simpler measures of exploration success in the United States since World War II: (1) gross discovery values (by five-year intervals) and (2) success ratios (gross discovery value/total cost of exploration). This analysis is an extension of Rose (1982), which considers mineral discoveries in the United States up to about 1980. Here an expanded list of discoveries (table 10-1) is examined, reflecting discoveries through 1982, production decisions through 1983, and a few additional pre-1980 discoveries.

Before proceeding to the estimates of exploration success, several terms used in compiling table 10-1 need to be defined. First, as in Cranstone (in this volume), a *discovery* is defined here as "a mineral deposit sufficiently attractive to have warranted the expenditure necessary to establish its tonnage and grade." Some judgment is necessary in deciding whether the tonnage and grade are established accurately enough for a deposit to qualify as a discovery, but if specific tonnage and grade figures are quoted by a major company, the deposit is usually included. In general, deposits with a gross value of less than $100 million are not considered, except for some minor metals. Such deposits are excluded for two reasons: (1) data on smaller deposits are much more difficult to obtain, and (2) smaller deposits appear to be of little or no significance in evaluating the overall success of metallic mineral exploration. In other words, most of the value is accounted for by the large deposits. Because of this size limit, table 10-1 is not equivalent to lists for Canada, where the government has collected more detailed data over the years.

The table includes some deposits that are not currently minable at a profit, but all are at least major resources that could come into production if metal prices rose or costs fell. Some of these discoveries are extremely large (for example, the Duluth, Minnesota, gabbro copper-nickel deposit discovered in 1958; the Stillwater, Montana, platinum deposit, 1975; and the Mount Tolman, Washington, molybdenum deposit, 1978). Several may become major producers in the future. They are included because tonnages and grades are at least approximately defined and because they are large accumulations of metal that help define an upper limit on future production costs.

Second, the definition of *discovery date* also follows the usage of Cranstone (in this volume): i.e., it is "the year in which the first drill hole intersected a mineral zone that was recognized relatively soon thereafter to be part of a mineral deposit of economic interest." The discovery dates listed in table 10-1 are based largely on published data; thus, the true discovery dates may be earlier than the quoted dates, but only by a year or two at most. The table is intended to include all large discoveries from 1940 through 1982.

Third, the term *metallic mineral deposits* as used in table 10-1 *excludes* uranium and iron. Uranium is excluded because good data are already available for uranium exploration success (see, for example, Lieberman, 1976; U.S. Department of Energy, Energy Information Administration,

Table 10-1. Discoveries and Estimated Metal Values of Major Metallic Mineral Deposits (Except Uranium and Iron) in the United States, 1940–82

Discovery date	Name and state of deposit	Company[a]	Type of deposit[b]	Metal[c]	Estimated gross metal value (billions of constant 1981 U.S. $)[d],[*]	Sum of metal values per 5-year period (billions of constant 1981 U.S. $)[e],[*]
1940	Lucky Friday, Idaho	Hecla	VR	Pb, Ag, Zn	1.0[*]	
1941	Yellow Pine, Idaho	Bradley (USGS)	VR	W, Sb, Au	0.8[*]	
1942	Castle Dome, Ariz.	Miami (USGS)	PC	Cu	0.5[*]	
1943	San Manuel, Ariz.	Magma Copper (USBM)	PC	Cu	10.0[*]	1940–44 period: 12.3[*]
1947	Indian Creek, Mo.	St. Joe	MV	Pb	1.2[*]	
1947	Shullsburg, Wis.		MV	Zn	?[*]	
1949	Copper Cities, Ariz.	Miami	PC	Cu	0.9[*]	1945–49 period: 2.1[*]
1950	Jefferson City, Tenn.	NJZinc	MV	Zn	0.3[*]	
1950	Silver Bell, Ariz.	Asarco	PC	Cu	1.4[*]	
1950	White Pine, Mich.	Copper Range	SC	Cu	5.0[*]	
1951	Yerington, Nev.	Anaconda	PC	Cu	1.8[*]	
1951	Pima–Mission, Ariz.	Pima–Asarco	PC	Cu	8.1[*]	
1952	Flat Gap, Tenn.	NJZinc	MV	Zn	0.22[*]	
1952	New Market, Tenn.	Asarco	MV	Zn	1.0[*]	
1954	Immel, Tenn.	Asarco	MV	Zn	0.8[*]	1950–54 period: 18.6[*]
1955	Esperanza, Ariz.	Duval	PC	Cu	0.95[*]	
1955	Viburnum, Mo.	St. Joe	MV	Pb	3.0[*]	
1957	San Xavier, Ariz.	Asarco	PC	Cu	1.74[*]	
1957	Questa, N. Mex.	Molycorp	PM	Mo	12.0[*]	

(Continued)

Table 10-1 (continued)

Discovery date	Name and state of deposit	Company[a]	Type of deposit[b]	Metal[c]	Estimated gross metal value (billions of constant 1981 U.S.$)[d],*	Sum of metal values per 5-year period (billions of constant 1981 U.S.$)[c],*
1958	Continental, N. Mex.	U.S. Smelting	SK	Cu	1.1*	
1958	Tyrone, N. Mex.	Phelps Dodge	PC	Cu	4.0*	
1958	Safford (Kennecott), Ariz.	Kennecott	PC	Cu	8.0	
1958	Fletcher, Mo.	St. Joe	MV	Pb	3.0*	
1958	Duluth (gabbro), Minn.	Inco	MS	Ni, Cu	33.0	
1959	Palo Verde, Ariz.	Banner	PC	Cu	1.5*	
1959	Glacier Peak, Wash.	Kennecott	PC	Cu, W	0.4	
1959	Burgin, Utah	Kennecott	VR	Pb, Ag, Zn	0.5*	
1959	Young, Tenn.	Asarco	MV	Zn	0.8*	1955–59 period: 70.0 (28.6*)
1960	Babbitt, Minn.	Kennecott	MS	Cu, Ni	4.8	
1960	Christmas, Ariz.	Inspiration	SK	Cu	1.2*	
1960	Mineral Park, Ariz.	Duval	PC	Cu	1.8*	
1960	Brushy Creek, Mo.	St. Joe	MV	Pb	2.0*	
1960	Buick, Mo.	Amax	MV	Pb, Zn	4.0*	
1960	Ruby Creek, Alaska	Kennecott	VR	Cu, Co	2.4	
1961	Safford (Phelps Dodge), Ariz.	Phelps Dodge	PC	Cu	5.8	
1962	Sierrita, Ariz.	Duval	PC	Cu, Mo	8.1*	
1962	Carlin, Nev.	Newmont	SG	Au	2.1*	
1962	Ozark Lead, Mo.	Kennecott	MV	Pb	1.9*	
1962	Magmont, Mo.	Cominco	MV	Pb	2.0*	
1962?	Hall, Nev.	Anaconda	PM	Mo	2.6*	
1962	Brady Glacier, Alaska	Newmont	MS	Ni, Cu	5.0	
1963	Sacaton, Ariz.	Asarco	PC	Cu	0.9*	
1963	Blue Hill, Maine	Kerr Addison	VS	Cu, Zn	0.07*	
1963	Troy, Mont.	Kennecott-Asarco	SC	Ag, Cu	1.6*	
1964	Twin Buttes, Ariz.	Banner	PC	Cu, Mo	15.0*	1960–64 period: 61.3 (43.3*)

Year	Location	Company	Type	Commodity	Value
1965	Bluebird, Ariz.	Ranchers	PC	Cu	0.7*
1965	Kalamazoo, Ariz.	Quintana	PC	Cu, Mo	6.7*
1965	Copper Canyon, Nev.	Duval	SK-PC	Cu, Au	2.6*
1965	Henderson, Colo.	Amax	PM	Mo	17.0*
1966	Cortez, Nev.	Amex	SG	Au	0.5*
1968	Sanchez, Ariz.	Inspiration	PC	Cu	0.6
1969	Taylor, Nev.	Silver King	VR	Ag	0.2*
1969	Elmwood, Tenn.	NJZinc	MV	Zn	0.76*
1969	Nacimiento, N. Mex.	Earth Resources	SC	Cu	0.2*
1969	Carthage, Tenn.	Occidental	MV	Zn	2.0
					1965–69 period: 31.26 (28.66*)
1970	Helvetia, Ariz.	Banner-Anaconda	PC	Cu	4.0
1970	Copper Creek, Ariz.	Newmont	PC	Cu	2.0
1970	Vekol, Ariz.	Newmont	PC	Cu	1.2
1970	Red Mountain, Ariz.	Kerr McGee	PC	Cu	6.0
1970	Lakeshore, Ariz.	Hecla	PC	Cu	7.85*
1970	Florence, Ariz.	Conoco	PC	Cu	6.4
1970	Metcalf, Ariz.	Phelps Dodge	PC	Cu	5.6*
1971	Pinson, Nev.	Cordex-Rayrock	SG	Au	0.2*
1971	Flambeau, Wis.	Kennecott	VS	Cu	0.4
1971	Stillwater, Mont.	Anaconda (Amax)	MS	Ni, Cu	7.5
1973	Pinto Valley, Ariz.	Cities Service	PC	Cu	3.5*
1973	Copper Basin, Ariz.	Phelps Dodge	PC	Cu	2.0
1974	Cyprus-Johnson, Ariz.	Bagdad	SK	Cu	0.2*
1974	Pinos Altos, N. Mex.	Exxon	SK	Cu, Zn	0.4
1974	Sultan, Wash.	Brenmac	PC	Cu, Mo	1.0*
1974	Alamo, Nev.	Union Carbide	SK	W	?
1974	Delamar, Idaho	Earth Resources	EG	Ag	0.75*
1974	McDermitt, Nev.	Placer	EG	Hg	0.06*
1974	Cornucopia, Nev.	Standard Silver	EG	Ag, Au	0.5*
1974	Round Mountain, Nev.	Copper Range	EG	Au	4.4*
					1970–74 period: 54.0 (23.1*)
1975	Candelaria, Nev.	Occidental	EG	Ag, Au	0.5*
1975	Rhinelander, Wis.	Noranda	VS	Cu, Zn	0.1

(Continued)

Table 10-1 (continued)

Discovery date	Name and state of deposit	Company[a]	Type of deposit[b]	Metal[c]	Estimated gross metal value (billions of constant 1981 U.S. $)[d,*]	Sum of metal values per 5-year period (billions of constant 1981 U.S. $)[e,*]
1975	Thompson Creek, Idaho	Cyprus	PM	Mo	6.5*	
1975	Red Dog, Alaska	Cominco (USBM)	EP	Pb, Zn	18.0	
1975?	Arctic, Alaska	Kennecott	VS	Cu	9.0	
1975	Ambler River, Alaska	Anaconda	VS	Cu	?	
1975	Stillwater, Mont.	Johns Manville-Anaconda	MS	Pt, Pd	14.0?	
1975	Green Creek, Alaska	Noranda	VS	Zn, Pb, Ag	1.0	
1976	Rochester, Nev.	Asarco	EG	Ag	1.6	
1976	Quartz Hill, Alaska	U.S. Borax	PM	Mo	8.2	
1976	Ashland, Maine	Superior-Louisiana Land	VS	Cu, Zn	1.5	
1976	Jerritt Canyon, Nev.	FMC-Freeport	SG	Au	1.7*	
1976	Crandon, Wis.	Exxon	VS	Cu, Zn	5.0	
1976	Oracle Ridge, Ariz.	Continental	VR?	Cu	0.4	
1976	Casa Grande, Ariz.	Hanna-Getty	PC	Cu	7.0	
1976	Ortiz, N. Mex.	Goldfields	EG	Au	0.2*	
1977	Stonewall, Tenn.	NJZinc	MV	Zn	0.25	
1977	Gordonsville, Tenn.	NJZinc	MV	Zn	0.7*	
1977	Pine Grove, Utah	Phelps Dodge	PM	Mo	6.0	
1977	Mount Emmons, Colo.	Amax	PM	Mo	12.0	
1977	Beaver Creek, Tenn.	NJZinc	MV	Zn	0.12	
1978	Wulik, Alaska	Houston	EP	Pb, Zn, Ag	2.4	
1978	Kearsarge, Calif.	Pickands-Mather	EG	Au, Ag	0.4	
1978	Nye County, Nev.	UV Industries	PM	Mo	0.7	
1978	Mt. Tolman, Wash.	Amax	PM	Mo	18.0	
1978	Alligator Ridge, Nev.	Amselco	SG	Au	3.0*	
1979	Bald Mountain, Maine	Superior	VS	Cu, Zn	0.3	
						1975–79 period: 118.6 (12.6*)

1980	Blackbird, Idaho	Hanna–Noranda	O	Co, Cu	0.6
1980	Pierrepont, N.Y.	St. Joe	EP?	Zn	0.35*
1980	Maggie Creek, Nev.	Newmont	SG	Au	0.25*
1980	Hillsboro, N. Mex.	Quintana	PC	Cu	0.5*
1980	West Fork, Mo.	Asarco	MV	Pb	0.9
1980	McLaughlin, Calif.	Homestake	EG	Au	1.6
1980	West End, Idaho	Superior	EG	Au	0.1*
1980	Silver Peak, Idaho	Sunshine	EG	Ag	0.1
1980	Borealis, Nev.	Houston	EG	Au	0.1*
1981	Mercur, Utah	Getty	SG	Au	0.45*
1981	Mt. Hope, Nev.	Exxon	PM	Mo	5.0
1981	Golden Sunlight, Mont.	Placer Amex	EG	Au	0.6
1981	Escalante (silver), Utah	Ranchers	EG	Ag	0.4*
1982	Gold Quarry, Nev.	Newmont	SG	Au	4.0
1982	Rain, Nev.	Newmont	SG	Au	0.5
1982	Zaca, Calif.	California Silver	EG	Au, Ag	0.1
1982	Boulder Creek (Dcc), Nev.	Rayrock	SG	Au	0.15*
1982	Horse Canyon, Nev.	Placer Amex	SG	Au	0.1
				1980–82 period:	15.8 (2.30*)

Note: In general, deposits with gross metal value of less than $100 million are not included.

Sources: A. W. Rose, "Mineral Adequacy, Exploration Success and Mineral Policy in the United States," *Journal of Geochemical Exploration* vol. 16 (1982) pp. 163–182; P. Gilmour, "Grades and Tonnages of Porphyry Copper Deposits," pp. 7–35 in S. R. Titley, ed., *Advances in Geology of the Porphyry Copper Deposits* (Tucson, Ariz., University of Arizona Press, 1982); and various other sources, including U.S. Bureau of Mines, *Minerals Yearbook* (Washington, D.C., Government Printing Office, various issues), mining journals, and company annual reports.

a Acronym in parentheses after company name means that the deposit was discovered by USGS (U.S. Geological Survey) or USBM (U.S. Bureau of Mines), as indicated.

b VR = vein and replacement hydrothermal; PC = porphyry copper; MV = Mississippi Valley-type lead-zinc; SC = sedimentary copper; PM = porphyry molybdenum; SK = skarn; MS = magmatic copper-nickel-platinum sulfide; SG = sediment-hosted disseminated gold; VS = volcanogenic massive sulfide; EG = epithermal gold and silver (and mercury) except sediment-hosted disseminated gold; EP = exhalative lead-zinc-silver; O = other.

c Pb = lead; Ag = silver; Zn = zinc; W = tungsten; Sb = antimony; Au = gold; Cu = copper; Mo = molybdenum; Ni = nickel; Co = cobalt; Hg = mercury; Pt = platinum; Pd = palladium.

d Value of metal is based on estimated total production plus reserves, evaluated at the following prices: Pb, $0.40/lb; Ag, $12/oz; Zn, $0.40/lb; Au, $500/oz; Cu, $1/lb; MoS_2, $9/lb; Ni, $3/lb; Co, $10/lb; Hg, $400/flask; Pt, $600/oz; Pd, $200/oz.

*Values in this column followed by an asterisk (*) are for deposits that have produced.

e Value in parentheses = five-year total for deposits that have produced, if that sum differs from the overall five-year total.

1985). Iron is excluded mainly because meaningful market prices are not published, making it difficult to assign a value to the success of iron exploration. Moreover, iron ore discoveries in the United States since 1945 do not appear large enough to affect the results significantly.

Fourth, the estimates of the *gross value of metal* discovered at each deposit, shown in table 10-1, are based on 1981 metal prices and on data for tonnage and grade. In other words, costs of production are not incorporated in the calculations. These estimates of gross discovery values are subject to considerable error for at least three reasons: (1) the ultimate size of some deposits is undoubtedly underestimated, (2) some mineralized material included in estimates of tonnage and grade is not recovered because of less-than-complete recovery in extraction and processing, and (3) relative mineral prices change over time. The first two factors work in opposite directions, reducing to some extent their net impact on estimates of discovery values.

Gross Discovery Values

Figure 10-1 displays gross discovery values (constant 1981 U.S. dollars) for five-year periods from 1940 through 1982. If only deposits that have produced metal are taken into account, the value of discoveries peaked in the 1960–64 period. However, other deposits discovered in the 1970s and 1980s undoubtedly will come into production in the next few years; Red Dog, Wulik, and Green Creek (all in Alaska) seem the best candidates, and others are likely to do so. If about half of the discoveries made between 1975 and 1979 come into production in the near future, as was the case for discoveries made in 1960–64 and 1970–74, then the value of 1975–79 discoveries would exceed the value for 1960–64 (the highest period). Making this assumption, the value of discoveries from 1955 to 1980 averaged about $30 billion per five-year period, with no clear upward or downward trend over time.

The major discoveries from 1940 through 1982 are classified by type of company in table 10-2. As shown, major metallic mining companies account for the bulk of the discoveries in both number and gross value. Companies of the medium-sized metallic and nonmetallic mining types account for significant, but smaller, shares of the discoveries. Oil companies discovered nearly as many deposits as the medium-sized metallic mining firms but were relatively unsuccessful in bringing these deposits into production. Finally, small mining companies, stock-promotional firms, and individuals also discovered a significant share of the deposits. The nature of some of these small companies is not clear, but the following comments on two of the most successful are instructive.

Banner Mining Company started with a small, underground copper mine and a 300-ton-per-day mill near Lordsburg, New Mexico. Then, starting in 1950, the company shrewdly acquired part or all of three major

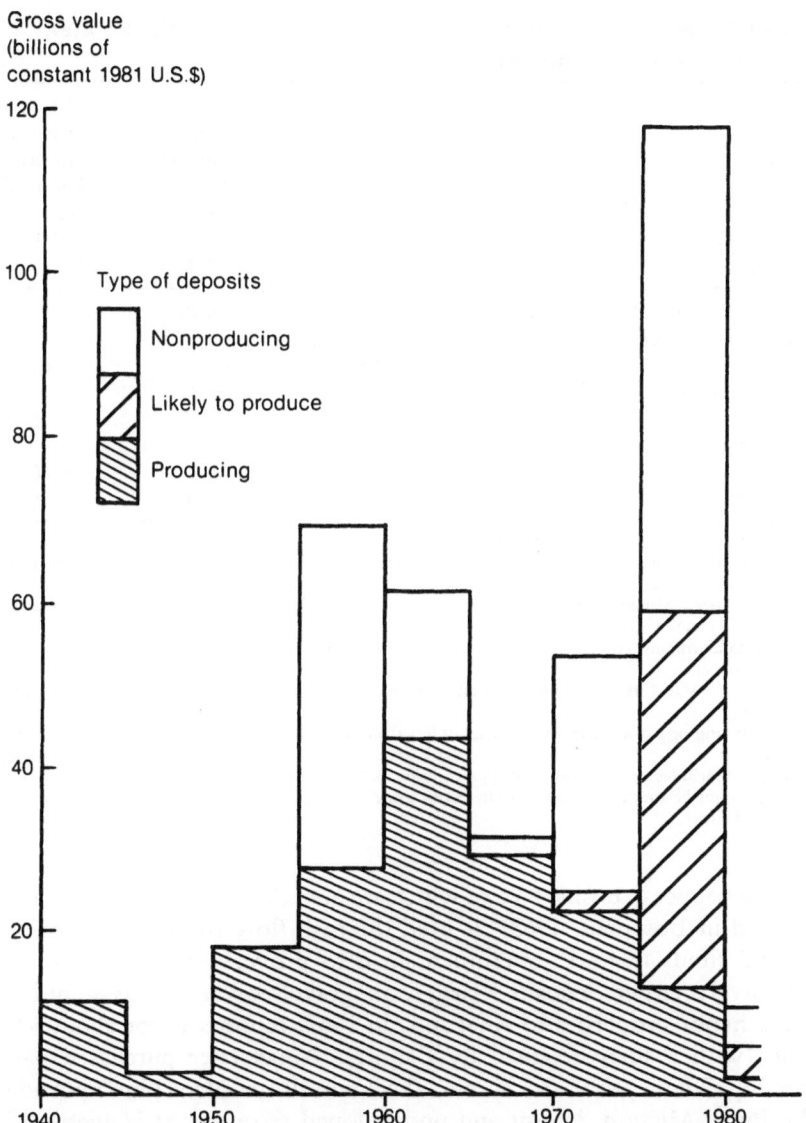

Figure 10-1. Gross metal value of major mineral discoveries (excluding uranium and iron) in the United States by five-year periods, 1940–82, at 1981 metal prices. The "likely to produce" category represents inferred value for producing deposits a few years in the future, assuming that 50 percent of total discoveries will be producing. *Source*: Table 10-1 in this chapter.

**Table 10-2. Major Metallic Mineral Discoveries in the United States,
1940–82, by Type of Company**

Type of company	No. of companies	No. of discoveries	Value, all discoveries (billions of constant 1981 U.S.$)	Value, produced discoveries (billions of constant 1981 U.S.$)
Major metallic mining[a]	17	68	226.10	80.01
Medium-sized metallic mining[b]	11	15	60.35	50.67
Nonmetallic mining[c]	5	8	42.35	15.15
Oil[d]	7	12	30.15	1.15
Small mining or stock[e]	10	15	32.20	26.60
Unclassified[f]	6	6	2.26	0.76

Sources: Table 10-1 in this chapter and authors' data.
[a]A major metallic mining company had sales exceeding $400 million in 1980 and a long-time position in U.S. metallic mining.
[b]A medium-sized metallic mining company had sales of less than $400 million in 1980 and a prior history of U.S. metallic mining.
[c]A nonmetallic mining company had its major interests in nonmetallics, including sulfur, prior to a metallics discovery.
[d]An oil company had its major interest in oil, gas, and possibly uranium before entering the metallic minerals field.
[e]A small mining or stock company had sales of less than $10 million and little or no previous mineral production.
[f]Unclassified includes six companies, all apparently small, for which adequate information on ownership and size could not be found. Controlling interest in some of these companies may be held by major or medium-sized firms.

porphyry copper deposits, two of which are now developed and producing (described in Bowman, 1963). It used the cash flow from the operating mine and funds from a government loan to prove, develop, and mine progressively larger deposits—largely in gravel-covered areas—until it sold the major properties to Anaconda in 1963, reportedly for about $20 million (with a share retained by Banner). Amax later purchased part ownership. These properties became the Twin Buttes Mine plus a portion of the Pima–Mission deposit and undeveloped resources at Helvetia (all in Arizona).

Another outstanding success was the discovery of the Kalamazoo, Arizona, ore body by Quintana Minerals. This ore body is actually the faulted half of the San Manuel, Arizona, porphyry copper ore body, originally drilled by the U.S. Bureau of Mines during World War II and later developed by Magma Copper (a subsidiary of Newmont Mining Company) with a government loan in 1952. Because of the complex geology and extensive cover in the area, the Kalamazoo portion was not apparent to geologists working in the region. A consultant, J. D. Lowell, developed a valid geologic hypothesis in the early 1960s and pressed for the 3,000-to 4,000-foot drill holes needed to discover the ore body, which eventually was sold to San Manuel Copper for a reported $27 million.

Despite these and other successes, small companies apparently play a smaller role in mineral discoveries in the United States than in Canada, where securities laws make it much easier to raise capital in stock markets. In Canada, 62 percent of the discoveries between 1951 and 1974 were made by small mining companies (Snow and Mackenzie, 1981). In the United States, it is unlikely that small companies would account for as large a share of discoveries even if discoveries valued at less than $100 million—not included here—were counted. Nevertheless, table 10-1 suggests that small companies accounted for a higher proportion of U.S. discoveries in recent years than previously. In part this results from more complete data in recent years, but it also is believed to be an actual trend. Small gold deposits, a major focus of U.S. exploration between the late 1970s and mid-1980s, can be discovered and developed with relatively small expenditures and pay back their investment in one to two years.

In table 10-3, which classifies discoveries by geologic deposit type, porphyry copper, porphyry molybdenum, and magmatic copper-nickel-platinum deposits together account for more than two-thirds of the gross value of all discoveries. The high value given for the magmatic sulfides is based completely on unmined discoveries, and thus this class of discoveries is of uncertain economic significance. As the table also shows, a large number of porphyry copper deposits were discovered in the 1950s, 1960s, and 1970s; discoveries declined after the early 1970s. Mississippi Valley lead–zinc deposits have been a continuing target of exploration, with some success in Missouri and Tennessee. Exhalative lead–zinc, sediment-hosted gold, and epithermal gold deposits only became popular targets in the 1970s but are currently a major focus of exploration because of their precious metal content and—in the case of exhalative deposits—high grade coupled with significant tonnage. Volcanogenic massive sulfides, a major exploration target in Canada since the 1950s, received increased attention in the United States during the 1970s, with some success.

Success Ratios

Going beyond mere tabulation of gross discovery values, success ratios incorporate exploration expenditures into estimates of exploration success. Data on U.S. expenditures for metallic mineral exploration are more limited than are these data for a number of other countries, especially Australia and Canada. There is no single source providing consistent data on U.S. expenditures from the 1950s to the 1980s. Various estimates are presented in table 10-4 and illustrated in figure 10-2 on p. 346 (expenditures are shown in constant in 1982 U.S. dollars for comparison with data in other

Table 10-3. Classification of Major Metallic Mineral Discoveries in the United States, 1940s through 1980s, by Deposit Type
(number of discoveries)

Type of deposit	1940s	1950s	1960s	1970s	1980s	Total	Value of metal (billions of constant 1981 U.S.$)
PC (porphyry copper)	3	9	8	11	1	32	125.7
PM (porphyry molybdenum)	0	1	2	4	1	8	82.0
SK (skarn)	0	1	2	3	0	6	5.5
VR (vein and replacement hydrothermal)	2	1	2	1	0	6	5.3
VS (volcanogenic massive sulfide)	0	0	1	8	0	9	17.4
EP (exhalative lead–zinc–silver)	0	0	0	3	1	4	21.0
MV (Mississippi Valley-type lead–zinc)	2	7	6	4	1	20	25.0
SC (sedimentary copper)	0	1	2	0	0	3	6.8
MS (magmatic copper–nickel–platinum sulfide)	0	0	2	1	0	3	60.0
SG (sediment-hosted disseminated gold)	0	0	2	3	6	11	12.8
EG (epithermal gold and silver [and mercury] except sediment-hosted disseminated gold)	0	0	0	7	7	14	11.4

Sources: Table 10-1 in this chapter and authors' data.

chapters in this volume, whereas gross discovery values and success ratios are given at 1981 mineral prices and valued in constant 1981 U.S. dollars).

Despite the obvious discrepancies among data series in table 10-4, two broad secular trends can be identified: (1) generally rising expenditures from the late 1950s to about 1980 and (2) falling expenditures from the early to middle 1980s. These broad trends mask year-to-year fluctuations in exploration expenditures that are not uncommon (see Eggert, in this volume). The major differences among the estimates of expenditures shown in the table result from differences in definitions of exploration, in mineral universes, and in geographic universes. Some data, for example, include exploration for uranium whereas others do not; in the late 1970s, expenditures for uranium exploration apparently exceeded those for other metals (compare table 10-5 on p. 347 with figure 10-2). Some, but not all, data include foreign exploration by U.S. companies. Table 10-4 elaborates on the various expenditure estimates.

Figure 10-3 (p. 348) and table 10-6 (p. 349) display success ratios calculated by dividing exploration expenditures into gross discovery values

Table 10-4. Estimates of Mineral Exploration Expenditures in the United States Between 1955–59 and 1984 (millions of U.S. dollars per year)

Sources and comments	Year or period	Exploration expenditures		
		Nominal U.S.$	Constant 1982 U.S.$[a]	Includes foreign exploration by U.S. companies	Includes uranium and nonmetals
Amax, in Committee on Mineral Resources and the Environment (1975)[b]	1955–59	35	121	No	Probably
	1960–64	55	172		
	1965–69	90	247		
Wargo (1973): sum of 21 companies	1968–71	110	268	Not indicated	Not indicated
Barber (1981): 12 major U.S. companies	1972	100	215	Yes	Yes
	1973	130	263		
	1974	150	278		
	1975	170	287		
	1976	180	285		
	1977	190	282		
	1978	210	291		
	1979	310	394		
	1980	410	478		
Barber (1981): 12 major U.S. companies	1979	200	254	No	Yes
	1980	240	280		
U.S. Dept. of Commerce (1979)	1977	75–110	111–149	Probably not	Yes
Brown (1983): 30 companies exploring in United States; based on published data, inquiries, estimates	1961	60	192	Probably not	Yes
	1965	80	237		
	1970	140	333		
	1975	210	354		
	1980	390	455		

(Continued)

Table 10-4 (continued)

Sources and comments	Year or period	Nominal U.S. $	Constant 1982 U.S. $[a]	Includes foreign exploration by U.S. companies	Includes uranium and nonmetals
				Exploration expenditures	
Emerson and Ivosevic (1984); basic data from Brown (1983); more details in Schreiber and Emerson (1984)	1961	60.9	195.2	Probably not	Yes
	1962	66.4	208.2		
	1963	71.6	221.0		
	1964	81.0	246.2		
	1965	86.3	255.3		
	1966	94.1	268.9		
	1967	103.5	288.3		
	1968	112.6	298.7		
	1969	129.3	324.9		
	1970	162.8	387.6		
	1971	166.6	375.2		
	1972	150.1	322.8		
	1973	192.6	389.1		
	1974	227.6	421.5		
	1975	220.0	371.0		
	1976	235.2	372.7		
	1977	271.2	403.0		
	1978	285.3	395.2		
	1979	327.2	416.3		
	1980	440.0	513.4		
	1981	484.1	515.0		
	1982	386.0	386.0		
	1983	314.7	303.2		
Preston (1960), from *Census of Mineral Industries*; nonproducing sites only	1954	14.9	56.7	Probably not	No

Preston (1960): 7 companies; includes development	1954	15	57	No	Probably not
U.S. Bureau of the Census (1968, 1973, 1978)c	1967	27	75	Probably not	No
	1972	116	250		
	1977	264	392		
This study: unpublished data on U.S. metals exploration for 6 companies, and published total mineral exploration data for 6 other firms	1980	190	222	In part	In part
This study: extrapolation of 1980 estimates, based on incomplete data from 6 to 12 companies	1981	260	277	In part	In part
	1982	275	275		
	1983	180	173		
Barber and Muessig (1984, 1985, forthcoming): total metallic mineral exploration in United States by U.S. and Canadian companies	1980	208	243	No	No
	1981	306	326		
	1982	258	258		
	1983	264	255		
	1984	194	180		

[a]Adjusted for inflation using the U.S. GNP implicit price index (1982 = 100).

[b]According to Pierce Parker, former president of Amax Exploration (personal communication, 1984), these estimates were based partly on counts of exploration personnel (identified from membership lists of the Society of Mining Engineers of the American Institute of Mining, Metallurgical, and Petroleum Engineers) multiplied by factors of dollars per explorationist.

[c]For 1968: Total capitalized and expensed mineral development and exploration plus mineral rights and geologic expenditures at nonproducing establishments for copper, lead, and zinc, plus the same expenditures at all establishments for other metals except uranium.

For 1973: Expensed mineral development and exploration expenditures plus mineral rights and geologic expenditures, minus uranium. Does not include capitalized successful exploration; probably includes some development at producing mines.

For 1978: Expensed mineral exploration and development, including land and rights, plus capitalized land and mineral rights, minus uranium. The data account for unsuccessful exploration, probably include some expensed development at operating mines, and do not include successful projects except land costs.

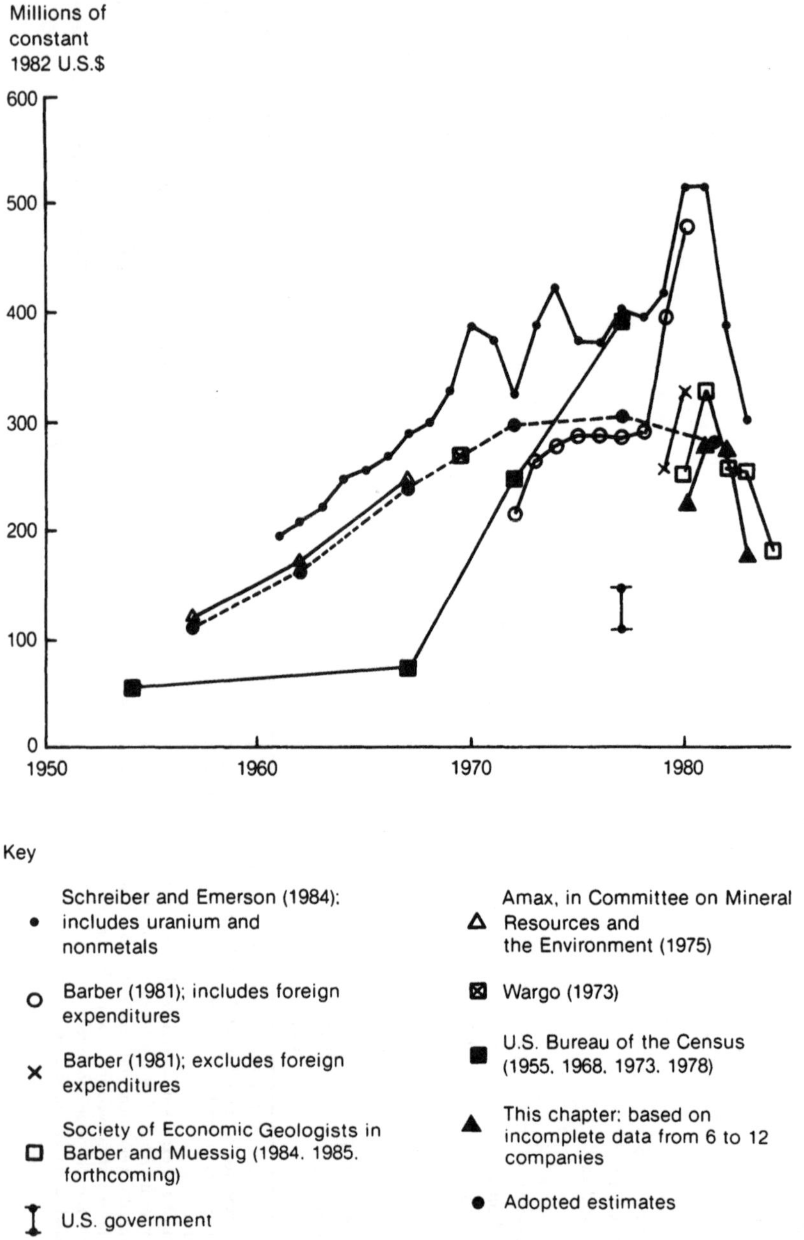

Millions of
constant
1982 U.S.$

Key

Schreiber and Emerson (1984):
• includes uranium and
nonmetals

○ Barber (1981); includes foreign
expenditures

✕ Barber (1981); excludes foreign
expenditures

Society of Economic Geologists in
□ Barber and Muessig (1984. 1985.
forthcoming)

⊥ U.S. government

Amax, in Committee on Mineral
△ Resources and
the Environment (1975)

⊠ Wargo (1973)

U.S. Bureau of the Census
■ (1955. 1968. 1973. 1978)

This chapter: based on
▲ incomplete data from 6 to 12
companies

● Adopted estimates

Figure 10-2. Estimated expenditures for mineral exploration in the United States versus time, 1954–84. The adopted estimates (dashed line) represent the authors' best estimates of the expenditures for metallic minerals, except uranium, in the United States through the stage of discovery but not including development; the authors based their estimates on the other data presented in this figure. *Source*: Table 10-4 in this chapter.

Table 10-5. Exploration Expenditures and Drilling Footage for Uranium in the United States, 1948–84

Year	Expenditures[a] (millions of U.S.$)		Drilling footage (millions of ft)
	Nominal	Constant 1982[b]	
1948	N.A.	N.A.	0.17
1949	N.A.	N.A.	0.36
1950	N.A.	N.A.	1.10
1951	N.A.	N.A.	1.08
1952	N.A.	N.A.	1.36
1953	N.A.	N.A.	3.65
1954	N.A.	N.A.	4.05
1955	N.A.	N.A.	5.27
1956	N.A.	N.A.	7.29
1957	N.A.	N.A.	7.35
1958	N.A.	N.A.	3.76
1959	N.A.	N.A.	2.36
1960	N.A.	N.A.	1.40
1961	N.A.	N.A.	1.32
1962	N.A.	N.A.	1.48
1963	N.A.	N.A.	0.88
1964	N.A.	N.A.	0.97
1965	N.A.	N.A.	1.16
1966	8.4	24.0	1.80
1967	24.8	69.1	5.44
1968	53.4	141.6	16.20
1969	58.7	147.5	20.50
1970	52.2	124.3	18.00
1971	41.2	93.2	11.4
1972	32.4	69.7	11.8
1973	49.5	100.0	11.7
1974	79.1	146.5	14.7
1975	122.0	205.7	15.7
1976	170.7	270.5	20.4
1977	258.1	383.5	28.0
1978	314.3	435.3	29.0
1979	315.9	401.9	28.1
1980	267.0	311.6	19.6
1981	144.8	154.0	10.9
1982	73.6	73.6	4.2
1983	38.9	37.5	2.1
1984	26.5	24.5	2.3

Note: N.A. = not available.

Sources: W. L. Chenoweth, *Industry Exploration Activities*, U.S. Department of Energy Open File Report GJO-108(79) (Washington, D.C., U.S. Government Printing Office, 1979); L. R. Sanders, *Uranium Exploration Expenditures in 1980 and Plans for 1981–1982*, U.S. Department of Energy Open File Report GJO-103(81) (Washington, D.C., U.S. Government Printing Office, 1981); U.S. Department of Energy, Energy Information Administration, *Uranium Industry Annual 1984* (Washington, D.C., U.S. Government Printing Office, 1985).

[a]About 15 percent of this expenditure appears to be for development activities.
[b]Adjusted for inflation using the U.S. GNP implicit price index (1982 = 100).

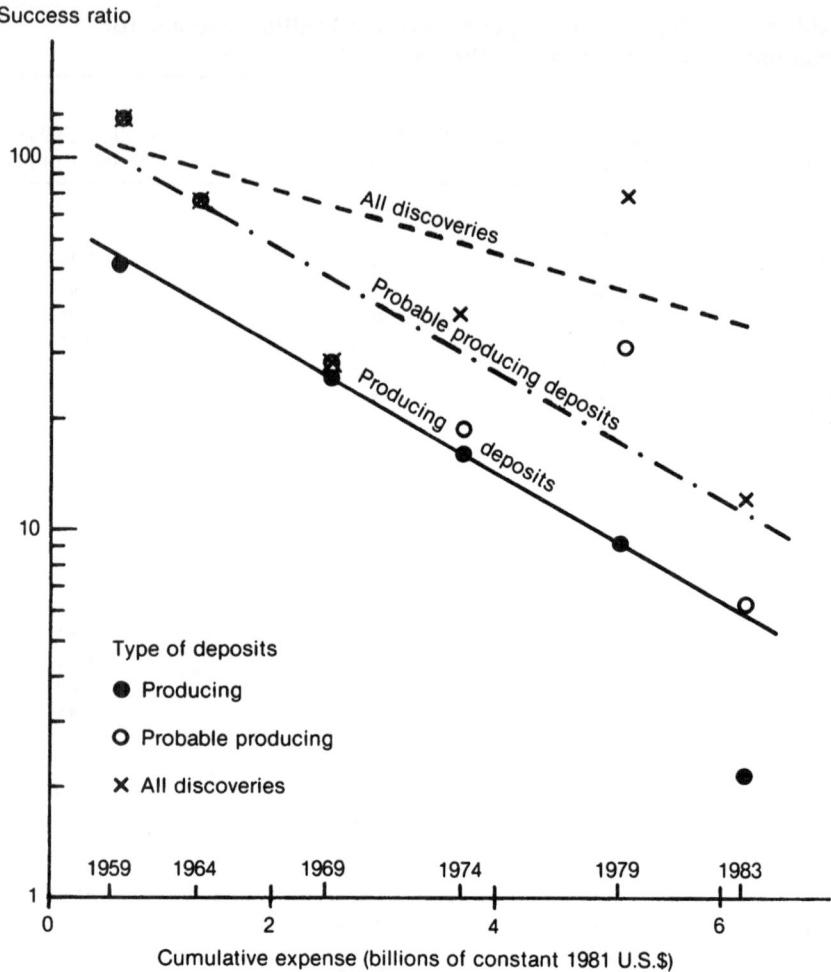

Figure 10-3.   Success ratios (inflation-adjusted value of discoveries/total cost of exploration in constant 1981 dollars) versus cumulative exploration expense (in constant 1981 dollars) for metallic minerals in the United States, 1955–83, and inferred trends with time. The 1983 point was not considered in estimating the trends. *Source*: Table 10-6 in this chapter.

per time period from 1955–59 to 1980–83 (with both expenditures and values expressed in constant 1981 U.S. dollars). The existence and steepness of any trend in figure 10-3 clearly depend on how one evaluates deposits that have not come into production. Considering only deposits that have produced metal, the success ratio decreases by a factor of about 10 over the period. This decrease is similar in magnitude to decreases observed for oil (Menard and Sharman, 1975) and uranium (Lieberman, 1976). However, as noted previously, at least some additional deposits are

Table 10-6. Estimated Exploration Expenditures in the United States and Success Ratios for Five-Year Periods, 1955–83

Period	Average expenditures per year (millions of current U.S.$)[a]	Average expenditures (millions of constant 1981 U.S.$)[b]	Cumulative expenditures (millions of constant 1981 U.S.$)[c]	Success ratio[d]
1955–59	35	108	540	53 (130)
1960–64	55	156	1,320	56 (78)
1965–69	90	226	2,450	26 (28)
1970–74	140	278	3,645	17 (38) [19]
1975–79	210	290	5,095	9 (81) [41]
1980–83	260	263	6,147	2.1 (12) [6]

[a]Based on data from A. W. Rose, "Mineral Adequacy, Exploration Success and Mineral Policy," *Journal of Geochemical Exploration* vol. 16, pp. 163–182, and on estimates discussed in this chapter.
[b]Data of the preceding column converted to 1981 dollars using the GNP implicit deflator. U.S. Department of Commerce (DOC), Office of Business Economics, *National Income and Product Accounts of the United States, 1929–1945* (Washington, D.C., U.S. Government Printing Office, 1966); U.S. DOC, Bureau of Economic Analysis, *Business Conditions Digest* (November 1979), p. 84; ibid. (February 1981), p. 84; U.S. DOC, Bureau of Economic Analysis, *Survey of Current Business* vol. 63, no. 7 (1983), p. 80.
[c]Cumulation of data for yearly deflated expenditures.
[d]Success ratio = value of metal discovered (table 10-1 in this chapter) divided by exploration expenditure for the period. Values at left in the column are for producing deposits, values in parentheses for all discoveries, and values in brackets for deposits likely to be producing in the next few years.

likely to come into production in the next five years or so. If the proportion is similar to that estimated in figure 10-1, then the success ratios are higher, although declining at a similar rate. Less of a decrease is possible if a larger proportion of nonproducing recent discoveries are included.

The absolute values of the success ratios range from about 20 to 100 if at least half of the nonproducing discoveries are included. Expressed differently, the cost of discovering metallic minerals has been some 1 to 5 percent of gross value. This range compares well with estimates of the costs of finding gold in the United States between 1960 and 1980: 2 to 8 percent of gross value, or $8 to $30 per ounce (Brown, 1983). As noted by Brown (1983) and others, success ratios vary significantly from company to company. Unfortunately, the data are not available for calculating success ratios for individual companies. (The data are also insufficient for calculating success ratios for groups of companies and specific geologic deposit types, as in the previous subsection on gross discovery values.)

Factors Affecting Exploration Success

Although many factors presumably influence the level of exploration success—including the quality of undiscovered deposits, the availability of

geologic information, the existence of geochemical and geophysical techniques for finding particular deposit types, and, to some extent, luck—
two factors appear to have been very important over the last decade or
so: (1) geologic models of idealized deposit geology and (2) the nature of
exploration planning and management. Although these factors have not
necessarily been the most important determinants of historical trends in
exploration success, they are worthy of more detailed attention than is
often found in discussions of exploration.

Geologic Models

Beginning in the 1950s and increasingly in the 1960s, 1970s, and 1980s,
exploration has been oriented toward specific geologic deposit types, such
as those listed in table 10-3. Geologic models of idealized deposit geology
have been developed as the basis for reconnaissance exploration and follow-
up of favorable indications. In essence, the models predict where ore should
occur. Previously, mineral exploration was much less scientific and systematic; mineral occurrences found by general prospecting tended to be
evaluated on an individual basis and were compared to other deposits only
after discovery.

As an example, the model for porphyry copper deposits includes the
following components:

1. Occurrence in a granitic porphyry stock and in adjacent sedimentary
and igneous country rocks.

2. Association with andesitic volcanics, developed in a subduction-zone
environment.

3. The presence of copper, molybdenum (richest within deposits formed
in continental environments), gold (richest in deposits formed in island
arc environments), and silver; these by-products may be useful geochemical guides as well as being economically important. In addition, lead, zinc,
silver, manganese, and other metals may occur in small deposits or geochemical anomalies around the periphery of the copper deposits.

4. Dissemination of the copper sulfides and accompanying pyrite in
very large volumes of rock, and possible extension of the pyritized rock
to several kilometers from the deposit, forming a halo that increases the
exploration target size. In addition to visual observations, the sulfides can
be detected by the induced polarization (IP) geophysical method, a new
technique developed to assist in exploration for these deposits, especially
where a cover of younger rocks conceals the sulfides.

5. Extensive alteration of the host rocks to clay, mica, feldspar, and
other minerals, extending out as much as several kilometers from the
deposit. The alteration is zoned, often forming a bulls-eye that can guide
exploration to the ore. Accessory magnetite is destroyed in most alteration

but is added in skarn zones if limestones are present, leading to magnetic anomalies.

6. Extensive shattering of the host rock, commonly in one or two particular directions that coincide with regional lineaments and swarms of igneous dikes that may assist in locating favorable targets.

The application of this model allows detection and efficient exploration of deposits that are concealed by younger rocks, which cover two-thirds of the favorable region in southwestern United States, and leads to much greater potential for discovery. Similar models now exist for all of the deposit types listed in table 10-3 as well as for other types (Ohle and Bates, 1981; Cox, 1983; Eckstrand, 1984). The development and refinement of these models are now a major focus of effort in exploration by major companies (Wilson, 1982; Bailly, 1979; Anderson, 1982).

Exploration Planning and Management

The year-to-year exploration activities of mining groups reflect the decisions and style of their exploration managers and interactions of these individuals with others both within and outside their companies. In 1982, the authors conducted extensive personal interviews with exploration managers of three large companies and telephone interviews with six other managers. The interviews indicated that there are two basic styles of operation in these companies. While the following description of the two types is generalized and does not exactly fit any one company, it is nevertheless useful in understanding how exploration decisions are made.

One style of operation is characterized by an exploration manager who, as a strong individual leader, tends to dominate the planning of exploration. He probably makes the selection of commodities himself, largely on a qualitative basis, with only incidental input from formal commodity and market analysis. A strong manager tends to have a wide range of contacts in the minerals industry and to rely on informal news and highly competent consultants and friends, in addition to his own ideas, for recognizing new possibilities and trends. The geologic potential for discovery and development of new types of deposits is given considerable weight in the manager's selection of commodities, and market analyses tend to be informal or to proceed from geologic considerations. In addition to his strong views on how exploration should be done, such a manager usually has access to the company president or chairman and calls upon these officials to ratify his decisions when necessary.

This type of management exists in several oil as well as mining companies with annual exploration budgets exceeding $15 million. And, although that amount exceeds the $3-million to $8-million budget level proposed for efficient exploration within a "hunting group" (Snow and Mackenzie, 1981), each of these firms allows for significant creative input from explorationists below the level of exploration manager, perhaps counterbalancing the problems of size.

In contrast to the style of operation in which a strong manager dominates exploration planning, under "team management" the planning is conducted by a group of specialists. The manager may act as coordinator or leader of the team, but with only slightly more influence than the other members. It is the team of specialists, which is often attached to a planning group at company headquarters rather than to the exploration department, that studies and recommends commodities. Although the exploration manager may suggest a list of commodities for study, the staff report developed by the specialists is the key to action. The commodity appraisal is commonly made before the development of geologic ideas for types of deposits and the selection of countries to be explored. A long-range plan (for five or even ten years) may be developed and updated by the planning group. The exploration manager in this type of structure generally does not have or use as much access to top company officers for making or confirming decisions. Team management is employed in both oil and mining companies, but more commonly in large companies.

In choosing and evaluating countries in which to explore, the two types of management operate along the same lines as described above. Strong managers tend to make decisions based on their own experience in the countries or on that of friends, consultants, and company personnel, whereas team management tends to assemble a group of specialists to evaluate and report on prospective countries or to organize a permanent group within the company for that purpose.

Companies with strong exploration managers also tend to have forceful presidents or chairmen, who generally have a higher level of involvement with the exploration department than do their counterparts in team management companies, although their input is appropriately broad in most instances. In several companies with the first style of operation, top officials suggest new commodities and countries; in companies of the second type, top management seems willing to let the exploration departments and evaluation groups choose the commodities and countries and to exercise only financial management.

Among the nine major companies covered by the interviews, there appears to be a trend toward the team management style; several of these companies had initiated organized evaluation groups during the preceding three years. Many but not all of the large oil companies that have entered the minerals business or purchased mining companies have team management, although the company most clearly in the strong-manager group was a moderate-sized oil company. The old-line mining companies were more likely to have strong managers, but the operations of several such companies were intermediate, with neither a strong manager nor a well-developed team. In several companies, team management was adopted as part of a plan to diversify, but two of the companies with strong managers were diversifying as well. However, the increasing number of small companies recently initiating gold exploration, together with cutbacks in both the number of major companies and the level of expenditures at remaining companies, may indicate an industry trend away from team management.

While it would be of interest to evaluate the success of the two types of management, the interviews yielded detailed data on discoveries and exploration expenditures mainly for the companies with strong managers, probably because these companies' officials were more willing to be interviewed. It is apparent that companies with both types of management have made multiple discoveries and can be successful. However, a tentative conclusion is that the companies with strong managers are probably more successful. Indeed, all of them have made at least two major discoveries since 1970 that have come into production or seem likely to do so within a few years. In contrast, although all of the team management companies have made discoveries since 1970, the number likely to come into production soon appears to be zero or one per company, even though several of these companies have exploration budgets similar to those of the larger companies with strong managers.

It appears that the greater success of the companies with strong managers is attributable to their being quicker to seize new opportunities, more entrepreneurial, and better able to satisfy most other requirements of successful exploration groups as expressed by Snow and Mackenzie (1981) and Bailly (1979). The disadvantage appears to be that if the exploration manager is not competent enough, the whole effort may fail.

In addition to the apparent trend toward team management, another trend that has important implications for future exploration is the tendency to evaluate discoveries not mainly by estimated return on investment, but primarily by comparing their estimated cost of production per unit of output with the same costs for the existing array of deposits. The severe inflation of several years ago and unusually depressed metal prices relative to long-term averages, which have obviously impressed management with the difficulty of estimating meaningful future prices for minerals, are the reasons for which return on investment cannot be calculated with any assurance. The planning of exploration to discover deposits with relatively low cost of production relative to that for existing deposits was noted as a company goal by all managers interviewed in depth and by several others. Similarly, when a deposit was found, this criterion was given heavy weight in deciding on development.

A related change in recent years has been the reorientation of exploration to geologic types of deposits that will produce higher and more certain profit, even though these are smaller than other types. For example, volcanogenic massive sulfides are preferred as an exploration target compared to larger but lower-grade porphyry copper deposits, and the low-grade gold deposits in Nevada are also desirable, commonly having short payout times and low capital cost.

The Episodic Nature of Exploration

From a broader, more historical perspective, exploration and discovery appear to be episodic in nature. Copper is an instructive example. Although data on exploration expenditures for copper are not available, data on discoveries appear in figure 10-4, presumably reflecting the general pattern of exploration effort.

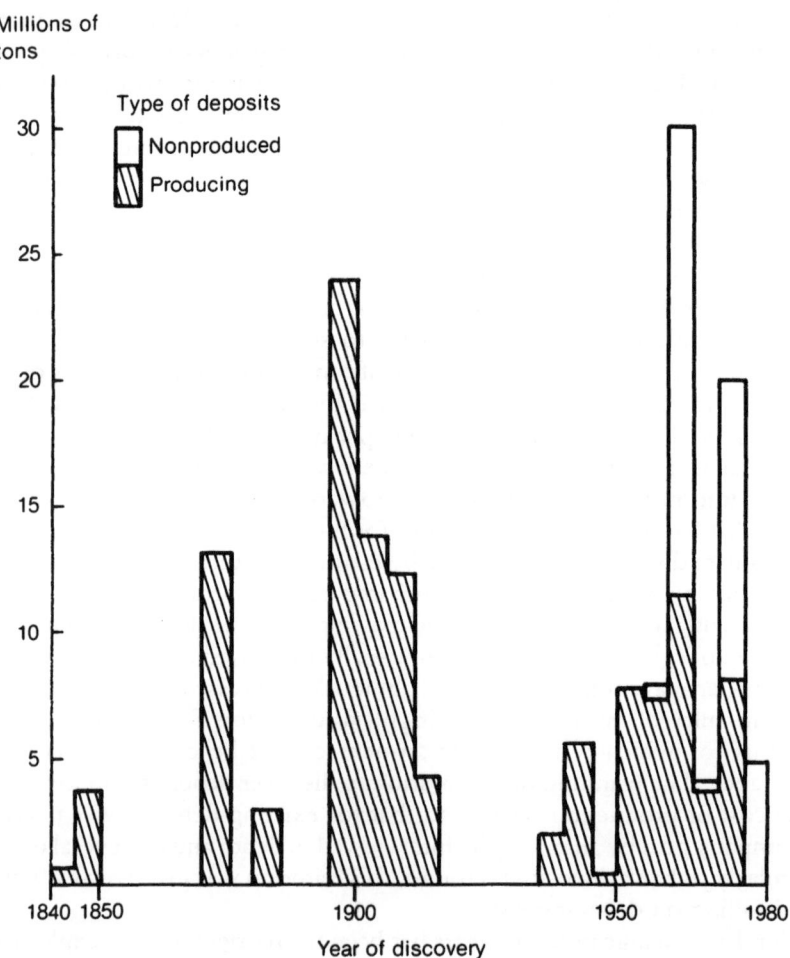

Figure 10-4. An example of the episodic nature of mineral discovery: major copper resources discovered in the United States, 1840–1980. *Source*: Data, with minor additions, from A. W. Rose, "Mineral Adequacy, Exploration Success and Mineral Policy in the United States," *Journal of Geochemical Exploration* vol. 16 (1982) pp. 163–182.

Initial deposits of native copper discovered in about 1845 in northern Michigan required only simple crushing and gravity separation to obtain the metal. Grades were apparently several percent copper. Geologically, the ores occurred in relatively simple flow tops and thin conglomerates. No further major discoveries were made until 1870, when an important deposit was found at Butte, Montana. The Butte ores (at this period) were in very high grade veins (5 to 20 percent copper) with significant amounts of silver and some gold. The veins had complex structural relations that spurred the development of geologic techniques for predicting extensions, but the grade was high enough to support relatively expensive underground mining.

In 1899 Daniel Jackling recognized the potential for low-cost, open-pit bulk mining of relatively low grade "porphyry copper" ore at Bingham, Utah (Parsons, 1957). The ore had a grade of 2 percent copper, but copper minerals were relatively uniformly distributed through a large mass of "porphyry," a granitic igneous rock. Jackling was initially ridiculed because the grade was similar to the tailings (waste after processing) at Butte, but by mining and processing large tonnages by new methods, he achieved competitive costs. The advent of the age of electricity had furnished increased demand for copper, and the success at Bingham, coupled with increased demand, led to the discovery and development of seven other porphyry copper deposits in the western United States and two in Chile between 1900 and 1915. The extremely high grade nonporphyry deposit at Kennecott, Alaska, also was discovered in 1900 and put into production in 1911.

Although production at Bingham was started on the basis of a total of about 12 million tons of 2 percent copper ore, it was readily apparent that very much larger tonnages of lower-grade ore were present. By increasing the scale of operations and by applying the new process of flotation to separate the copper minerals from useless gangue, profitable grades dropped to about 1 percent in 1925 and 0.8 percent in 1960. The total production of Bingham to 1972 was 1.24 billion tons of 0.91 percent copper plus valuable by-products of 0.036 percent molybdenum, 0.0064 ounces per ton of gold, and 0.058 ounces per ton of silver, with a total value of $35 billion at 1981 metal prices. In addition, 1.7 billion tons with 0.71 percent copper still remained in 1972 (Gilmour, 1982). Similar decreases in profitable grade were accomplished at other deposits, although tonnages of ore are generally smaller, in the hundreds of millions of tons, as illustrated in figure 10-5.

The major copper companies (Kennecott, Phelps Dodge, Anaconda, and Miami Copper) were organized by consolidations of ownership between 1910 and 1925. In recent years, porphyry copper deposits have accounted for about 80 percent of U.S. copper production and 40 percent of world copper production.

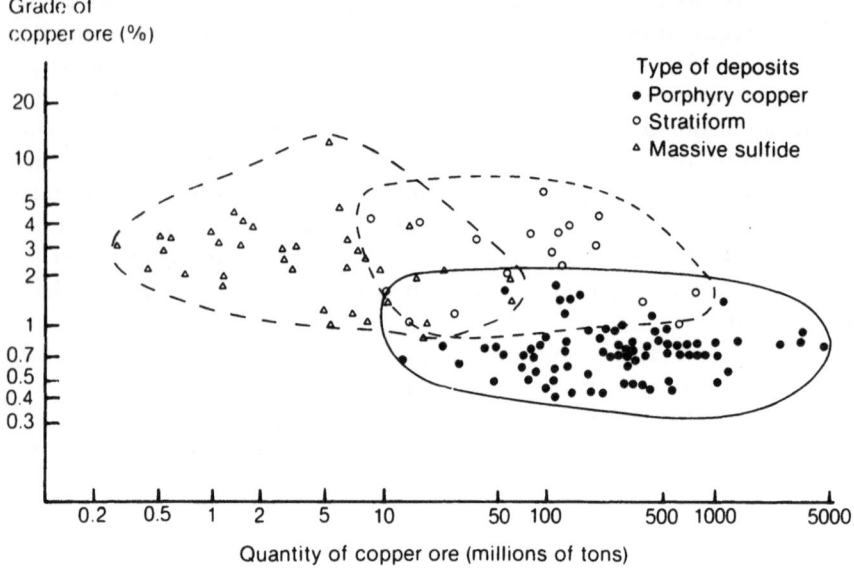

Figure 10-5. Tonnage and grade of major types of copper deposits in the world. *Source*: Adapted from Committee on Mineral Resources and the Environment, Commission on Natural Resources, National Research Council, *Mineral Resources and the Environment* (Washington, D.C., National Academy of Sciences, 1975, fig. 4, p. 167).

Through technological improvement and expansion, the demand for copper was satisfied until about 1950. Very little exploration for copper was carried out in the 1915–45 period, and of the few domestic deposits that were found, only the Bagdad deposit in Arizona was put into production, by an independent company. The large copper consumption during World War II probably accelerated a tightening of supplies. In the early 1950s several companies that had not been major producers (Asarco, Pima Mining, Copper Range, and Magma Copper) responded to increasing demand by discovering and opening new mines, mostly of the porphyry copper type, and the major copper producers greatly stepped up exploration. Government loans and price-support agreements during the Korean War also promoted development. This exploration "boom" for porphyry copper deposits lasted until about 1974 and resulted in discovery of about thirty porphyry copper deposits in the western United States and many others in Canada, Mexico, Panama, Ecuador, Peru, Chile, Argentina, the Philippines, Papua New Guinea, Yugoslavia, Iran, and elsewhere.

This successful period of copper exploration ended in the early 1970s, because the price of copper no longer justified the very high capital expense of constructing the large mining and processing facilities. The halt in exploration reflected an adequate supply of copper available from existing deposits, a slowing in the growth of the world economy, and a prior, extremely successful period of exploration. Exploration for copper is now largely limited to the search for the smaller, but higher-grade, massive sulfide deposits, which tend to contain high values in zinc and precious metals as well as copper.

A major question is whether exploration for copper will recover in the near future, as the economy improves, or will remain at a low ebb. Table 10-1 and the equivalent table in Rose (1982) indicate that nonproducing discoveries contain about 50 million tons of copper. At least 40 million tons remain in producing deposits. Given the consumption of about 2 million tons of copper per year in the United States and questionable growth in consumption, this reserve would last for about forty-five years. These data suggest that another period of little exploration for copper is at hand unless world demand and price increase to the extent that the United States becomes a major exporter of copper. In the meantime, discussions with managers of exploration companies suggest that they are interested only in deposits of markedly higher grade and lower production cost than are associated with the classical porphyry copper.

Similar conclusions about the episodic nature of exploration appear valid for other metals. The discovery of the Climax deposit in Colorado in about 1915 supplied the molybdenum market until about 1957, when the deposit at Questa, New Mexico, was found. As indicated in table 10-1, seven major discoveries of molybdenum were made between 1965 and 1981. The total reserves in these deposits amount to at least 4 million tons of MoS_2 (molybdenum disulfide), which would supply U.S. consumption for forty years, even without the continuing supply of by-product molybdenum from porphyry copper deposits and the remaining reserves at the four producing deposits.

For iron, a major worldwide period of exploration between 1946 and 1960 was so successful that little or no exploration is now being done for this commodity. The technological developments in taconite mining and processing were an important part of this success, as was the recognition of a very large type of deposit, the Superior-type ores of middle Precambrian age. Since 1970 the major iron companies have largely diversified their mineral exploration staffs to ferroalloys.

For gold, a similar pattern of episodic exploration is evident, although in this case reflecting abrupt changes in price plus recognition of a new ore type—large tonnages of "no see-um" gold (as at Carlin, Nevada) that was missed by early prospectors. The United States is currently in the midst of an exploration boom after a period of negligible exploration between 1940 and 1970.

For uranium, exploration from 1940 to 1957 supplied adequate reserves for nuclear weapons. Development of nuclear power plants led to another boom between 1967 and 1980, largely terminated by major high-grade discoveries in Canada and Australia, along with decreased demand. Uranium exploration is now dormant.

The episodic character of exploration for other metals is less clear, but the evidence suggests similar patterns. The episodes appear to be caused by several factors:

1. The very large size of some discoveries or deposit types (Superior-type iron formations, porphyry copper and molybdenum deposits, unconformity uranium deposits), bringing about drastic changes in supply as a result of one or a few discoveries.

2. The successful exploitation of exploration models and consequent discovery of many deposits once a new type of ore is identified. The porphyry copper and porphyry molybdenum models are good examples, but many others have been in use in recent years (Cox, 1983).

3. Increases in demand and price created by new uses, together with recognition of tight supply conditions; decreases in demand and price resulting from obsolescence of uses and excess supply.

4. Technological changes—such as the development of flotation, and heap leaching of low-grade gold—which open up a new class of deposits. In a few cases, technological improvements in exploration, such as airborne magnetic methods for iron ore in the 1950s, have had a major impact.

The episodic patterns of metal exploration contrast greatly with the Gaussian pattern of petroleum discovery proposed by Hubbert (1974). However, the Hubbert curve deals with reserves rather than with exploration for new fields. Even in oil, episodic periods of surplus can be recognized (for example, during the 1930–60 period when the Texas Railroad Commission limited production to avoid sharp decreases in price resulting from major Texas discoveries). It appears that oil discovery within the United States is also episodic if the total reserves in supergiant fields are counted in the year of their discovery, as is done here for metals, rather than during development of the fields.

If the stock of reserves declines to the point of inadequacy or perceived inadequacy for supplying consumption and the real metal price increases markedly, the history of exploration suggests that explorationists are stimulated to find new deposits and to search for and find new types of deposits. The discoveries of porphyry copper, porphyry molybdenum, Carlin gold, sandstone and unconformity uranium, and copper-nickel-cobalt-bearing manganese nodules in the deep sea appear to illustrate this pattern. Exploitation of the ideas generated by a new deposit type may lead to a glut of reserves and near cessation of exploration. The price of the commodity may increase initially, but the eventual price is likely to depend on production technology as well as on supply–demand relations.

Sherwood (1984) examined the exploration–production price patterns discussed above and pointed out that the ability to produce mineral commodities nearly always exceeds demand, thus ensuring low prices except during short periods of disruption or increased demand. He suggests that mineral development depends on the ability and willingness of industry to finance additional exploration and development, not on the ability to find deposits.

In the past, many observers of the mineral scene apparently assumed that increasing demand for mineral commodities would lead to constantly increasing levels of mineral exploration. The historical record does not justify this assumption. More likely, the rate of mineral exploration for individual commodities is inherently cyclical, especially if the ore bodies being mined are relatively small, with a grade tens to hundreds of times that present in average rocks, and if their discovery involves a large component of qualitative geologic knowledge and risk capital. The total metallic exploration effort, which is the sum of efforts for individual commodities, may smooth some of the peaks and valleys, but large fluctuations appear to remain. This conclusion appears to have major implications for long-range resource planning, exploration, research, and education.

Conclusions

Very little useful data exist on the expenditures for and the success of exploration in the United States. Companies tend not to release explicit data, and figures collected by the U.S. government are incomplete and inappropriate for evaluation of exploration. Based on incomplete data, it can be said that annual expenditures for metallic mineral exploration in the United States increased by a factor of about 2.5 in constant dollars over the 1955–83 period, despite apparently falling in 1982 and 1983.

The gross value of metallic mineral discoveries in the United States averaged about \$6 billion annually (in 1981 dollars) between 1955 and 1980, but showed no clear trend, although large short-term fluctuations make discerning a trend very difficult. Large oil companies entering metallic mineral exploration appear to have been less successful than major metallic and nonmetallic mining companies and small mining and oil companies.

The success ratio (gross value of discoveries/cost of all exploration) is difficult to evaluate for the last ten years or so because of uncertainty as to how many recent discoveries will actually come into production. The ratio appears to decrease with time.

Exploration planning in recent years has been organized around specific geologic types of deposits considered to have favorable economic and exploration characteristics. An idealized set of geologic and exploration attributes guides exploration for each deposit type. This approach has been responsible for groups of discoveries over periods of ten years or so and for considerable increases in reserves.

Management and planning styles in exploration groups range from the "strong manager" approach at one extreme to the "team management" approach at the other. There has been a trend toward team management, even though some of the most successful exploration groups are organized around strong managers.

Because of increased uncertainties surrounding future prices and demand, new discoveries now tend to be evaluated more in terms of estimated costs relative to existing producers than in terms of return on investment. There also is a trend toward exploration for smaller, often higher-grade deposits with smaller capital investments and shorter payback periods than for many larger, lower-grade deposits.

Exploration for copper, molybdenum, iron, uranium, and gold has been episodic over the past eighty years or so, with spurts of activity for five to fifteen years followed by periods of little activity. Causes of episodic exploration include discovery of one or a few very large deposits that drastically change the reserves and costs of production, successful exploitation of exploration models that leads to the discovery of many deposits over a short period, increases in demand caused by new uses of metals, decreases in demand caused by obsolescence, and technological change in methods of mineral processing.

References

Anderson, J. A. 1982. "Gold—Its History and Role in the U.S. Economy and the U.S. Exploration Program of Homestake Mining Co.," *Mining Congress Journal* vol. 68, no. 1, pp. 51–58.

Bailly, P. A. 1979. "Managing for Ore Discoveries," *Mining Engineering* vol. 31, pp. 663–671.

Barber, G. A. 1981. "Foreign Exploration—Pros and Cons," *Mining Congress Journal* vol. 67, no. 2, pp. 20–23.

Barber, G. A., and S. Muessig. 1984. "Minerals Exploration Statistics for the Years 1980, 1981, and 1982," *Economic Geology* vol. 79, pp. 1768–1776.

———. 1985. "1983 Minerals Exploration Statistics: United States and Canadian Companies," *Economic Geology* vol. 80, pp. 2060–2066.

———. Forthcoming. "1984 Minerals Exploration Statistics: United States and Canadian Companies," *Economic Geology*.

Bowman, A. B. 1963. "History, Growth and Development of a Small Mining Company," *Mining Engineering* vol. 15, no. 6, pp. 42–49.

Brown, W. K. 1983. "Exploration for Gold: Costs and Results." Preprint 83-142 (New York, Society of Mining Engineers of the American Institute of Mining, Metallurgical, and Petroleum Engineers).

Chenoweth, W. L. 1979. *Industry Exploration Activities*, U.S. Department of Energy Open File Report GJO-108(79) (Washington, D.C., Government Printing Office).

Committee on Mineral Resources and the Environment, Commission on Natural Resources, National Research Council. 1975. *Mineral Resources and the Environment* (Washington, D.C., National Academy of Sciences).

Cook, D. R. 1983. "Exploration or Acquisition: The Options for Acquiring Mineral Deposits," *Mining Congress Journal* vol. 69, no. 21, pp. 10–12.

Cox, D. P., ed. 1983. *U.S. Geological Survey—INGEOMINAS Mineral Resource Assessment of Columbia: Ore Deposit Models*, U.S. Geological Survey Open File Report 83-423 (Washington, D.C., Government Printing Office).

Eckstrand, O. R. 1984. *Canadian Mineral Deposit Types—A Geological Synopsis*, Economic Geology Report 36 (Ottawa, Geological Survey of Canada).

Emerson, M. E., and S. W. Ivosevic. 1984. "Current Approach to U.S. Minerals Exploration," *Mining Engineering* vol. 36, no. 4, pp. 345–349.

Gilmour, P. 1982. "Grades and Tonnages of Porphyry Copper Deposits," pp. 7–35 in S. R. Titley, ed., *Advances in Geology of the Porphyry Copper Deposits* (Tucson, Ariz., University of Arizona Press).

Hubbert, M. K. 1974. "U.S. Energy Resources, A Review of 1972, Pt. 1," in *A National Fuels and Energy Policy Study*. 93rd Cong., 2d sess., Senate Committee on Interior and Insular Affairs, Serial No. 93-40(92-75) (Washington, D.C., Government Printing Office).

Lieberman, M. A. 1976. "United States Uranium Resources—An Analysis of Historical Data," *Science* vol. 192, pp. 431–436.

Menard, H. W., and G. Sharman. 1975. "Scientific Uses of Random Drilling Models," *Science* vol. 190, pp. 337–343.

Ohle, E. L., and R. L. Bates. 1981. "Geology, Geologists and Mineral Exploration," pp. 766–774 in *Economic Geology 75th Anniversary Volume* (El Paso, Tex., Economic Geology Publishing Co.).

Parsons, A. B. 1957. *The Porphyry Coppers in 1956* (New York, American Institute of Mining, Metallurgical, and Petroleum Engineers).

Preston, L. E. 1960. *Exploration for Non-Ferrous Metals* (Washington, D.C., Resources for the Future).

Rose, A. W. 1982. "Mineral Adequacy, Exploration Success and Mineral Policy in the United States," *Journal of Geochemical Exploration* vol. 16, pp. 163–182.

Sanders, L. R. 1981. *Uranium Exploration Expenditures in 1980 and Plans for 1981–1982*, U.S. Department of Energy Open File Report GJO-103(81) (Washington, D.C., Government Printing Office).

Schreiber, H. W., and M. E. Emerson. 1984. "North American Hardrock Gold Deposits: An Analysis of Discovery Costs and Cash Flow Potential," *Engineering and Mining Journal* vol. 185, no. 10, pp. 50–57.

Sherwood, D. H. 1984. "Is There a Future for Mining Lawyers." Manuscript (Denver, Colo., Sherman and Howard).

Snow, G. G., and B. W. Mackenzie. 1981. "The Environment of Exploration: Economic, Organizational and Social Constraints," pp. 871–896 in *Economic Geology 75th Anniversary Volume* (El Paso, Tex., Economic Geology Publishing Co.).

U.S. Bureau of the Census. 1955. *Census of Mineral Industries 1954* (Washington, D.C., Goverment Printing Office).

———. 1968. *Census of Mineral Industries 1967* (Washington, D.C., Government Printing Office).

———. 1973. *Census of Mineral Industries 1972* (Washington, D.C., Government Printing Office).

———. 1978. *Census of Mineral Industries 1977* (Washington, D.C., Government Printing Office).

U.S. Department of Commerce, Office of Business Economics. 1966. *National Income and Product Accounts of the United States, 1929–1945* (Washington, D.C., Government Printing Office).

U.S. Department of Commerce, Bureau of Economic Analysis. 1979. *Business Conditions Digest* (November), p. 84.

———. 1981. *Business Conditions Digest* (February) p. 84.

———. 1983. *Survey of Current Business* vol. 63, no. 7, p. 80.

U.S. Department of Energy, Energy Information Administration. 1985. *Uranium Industry Annual 1984* (Washington, D.C., Government Printing Office).

Wargo, John C. 1973. "Trends in Corporate Mineral Exploration Expenditures 1968–71," *Mining Engineering* vol. 25, no. 5, pp. 43–44.

Wilson, John C. 1982. "Assessment of Metallic Mineral Resource Potential," pp. 93–113 in Committee on Mineral Resource Evaluation, National Research Council, *Assessing Mineral and Energy Resource Potential* (Washington, D.C., National Academy Press).

11

Economic Productivity of Base Metal Exploration in Australia and Canada

BRIAN MACKENZIE
ROY WOODALL

The economic health of both Australia and Canada depends to a significant extent on the mining sector of the respective economies. The viability of the mining sector, in turn, depends to a large degree on the success of mineral exploration. And, while luck may be a short-term determinant of exploration success, the importance of the mining sector in each of these countries is too great for its long-term survival to be left to chance. Thus the importance of a greater awareness of the economic nature of mineral exploration among exploration geologists, mining company executives, and government officials responsible for mineral policy. With better understanding of exploration investment, decisions affecting the mining sector can be made rationally. Knowledge of the economic productivity of mineral exploration—the subject of this chapter—is one aspect of that understanding.

The mining cycle proceeds from exploration through development to production. Investment starts at the exploration stage, when there is not only a long period of cash outflow but a high risk of total loss through failure to make an economic discovery. For investment in mineral exploration to be economically attractive, the expected value at the very beginning of the mining cycle should be positive. Also, the rate of return on the investment should be substantially greater than the current interest rate to preserve the value of the money invested and to reward the investor for the high risk being taken. This risk can be assessed as the probability of failure to make an economic discovery or, alternatively, as the level of exploration expenditure required to achieve an acceptable confidence level of success.

Historically, Australia and Canada have displayed similar patterns of mineral exploration and development. These overall patterns suggest that the economics of base metal exploration in the two countries may also be similar. The purpose of this chapter is to examine that premise and to explain, in the light of historical evidence, similarities and differences in base metal exploration in the two countries.

The chapter compares the economic productivity of base metal exploration in Australia during the period 1955 through 1978 with a similar assessment made in Canada for the period 1946 through 1977. The comparison is based on historical experience evaluated in the context of current conditions. Mineral exploration productivity is measured by the relationship between real exploration expenditures and the value of economic deposits discovered as a result of that exploration.[1] Changes in the economic productivity of exploration over time reflect the effects of depletion, which include changes in the endowment characteristics of discoveries, and advances in geologic concepts and exploration technology.

The approach applied in this chapter is based on the authors' confidence in the value of providing, insofar as is practically possible, a factual economic basis for mineral exploration; of assessing exploration performance in terms of conventional measures of economic value; and of studying the past as a practical guide to planning for the future.

Methodology

The Mineral Supply Process

The mineral supply process consists of a series of activities by which minerals are converted from unknown geologic resources to marketable commodities. As portrayed in figure 11-1 and described below, the major phases of activity in the process are (1) exploration, (2) development, and (3) production. The basic stimuli of this process are the physical occurrence of mineral deposits in nature and the demand for mineral commodities in the economy. A positive response from both exploration geologists and market researchers to particular geologic and market factors combine to guide the selection of environments for exploration.

[1]The term *productivity* encompasses a wide variety of empirical relationships between the inputs and outputs of productive activities. *Economic productivity*, as used in this chapter, measures the efficiency with which all inputs, including human, financial, and technological resources, are applied in mineral exploration to discover and delineate economic mineral deposits. Thus, the definition adopted here differs from that commonly used in the popular press, which concerns labor productivity, or the relationship between the human resource input and production, as expressed by indices such as output per person employed or metric tons per manshift. For a more detailed discussion of the concept of productivity, see Fabricant (1968).

Mineral exploration, the first phase of the mineral supply process, is a sequential, information-gathering process consisting of two stages: primary exploration and delineation. During primary exploration, potentially favorable areas of land are selected within an environment of interest and then are subjected to a series of geological, geophysical, and geochemical tests. Successful primary exploration results in the discovery of one or

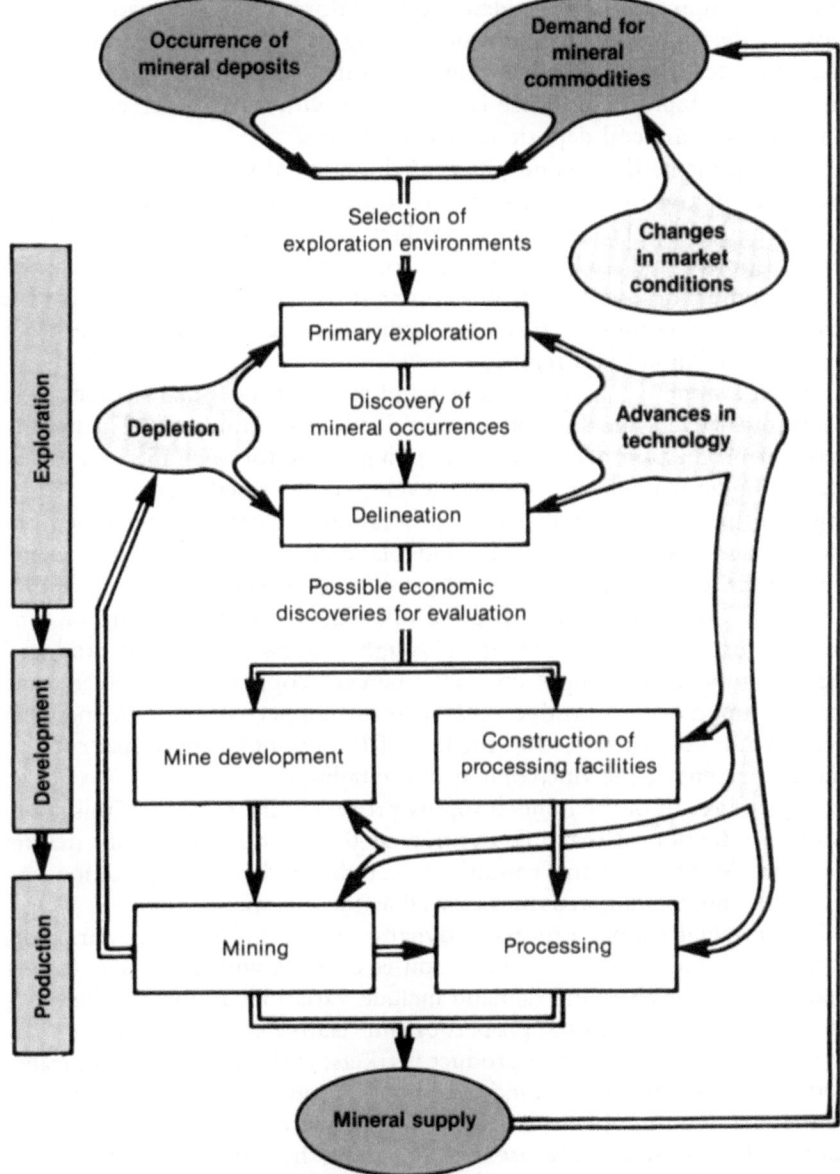

Figure 11-1.   The mineral supply process.

more mineral occurrences, although at this stage the ultimate size and value of any occurrence is unknown.

The mineral occurrences discovered provide justification for delineation, the second stage of exploration, during which information is provided for estimating the size, grade, and physical characteristics of the discovery that are relevant to mineral or metal recovery. The objective of delineation is the identification of possible economic discoveries for evaluation. When the delineation stage has yielded enough information on a particular deposit, a decision is made about developing the deposit to production. If the characteristics of the delineated deposit warrant mine development, the mineral exploration phase is completed with what is perceived to be an economic mineral deposit as its end result.

Development, the second phase of the mineral supply process, establishes productive mining and mineral processing capacity. Since processing is required to upgrade the mine product to a metal concentrate that can be transported and sold, the construction of processing facilities is carried out in conjunction with mine development. A mill may be installed at the mine site, or a common processing facility may be used to treat ores from a number of mines in a region.

When a mine has been developed and related processing facilities have been constructed, the third phase—production—commences. The mining stage of production may include stripping waste for open pits, preparing stopes (underground excavations), developing ore reserves, drilling, blasting, transporting materials to the processing facilities, filling mined-out stopes, and associated technical and planning services. For base metal operations, the processing stage of production usually includes crushing, grinding, flotation, drying, disposal of tailings, and the loading of concentrate products for shipment. Through processing, the ore produced from a mine which might contain 2 percent copper and 5 percent zinc may be upgraded to produce a 25 percent copper concentrate and a 55 percent zinc concentrate, with the loss of 10 percent of the copper content and 20 percent of the zinc content in the tailings.

The end result of the mineral supply process in the context of this study is the production of metal concentrate at the mine site to satisfy market demand. Market demand conditions include product transportation and smelting and refining activities as well as the sale of the metal.

The mineral supply process is dynamic in several respects. First, the market demand for mineral commodities changes with time. The factors that account for changing demand include variations in the requirements of end uses; changes in the properties and relative costs of substitute materials; development of new product markets; and modifications in transportation, smelting, and refining conditions. Second, the physical exhaustion of mineral deposits, or depletion, is inherent in their exploitation. Thus, continual exploration is required just to sustain the existing level of mineral production. Furthermore, exploration, guided by geologic concepts and skills, is a systematic process in the long term, tending to detect first those deposits that are largest, of highest grade, closest to the surface, or closest to markets. Consequently, such deposits will, on average, be discovered, developed, and exhausted first, and deposits that are of lower quality, smaller, or harder to find come later. Thus, depletion causes the cost of mineral supply to rise over time. However, a third, offsetting dynamic force—advances in technology—is also at work. Such advances may include more efficient and extensive exploration techniques and improved mining and mineral-processing methods. Advances in technology act to reduce the cost of mineral supply. The combined effects of these market, depletion, and technological forces determine the economics of mineral supply over time.

Exploration Costs, Risks, and Returns

Mineral exploration cannot be justified as an end in itself. Thus, the economic productivity of exploration is determined by the costs, risks, and returns of the three-phase mineral supply process described above.

Economic mineral deposits are central to the mineral supply process. Consequently, the economic productivity of exploration can be measured by the relationship between the exploration expenditures required to find and delineate an economic deposit and the net return associated with its subsequent development and production.

In this chapter, the economic productivity of exploration is assessed on a before-tax, potential-value basis. All direct costs and revenues through the exploration, development, and production phases of mineral supply are included. Such an assessment represents the value of exploration to society.[2] The procedure for assessing the potential value of mineral exploration is described below, and involves a number of economic considerations associated with expected-value criteria and risk assessment.

Expected-Value Criteria

Expected value measures the average value that exploration yields in the long term, when the successes and failures associated with a very large (theoretically infinite) number of discoveries are considered. The expected value of exploration is derived from the time distribution of average cash flows for the discovery of an economic mineral deposit, as portrayed in figure 11-2. Because of time-value considerations, the average cash flow characteristics must be brought to a common point in time or spread

---

[2]This chapter does not deal with mining taxation and royalties and how they affect the productivity of exploration from the viewpoint of the firm. While this would more fully address the question of investment incentive to a mining company deciding on new or continuing exploration programs, the procedure used reflects the productivity of exploration from the viewpoint of society as a whole.

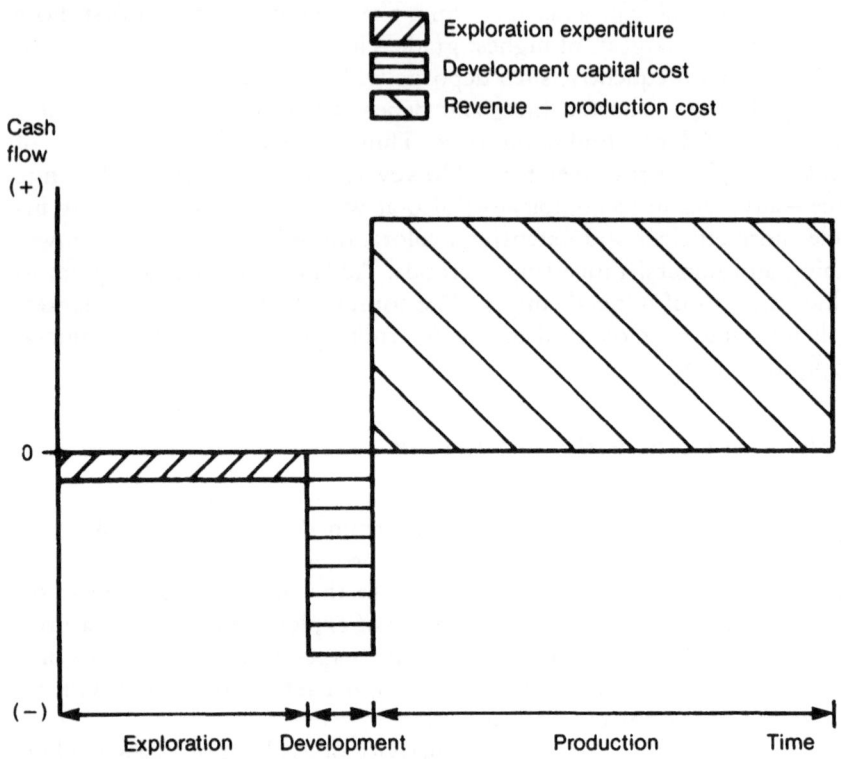

Figure 11-2. **Time distribution of average cash flows for an economic mineral deposit.**

evenly over a common period of time to make a valid assessment. The following two measures of expected value are applied in this study.

1. Expected Value per Economic Discovery

An exploration expenditure E is required on average for the discovery and delineation of an economic deposit, yielding an average net return R. Thus:

$$EV = R - E$$

where EV = expected net value per economic discovery, R = average net return from development and production associated with the discovery of an economic mineral deposit, and E = average exploration cost required to find and delineate an economic deposit.

To make a valid comparison, R and E are expressed as discounted values at the start of the exploration phase, using a cost-of-capital estimate for investment funds as the discount rate.

The higher the EV assessment, the more attractive exploration is perceived to be in the long term. Usually an expected value per economic discovery of greater than zero is regarded as a necessary condition for investment. A negative EV signifies that exploration is an unacceptable investment in the long term, with more money going "into the ground" in discounted terms than coming out.

The expected value per economic discovery is analogous to the criterion of net present value generally used in evaluation practice. As such, this measure reflects the intrinsic economic value of mineral exploration. Thus, EV comprises both size and profitability components.

2. Expected Rate of Return

The expected discounted cash flow rate of return on investment may also be evaluated based on the time distribution of average cash flows for an economic deposit, as depicted in figure 11-2. By definition, *rate of return* is the discount rate that equates the present value of the positive cash flows with the present value of the investment. In economic terms, *rate of return* is the average percentage annual return that exploration is expected to yield over the life of the mineral supply cycle. Using this method, the minimum acceptable condition for long-term exploration investment is an expected rate of return equal to the cost of capital.

Expected values are assessed in this study on the basis of overall exploration activity and reflect the average performance of all organizations that have undertaken exploration in a country or region. Obviously there are benefits to be gained from the application of superior skills and selectivity in exploration. In exploration environments characterized by unacceptable overall expected values, the selection of opportunities of above-average merit is a necessary condition for long-term exploration success. This can only result from the application of superior exploration skills, developed within a mining company over time through continued exposure to particular environments of interest.

Risk Considerations

Three fundamental types of risk are associated with the realization of exploration expectations. Individually and collectively these risks present challenges to the long-term profit, survival, and growth of organizations active in mineral exploration. They are

1. The risk caused by the sensitivity of the economic productivity of exploration to metal price uncertainties;

2. The risk related to the uncertainty of the return, given an economic discovery, due to geologic variability among economic deposits;

3. The risk associated with the discovery of economic mineral deposits.

The first type of risk is associated with the materials market for mineral commodities. There is typically a high level of uncertainty associated with the forecasting of short-term fluctuations and long-term trends in mineral market prices. Exchange rates are one component of these price uncertainties. The economics of mineral exploration is highly sensitive to prices. Flexibility is required in the exploration planning process to contend with unexpected changes in market conditions, which are inevitable. Among the exploration strategies that may be adopted to address this risk is that of directing exploration toward polymetallic deposits.

The second type of risk is the variability of the return due to geologic factors, given the discovery of an economic mineral deposit. The downside risk and upside potential associated with the variability of geologic characteristics among deposits have important implications for exploration planning. The possibility that any exploration program might lead to a multi-billion-dollar discovery, although extremely unlikely, is no doubt a great motivator in mineral exploration. Such a giant target probably carries weight in exploration decisions beyond its actual numerical contribution to the expected value measures.

The third and most direct risk faced in mineral exploration is the discovery risk: the low probability—typically, a 1 to 2 percent chance—of an economic mineral deposit, given the discovery of a mineral occurrence.[3] The implications of discovery risk for the exploration organization should be assessed. Because this risk is so high, the application of limited corporate funds does not ensure the realization of expected values, and exploration resources may be expended without success. The organizational risk associated with exploration success can be quantified by applying the classical problem of the gambler's ruin, which, in probability theory, concerns a gambler with limited capital who wagers against a house with essentially unlimited resources. The gambler is ruined and the game terminates if at any point his or her capital balance falls to zero. The problem is to determine, for any specified set of conditions, the gambler's probability of ruin.

Applying this theory, the practical implication of discovery risk for a mining company is that there is a large difference between the average exploration cost required to find and delineate an economic deposit and the exploration funds required to ensure success.[4] For example, the exploration funds required for a 90 percent certainty of making at least one economic discovery are 2.3 times the average exploration cost associated with an economic discovery.[5]

---

[3]In distinction to the discovery of an *economic* mineral deposit, the discovery of a mineral occurrence represents a technical or geologic success. For example, it may be defined as indications of mineralization of potentially economic grades across minable widths obtained by drilling.

[4]Let P_1 = probability of discovering at least one economic deposit, A = exploration funds available or required over appropriate planning horizon, and E = average exploration cost required to find and delineate an economic deposit. Then, $P_1 = 1 - e^{-m}$, where $m = \dfrac{A}{E}$. Thus, $A = -E [\ln (1 - P_1)]$.

[5]$A = -E [\ln (1 - .90)] = -E [-2.3] = 2.3 E$.

Assessment of the Potential Value of Exploration

The *potential value of mineral exploration* is defined as the difference between the revenues realized from mineral production and all the costs required to realize that revenue, including an allowance for the cost of capital. Since the cost of capital is deducted, this potential value represents the increase in real wealth that results from investing in exploration rather than in some other economic activity.[6] Thus, potential value reflects both the quality of mineral resource endowment and the economic viability of exploration. It also measures the productive capability of mineral resources and represents what is available for sharing between industry and government before mining taxation considerations.

In this chapter, the expected value and risk characteristics of mineral exploration are evaluated on the basis of historical footprints. Two assumptions are necessary if the results are used in planning for the future. First, deposits yet to be found must resemble, in economic terms, those that have been found to date. Second, the cost of making a future discovery must be similar to the cost in the past.

Thus, assessment of the cash flow characteristics associated with mineral exploration is based on actual experience over a relevant historical time period. This information is then placed in the context of current outlook conditions.

In essence, the methodology (diagrammed in figure 11-3) follows these steps:

1. Total exploration expenditures are estimated for the historical time period of interest.

2. Significant deposits discovered as a result of these expenditures are classified by discovery date and listed for evaluation.

3. The cash flow characteristics of the development and production phases for each of these possible economic discoveries are evaluated on the basis of current outlook conditions.

4. Discoveries that, on evaluation, satisfy minimum conditions of size and profitability are considered to be economic deposits.

5. Cash flow characteristics of the development and production phases of all economic deposits are averaged.

6. Total exploration expenditures, which cannot in general be directly associated with the economic discoveries, are prorated across all economic deposits evaluated.

7. The estimate for the exploration phase is integrated with the average development- and production-phase characteristics to portray the time distribution of average costs and revenues for an economic deposit from the start of exploration to the end of production.

---

[6]The cost of capital is a real cost which affects the new wealth created by mineral exploration.

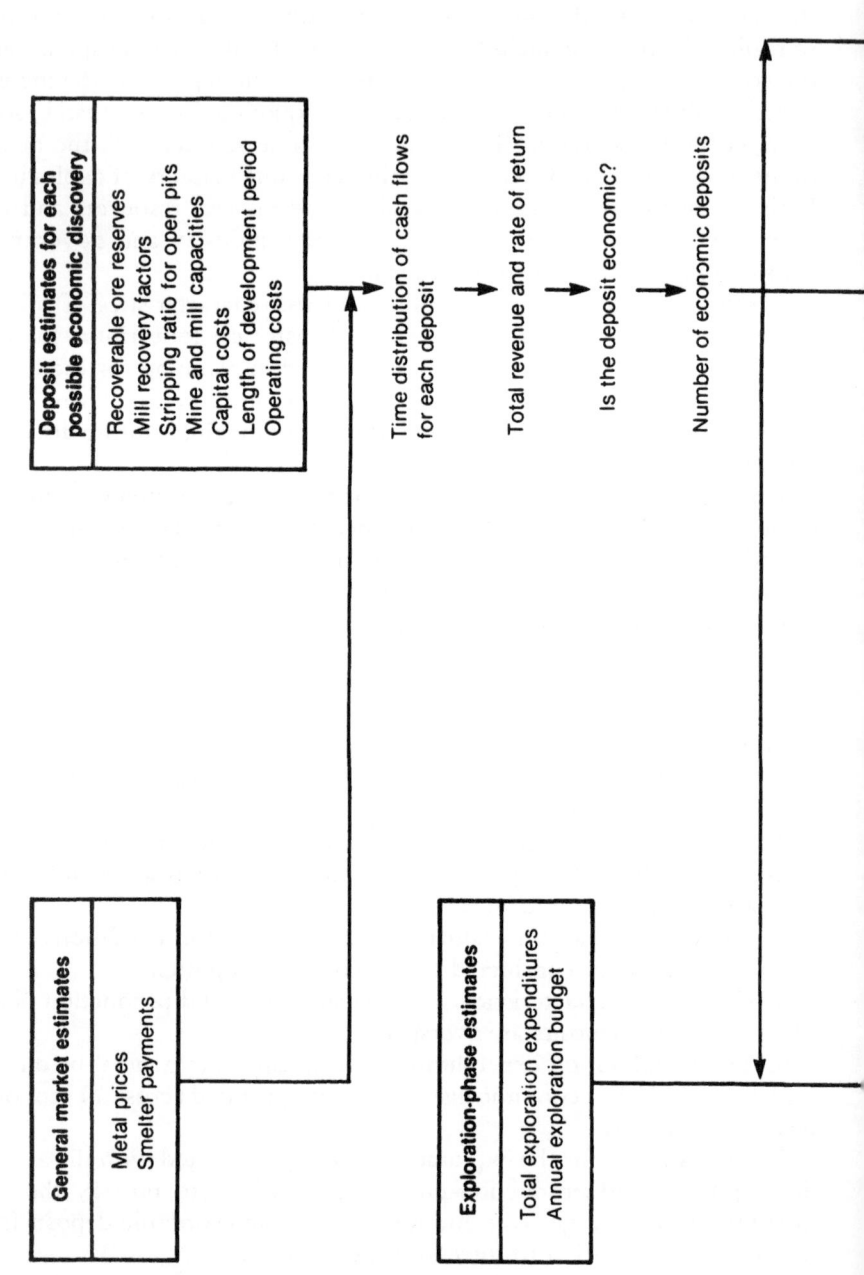

Deposit estimates for each possible economic discovery

Recoverable ore reserves
Mill recovery factors
Stripping ratio for open pits
Mine and mill capacities
Capital costs
Length of development period
Operating costs

Time distribution of cash flows for each deposit

Total revenue and rate of return

Is the deposit economic?

Number of economic deposits

General market estimates

Metal prices
Smelter payments

Exploration-phase estimates

Total exploration expenditures
Annual exploration budget

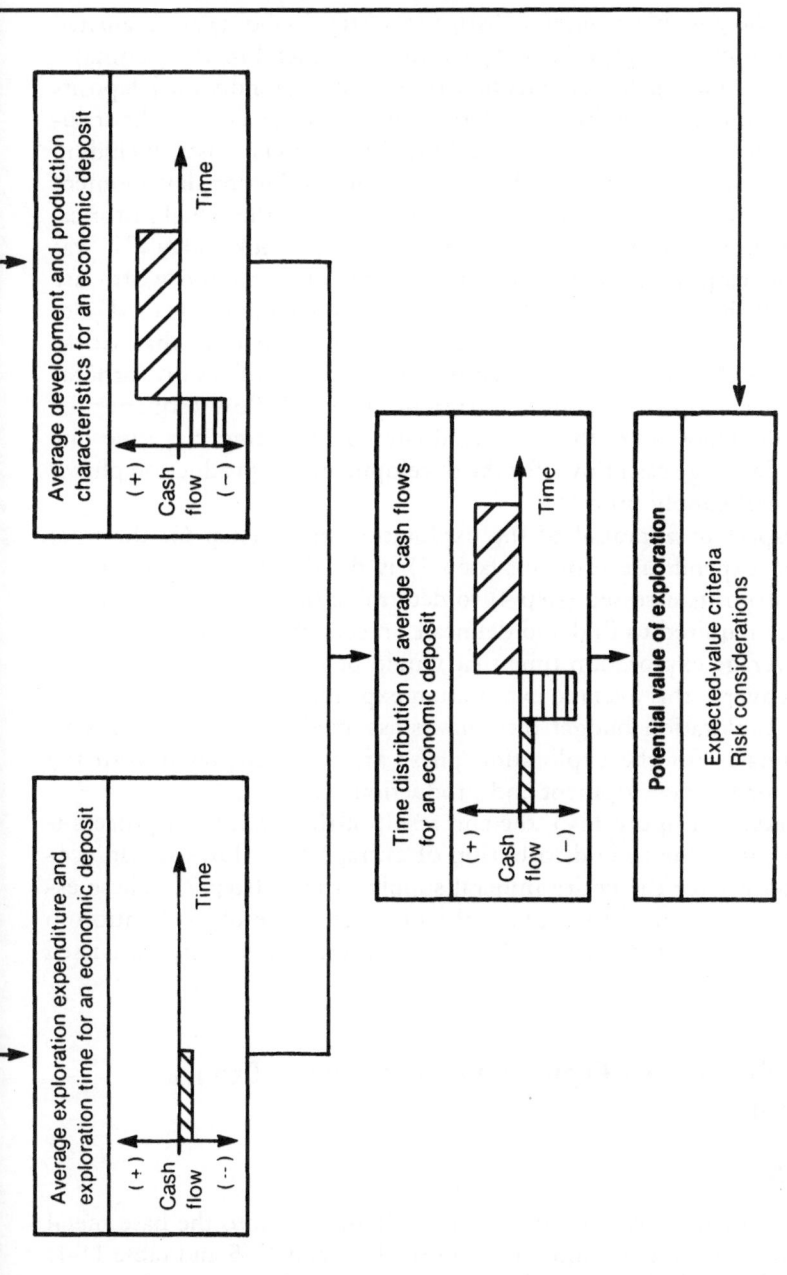

Figure 11-3. Procedure for evaluating the potential value of exploration.

8. Several expected-value criteria and risk measures, defined above, are derived from the assessments. These indicators reflect the economic productivity of mineral exploration.

More specifically, the development- and production-phase characteristics for each of the possible economic discoveries (step 2, above) are evaluated (step 3), as shown in figure 11-3, by combining general market estimates of metal prices and smelter payments with estimates for individual deposits of the following: recoverable ore reserves; mill recovery factors; the stripping ratio for open pits; mine and mill capacities; capital costs, including the working capital requirement; length of the preproduction development period; and operating costs. These estimates, based on the actual historical record, attempt to portray how each deposit would look today if it was awaiting development. A number of measures of economic worth are derived from the resulting cash flow distributions, including the total sales revenue generated and the rate of return. Those discoveries that satisfy minimum conditions for total revenue (size) and rate of return (profitability) are deemed to be economic (step 4). The cash flow estimates for all economic deposits are then averaged (step 5), resulting in a time distribution of average cash flows for the development and production phases of the mineral supply process.

With respect to appraisal of the exploration phase (step 6), the total exploration expenditure estimate (step 1) is divided by the number of economic deposits assessed (step 4) to determine the average exploration expenditure required to find and delineate an economic deposit. To evaluate the average exploration time that would be needed to make an economic discovery, this average exploration expenditure is then divided by an annual exploration budget rate that is assumed to be most efficient. These estimates for the exploration phase are then integrated with the appraisals for the development and production phases.

As portrayed in figure 11-3, the end result of this evaluation process is an assessment of the time distribution of average cash flows for an economic deposit over the entire mineral supply process (step 7). These estimates are then applied to appraise the potential value of exploration in terms of the expected value and risk indicators previously discussed (step 8).

Geologic Setting and Exploration Conditions in Canada and Australia

Geologic Setting

Canada and Australia can be conveniently subdivided into the base metal exploration environments shown in figures 11-4 and 11-5 and table 11-1. The Australian subdivision is by geologic age, whereas the Canadian

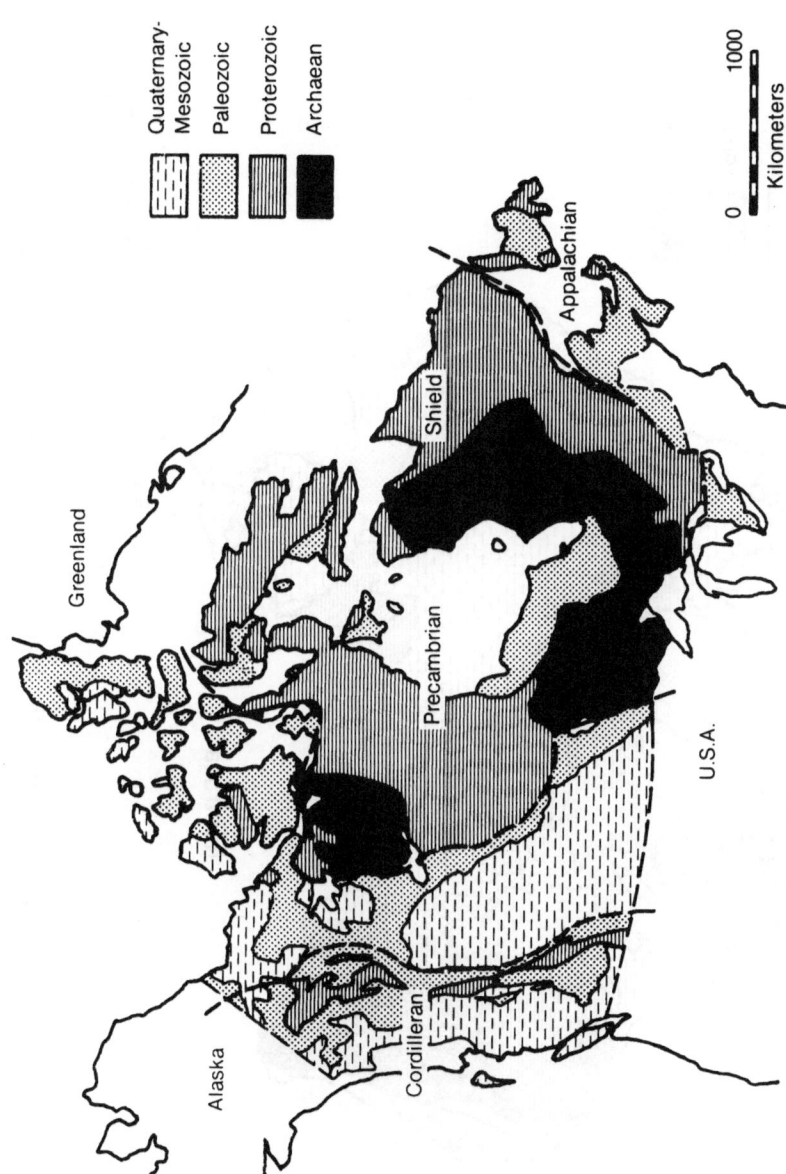

Figure 11-4. Geologic map of Canada. *Source:* Western Mining Corporation, Adelaide, Australia.

376

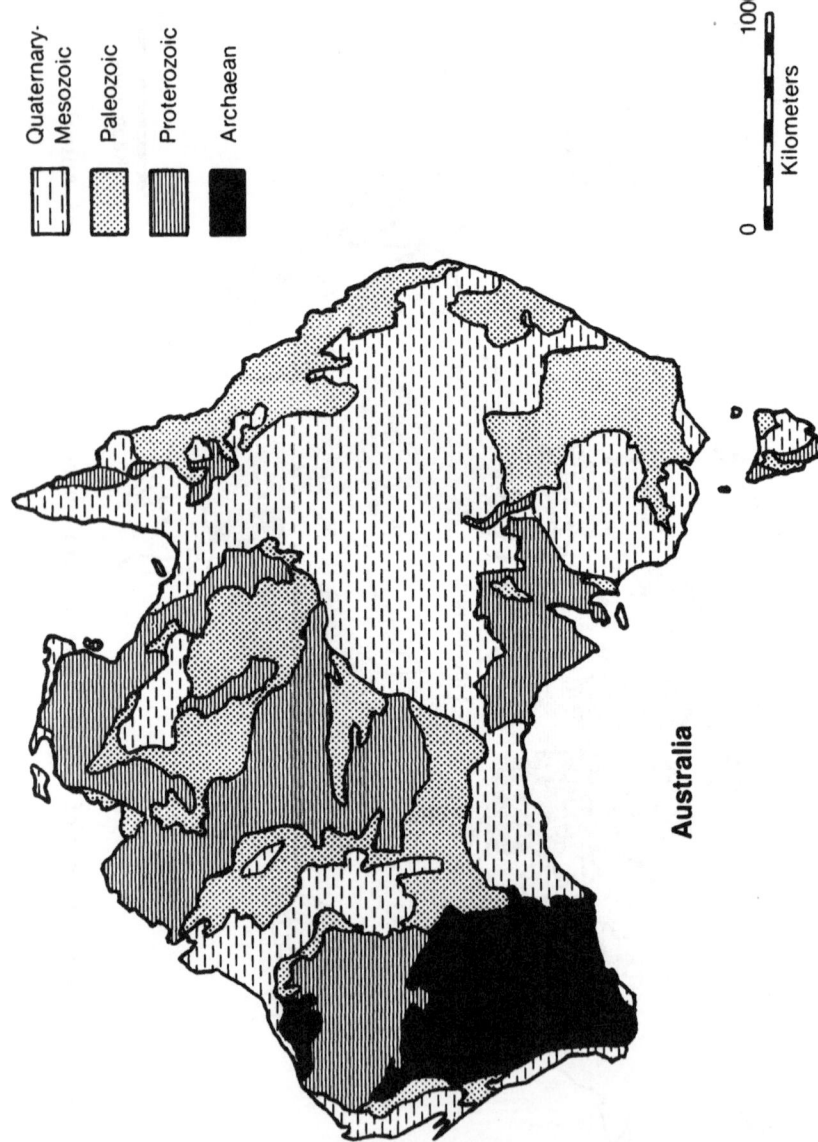

Figure 11-5. Geologic map of Australia. *Source:* Western Mining Corporation, Adelaide, Australia.

Table 11-1. Base Metal Exploration Environments of Australia and Canada

Age (millions of years)	Australian subdivision		Canadian subdivision	
2,500	Archean	Precambrian	Shield	Cordilleran
	Proterozoic			
570	Paleozoic		Appalachian	
225				
65				

breakdown is primarily by regions, which are then classified by age. Thus, the subdivisions of exploration environment used in the two countries are only partially comparable, reflecting the independent origins of the work that provides the basis for the comparison made in this chapter. Also, exploration expenditure data for Canada are only available by region.

The geologic evolution and epochs of mineralization in Canada and Australia were similar for more than 3 billion years. Prior to 2.5 billion years B.P. (before the present), belts of volcanic rocks and sediments derived from these rocks accumulated in areas now recognized as Archean provinces. Base metal deposits of copper and zinc, some containing gold, silver, and lead, formed at this time in the Shield region of central Canada and in Western Australia.

The Proterozoic era, which followed the Archean, extended from 2.5 billion to 570 million years B.P. During the early part of this era the earth's atmosphere became oxygenated, and sediments similar to those of the present formed, along with volcanic rocks. Base metal deposits of lead and zinc formed in western Canada, while deposits of lead-zinc, lead-zinc-copper, and copper-gold formed in Australia. During the Paleozoic era (570 million to 225 million years B.P.), deposits of lead, zinc, and copper formed both in the Cordilleran and Appalachian provinces of Canada (figure 11-4) and in rocks of the same age in eastern Australia (figure 11-5). Base metal mineralization continued in the sedimentary–volcanic environments of the Cordilleran province of Canada during the Mesozoic era (225 million to 65 million years B.P.), when important copper and copper-molybdenum deposits were formed.

Exploration Conditions

Mineral exploration is strongly influenced by surface and near-surface conditions. The deposits that outcrop were the first to be discovered by the relatively unsophisticated exploration of the early prospectors. But both Canada and Australia have extensive areas of concealed bedrock where the discovery of mineral deposits had to await the advent of drilling and more scientific prospecting. Use of these techniques began vigorously in Canada about 1946 and in Australia during the 1950s.

The type of bedrock concealment is very markedly different in each country. In Canada, it is largely in the form of extensive glacial debris and freshwater lakes beneath which the bedrock is essentially free of the effects of weathering. Concealed, sulfide-bearing base metal deposits can therefore exist close to the surface beneath this cover. In Australia, in contrast, extensive areas are covered with detritus and zones of highly weathered rock. In such areas the base metal sulfides are destroyed near the surface and to depths often in excess of 100 meters and now lie beneath a near-surface zone that is difficult to penetrate with electrical geophysical prospecting techniques. The black and shaded areas in figure 11-6 show the extent of this weathered zone of concealment on the Australian continent.

These different surface and near-surface conditions have had a marked influence on the nature, cost, and timing of effective scientific exploration. In Canada relatively rapid and cheap airborne electromagnetic prospecting methods became available in the late 1940s and early 1950s; they were effective because they could penetrate the glacial debris over large areas. In Australia these techniques were largely ineffective. The near-surface debris and weathered zones in Australia have a high electrical conductivity; consequently, effective geophysical mineral prospecting with electrical systems was delayed until more sophisticated systems of geophysical exploration evolved in the 1960s and 1970s. Even today much of Australia is difficult and costly to explore with electrical geophysical methods, and geochemical exploration is also difficult in these areas.

Evaluating the Cost of and Return to Exploration

The information required to assess the economic productivity of base metal exploration in Australia is derived from work completed for Western Mining Corporation (Mackenzie and Bilodeau, 1984). The Canadian part of the analysis uses a generally comparable data base and methodology that was developed at the Centre for Resource Studies at Queen's University in Kingston, Ontario. (For the most up-to-date description of the Canadian data base and methodology, see Mackenzie and Bilodeau [1982, chaps. 2 and 3] and Mackenzie [1985].)

The overall assessments of base metal exploration presented in this chapter are based on the twenty-four-year period 1955 through 1978 for Australia and on the thirty-two-year period 1946 through 1977 for Canada.[7]

_____

[7]The initial year of each period is the first for which statistics on exploration expenditures are available for the respective countries. As of this writing, the Canadian data base does not extend beyond 1977.

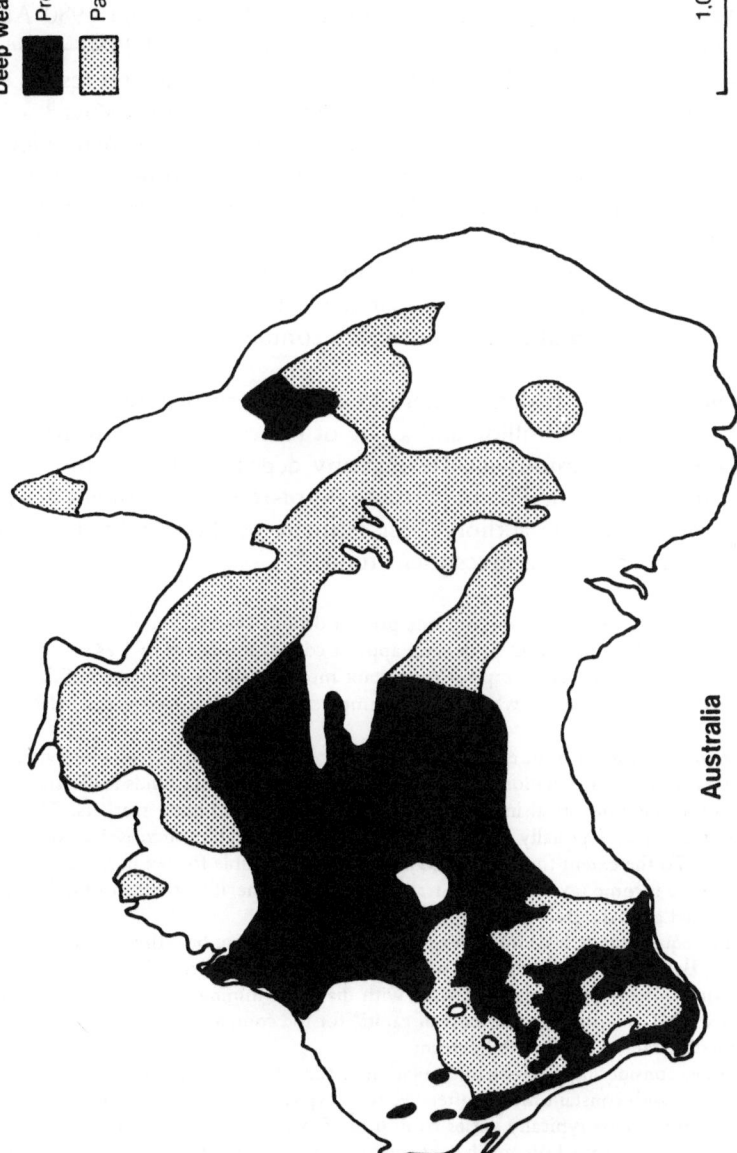

Figure 11-6. Weathered zones of concealment in Australia. *Source:* Western Mining Corporation, Adelaide, Australia.

While the 1955–77 period for Canada could have been used to match the
Australian time frame more closely, the longer period was chosen for two
reasons: (1) it seemed desirable to include more information in the comparison rather than less, and (2) the time periods selected correspond to
the modern era of exploration for concealed deposits, which began in
Canada after World War II and in Australia during the 1950s.

All money values in this chapter are expressed in constant 1980 Australian dollars (A$). The exchange rate prevailing in mid-1980—1.3412
Canadian dollars (C$) per Australian dollar—is used to convert results for
base metal exploration in Canada to Australian dollar equivalents.[8]

The measure of deposit size used throughout this study is total revenue
over the life of the mine (i.e., net smelter return price times recoverable
metal content valued at the mine site), which provides a valid economic
measure of deposit size as distinct from profitability. (The size of a deposit
can be measured in a number of ways. Ore reserve tonnage and recoverable
metal content are two indicators that neglect grades and prices. Such
criteria as cash flow and net present value combine size and profitability
considerations.)

An *economic deposit* is defined here as a discovery that realizes a total
revenue of at least $20 million and a rate of return of at least 10 percent.
The $20-million minimum ensures that any deposit of significance is in-
cluded in the evaluation. The 10 percent rate-of-return investment margin
has been selected as the authors' best estimate of the weighted average
cost of capital for mineral investment funds.[9]

[8]Most of the other chapters in this volume present data in 1982 U.S. dollars. To convert
1980 Australian dollars to 1982 U.S. dollars, apply a conversion factor of 1.24.

[9]The weighted average cost of capital in constant money terms is comprised of the costs
of individual sources of funds, which can be simply grouped into two types—debt and
equity.

The cost of debt capital is an explicit interest cost, nominated in current money terms,
that must be corrected for inflation. The constant money cost of debt funds is illustrated by
the 3 percent average historical interest rate that has prevailed in bond markets. This real
cost varies with time, especially when the general rate of inflation increases or decreases
unexpectedly. To the extent that interest payments are deductible for tax purposes, as they
are for corporate income tax in both Australia and Canada, the after-tax cost of debt funds
must be reduced accordingly.

The cost of equity funds is an implicit opportunity cost that can be estimated empirically.
For example, Ball (1981) evaluated a cost of equity capital for Northern Mining Corporation,
an Australian company primarily concerned with diamond-mining opportunities. He estimated a most-likely constant money cost of capital for the company's equity of 6.8 percent
within a possible range of 5.2 to 9.9 percent.

In summary, considering inflation and taxation effects, debt capital represents a low-cost
source of funds, with constant money after-tax costs typically in the range of 1 to 3 percent.
The cost of equity funds typically varies from 6 to 15 percent. The financial structure of a
mining company—basically how much debt and how much equity it utilizes—is then applied
to cost estimates for the individual sources to assess the weighted average cost of capital.
The 10 percent weighted average cost of capital assumed in this study should be viewed
within this framework.

Total Exploration Expenditures

Estimates of total exploration expenditures are required for each exploration environment and time interval of interest. The intention is to include all costs associated with searching for and delineating the possible economic discoveries of interest including, of course, expenditures for unsuccessful exploration. Total exploration expenditures for each environment and time interval of interest are difficult to estimate, mainly because government statistics are too aggregated to be directly applicable.

Total base metal exploration expenditures in Australia were estimated by

1. Deriving a series of exploration expenditures for the category *other metallic minerals on nonproducing leases* from published government statistics by year and by state and territory;

2. Estimating the percentage of these exploration expenditures spent within each exploration environment by state and territory and by year. (See table 11-1 for the classification of Australian base metal exploration environments.)

Appropriate adjustments are made to bring the exploration expenditures to a calendar-year basis and to convert current dollar statistics to constant 1980 Australian dollars.

Estimates of total base metal exploration expenditures in Canada are derived from annual data from the federal government agency Statistics Canada for metallic mineral exploration expenditures by province. To derive the estimates, it was necessary to deduct from these provincial totals the exploration expenditures for metallic minerals excluded from this study, particularly nickel, uranium, and gold. The estimates for provincial base metal expenditures were then grouped within the Appalachian, Shield, and Cordilleran exploration environments. (For a description of the method of estimation and sources of data used for base metal exploration expenditures in Canada, see Mackenzie and Bilodeau [1979], app. II.)

Total base metal exploration expenditures in Australia and Canada are summarized in table 11-2 by exploration environment and by time interval. Exploration expenditures total approximately A$1 billion in Australia and A$2 billion in Canada for the respective study periods.

Annual Exploration Budget Rate

Based on the authors' discussions with exploration personnel and observations of both successful and unsuccessful exploration at work, a $2.5-million annual exploration budget was chosen as an efficient level for an exploration organization. This size of budget is considered large enough to afford the best possible exploration skills and techniques and small enough to encourage a high degree of motivation and organizational ef-

Table 11-2.   Total Base Metal Exploration Expenditures in Australia (1955–78) and Canada (1946–77), by Exploration Environment and Time Period
(millions of constant 1980 Australian dollars)

Category	Australia			Canada		
	Environment	Years	Expenditure	Environment	Years	Expenditure
By exploration	Paleozoic		404.6	Appalachian		267.0
environment	Precambrian		596.5	Shield		955.4
	Archean		(184.0)	Cordilleran		726.4
	Proterozoic		(412.5)	Other		51.7
Total			1,001.1			2,000.5
By time period					1946–54	187.8
		1955–62	80.4		1955–62	519.8
		1963–70	339.3		1963–70	764.0
		1971–78	581.4		1971–77	528.9
Total			1,001.1			2,000.5

Note: Figures in parentheses = breakdown of Precambrian.
Source: B. W. Mackenzie and M. L. Bilodeau, *Comparison of the Economics of Base Metal Exploration in Australia and Canada.* Unpublished report prepared for Western Mining Corporation, Adelaide, Australia, 1983.

ficiency. Assessments are also carried out in this chapter for annual exploration budgets of $1.5 million and $4.0 million. Total exploration expenditures are not changed by these budget rate variants. All that is affected is the estimated average time required for an exploration organization to discover an economic deposit.

While opinions on the subject of budget size vary and the $2.5-million figure may be criticized, there is little doubt that the most efficient budget lies within the range of $1.5 million to $4.0 million per year. A group of five to eight geologists—"hunting band" size—with an annual budget in this range appears to be the most efficient and effective, retaining the characteristics of an entrepreneurial organization while avoiding the weaknesses of a bureaucracy. Bailly (1977) analyzed the relationship between exploration expenditures and the number of economic discoveries as a function of exploration budget size, considering metallic mineral exploration in Canada during the 1958–73 period. His results, converted here to 1980 Australian dollars, show that the $1.2-million to $6.3-million budget range is dramatically more efficient than larger or smaller sizes: mining companies with annual budgets above $6.3 million were found to account for 44 percent of all exploration expenditures but realized only 15 percent of the economic deposits discovered; the corresponding percentages for annual budget levels below $1.2 million are 29 and 10 percent, whereas these percentages at budget levels of $2.5 million to $6.3 million are 14 and 35 percent, respectively, and for $1.2-million to $2.5-million budgets they are 13 and 40 percent.

If its exploration budget is too small, the exploration organization cannot mount a modern, efficient, and competitive program. At the other extreme, companies with larger budgets typically develop organizations where procedures become formalized and bureaucratic styles develop. The inefficiencies of organizations with the larger budgets are clearly shown in Bailly's results.

A dilemma arises when a mining company wishes to commit large amounts to exploration yet remain efficient. This problem can be resolved by dividing exploration activity into a number of decentralized, semiautonomous units that are able to maintain the favorable organization characteristics with competitive budget and staff size. These units should be given broad guidelines by a central exploration group that also provides specialized, high-quality support, and the financial ability to sustain exploration activities.

Possible Economic Discoveries for Evaluation

In Australia, 77 base metal discoveries that could possibly satisfy the definition of economic deposit were selected for evaluation, and in Canada, 275 deposits were selected. In addition to the deposits discovered during the respective time periods studied for each country, discoveries preceding the first year of those periods are also included. The deposits are listed in appendixes 11-A and 11-B of this chapter for Australia and Canada, respectively. They are classified by exploration environment and by discovery date.

A discovery is considered to have been made when the deposit that ultimately forms the basis for mine development was first encountered by drilling. Deposits are treated individually if their discovery required essentially independent primary exploration programs. The intention is to include for both countries all known deposits of at least 200,000 metric tons that offer medium-term potential for development.

Table 11-3 shows the distribution of possible economic discoveries in Australia and Canada by exploration environment and by discovery date. Of the 77 base metal deposits in Australia, 39 were discovered during the 1955–78 period. In Canada, 210 of the 275 possible economic deposits were discovered during the 1946–77 period. While discoveries made prior to the start of the respective study periods cannot be included in the overall analysis of exploration productivity presented here, they do help provide a more complete picture of mineral endowment characteristics and the time trend of returns from economic discoveries.

General Market Estimates

General market estimates are made to evaluate the time distribution of revenues for each of the possible economic discoveries. Metal price is both

Table 11-3.   Number of Australian and Canadian Possible Economic
Deposits for Evaluation, by Exploration Environment and Time Period

Category	Australia			Canada		
	Environment	Years	Number	Environment	Years	Number
By exploration	Paleozoic		34	Appalachian		39
environment	Precambrian		43	Shield		139
	Archean		(7)	Cordilleran		88
	Proterozoic		(36)	Other		9
Total			77			275
By time period					Pre-1946	64
		Pre-1955	38		1946–54	45
		1955–62	7		1955–62	56
		1963–70	15		1963–70	63
		1971–78	17		1971–77	47
Total			77			275

Note: Figures in parentheses = breakdown of Precambrian.
Source: B. W. Mackenzie and M. L. Bilodeau, Comparison of the Economics of Base Metal Exploration in Australia and Canada. Unpublished report prepared for Western Mining Corporation, Adelaide, Australia, 1983.

the most important and the most uncertain variable in the assessment of cash flows; therefore, cash flows are evaluated as a function of price. An expected or mean price outlook is bounded by upper- and lower-limit prices for each mineral commodity. (The prices used are shown in table 11-4.) These are long-term average price estimates, the period of which is intended to correspond to the time frame of the three-phase mineral supply process being evaluated.[10]

The assessment of costs and revenues assumes that metal concentrate is produced at each possible mine. General net smelter return conditions for copper, lead, and zinc concentrates are used to convert the forecast metal prices to a revenue estimate for the sale of concentrate at the smelter.[11] Concentrate transportation charges are then deducted to work the smelter payment back to a mine site value.

[10]The basic U.S. dollar price estimates have been combined with exchange rate estimates to give the widest possible domestic price range. The correlation between metal prices and the exchange rate has not been considered. It would have the effect of increasing the U.S.$/ A$ and U.S.$/C$ exchange rates as a function of metal prices.

[11]The net smelter conditions for copper, lead, and zinc concentrates are based on the worldwide survey and analysis of approximately 200 actual smelter contracts by Lewis and Streets (1978). Based on discussion with marketing personnel, a number of modifications have been made to adapt the general results of Lewis and Streets (1978) to more specific Australian and Canadian conditions.

Table 11-4. Long-term Price Variants for Mineral Commodities
(1980 U.S. dollars)

Metal	Price basis	Prices		
		Lower limit	Expected	Upper limit
Bismuth	pound	2.60	3.15	7.35
Copper	pound	0.90	1.00	1.15
Gold	ounce	350	400	600
Lead	pound	0.26	0.37	0.42
Molybdenum	pound			
	concentrate	4.50	6.00	10.00
Silver	ounce	10.00	13.00	21.00
Uranium	pound U_3O_8	28.00	32.00	45.00
Zinc	pound	0.26	0.39	0.47
Exchange rates	U.S.$/A$	1.20	1.14	1.08
	U.S.$/C$	1.00	0.85	0.75

Source: B. W. Mackenzie and M. L. Bilodeau, *Comparison of the Economics of Base Metal Exploration in Australia and Canada*. Unpublished report prepared for Western Mining Corporation, Adelaide, Australia, 1983.

Estimates for Individual Deposits

Based on current economic and technological conditions, estimates for the following were made for each of the possible economic discoveries in Australia and Canada: recoverable ore reserves, ore capacity, preproduction period, preproduction capital cost, the cost of major modifications, sustaining capital, operating cost, metallurgical recoveries, concentrate grades, and concentrate transportation. These deposit estimates are based on actual historical data when available. Otherwise, generalized Australian and Canadian costing relationships are applied, reflecting an order-of-magnitude level of accuracy.

With respect to processing facilities, where more than one deposit of the same type exists in a region, deposits are evaluated on the basis of the use of a common mill facility. The overall regional mill capacity is specified, as is the distance between the deposit and the mill. The deposit is allocated a proportion of the mill capital cost based on relative capacities. The distance is used to estimate a road-haulage transportation cost for the ore. Additional road construction may be specified.

Table 11-5 presents average base metal deposit statistics in Australia and Canada by exploration environment. These figures provide typical deposit profiles.

Although Australian base metal deposits have somewhat smaller ore reserves on average than those in Canada, they have higher capital and operating cost requirements, higher grades, and total revenues that are

Table 11-5. Average Development Statistics for Base Metal Deposits in Australia and Canada, by Exploration Environment

Country and environment	Possible economic discoveries (number)	Recoverable ore reserves (millions of metric tons)	Annual mill capacity (thousands of metric tons)	Productive mine life (years)	Preproduction capital cost (millions of 1980 A$)	Unit operating cost ($/metric ton)	Unit revenue ($/metric ton)	Total revenue (millions of 1980 A$)
Australia	77	26	1,130	12	68	34	59	1,260
Paleozoic	34	14	830	12	54	33	54	547
Precambrian	43	36	1,361	12	78	35	64	1,829
Archean	(7)	7	517	11	55	29	39	231
Proterozoic	(36)	41	1,498	12	82	36	68	2,088
Canada	275	32	1,405	13	47	29	41	593
Appalachian	39	16	698	13	34	30	50	497
Shield	139	7	352	11	27	36	49	363
Cordilleran	88	83	3,458	16	85	19	26	1,020

Notes: Revenue estimates are based on expected metal prices and evaluated at the mine site. Figures in parentheses = breakdown of Precambrian. Environmental breakdowns for Canada do not include "other" regions.

Source: B. W. Mackenzie and M. L. Bilodeau, *Comparison of the Economics of Base Metal Exploration in Australia and Canada.* Unpublished report prepared for Western Mining Corporation, Adelaide, Australia, 1983.

typically twice as high as those of base metal deposits in Canada.[12] These distinctions between the two countries may result from differences in basic mineral endowment characteristics, the difficulty of exploration as reflected by the character of mineral deposits found to date, and economic conditions. The marked variations in revealed endowment characteristics among environments within Australia and Canada have important implications for exploration planning.

Critique of the Data Base

There are inherent limitations in the data base assembled for this study. First, possible economic deposits are classified on the basis of their discovery dates. However, the uncertain time lag between primary exploration expenditures and discovery means that exploration expenditures included in the assessment of cost characteristics for one time interval may result in discoveries being included in the following time interval. This is particularly important if it results in the exclusion of possible economic deposits from the overall time period being assessed. For this reason, several discoveries made shortly after the end of the respective Australian and Canadian study periods have been included in the evaluation.

Second, the completeness and reliability of data on individual deposits are a function of discovery date; that is, the data base for older mines is more comprehensive than that available for more recent discoveries. This is especially important with respect to estimates of ore reserves, since development and production typically proceed on the basis of partial delineation results. The objective was to include as complete an estimate as possible for the tonnage and grade characteristics of each deposit. Nevertheless, there is a tendency to undervalue cash flows, particularly for mines discovered most recently. This bias is most significant with respect to tonnage. The problem can be eased by periodic updating and improvement of the data base.

The importance of these two limitations in the data base to overall results depends on the length of the study period and the proximity of the end of the period to the present. The twenty-four-year and thirty-two-year time periods chosen for this study in Australia and Canada, respectively, are considered to be of sufficient duration to minimize the effects of these limitations. Furthermore, information has been assembled several years after the end of each period.

With respect to the estimates of exploration expenditures, there is a question as to what extent "overhead" costs are included in the totals.

[12]It may be noted in table 11-5 that the combination of average statistics does not give the average statistics of the combination. Thus, average recoverable ore reserves times average unit revenue does not equal average total revenue. Similarly, average recoverable ore reserves divided by average annual mill capacity does not give the average productive mine life. These inequalities are caused by correlations that exist between the variables being combined.

The instructions provided by the Australian and Canadian governments in their exploration questionnaires are poorly defined in this regard. In more recent years, the intention in Australia has been to include expenditures for other activities indirectly attributable to exploration. In Canada, the intention has been to have administrative, overhead, and head office expenditures separately reported and excluded from total exploration expenditures. Nevertheless, company practices appear to vary widely in allocating the overhead component of their exploration expenditures and reporting it on government questionnaires. Thus, there is considerable scope for improving the completeness and consistency of the uncertain estimates of exploration expenditures applied in this study.

It can be argued that the use of historical exploration expenditures results in overestimating the exploration costs that would be required under today's conditions. With the increasing effectiveness of exploration, as one argument goes, a lower level of exploration expenditure would be required now to find the deposits discovered in the past. However, depletion effects as well as advances in exploration technology act over time, and, thus, the costs of finding minerals may increase or decrease with time depending on both of these dynamic forces. This study considers these effects by examining time trends within the two major periods of study. Otherwise it assumes that the cost of making a future discovery will be similar to the cost in the past.

It is also argued that exploration expenditures are overestimated because they are a function of commodity prices and the perceived incentive to invest—that is, if there is less incentive to invest in exploration now than there was in the past, exploration expenditures will be correspondingly lower. While this is no doubt true, this study does not attempt to predict *future* exploration expenditure levels, but instead assesses the exploration expenditure required to achieve a particular result—the discovery and delineation of an economic mineral deposit.

The historical exploration expenditures estimated for Australia from 1955 through 1978 and for Canada from 1946 through 1977 resulted in the discovery of thousands of interesting mineral occurrences—and of 39 Australian and 210 Canadian deposits that qualify as possible economic discoveries. The number of discoveries that are counted as economic depends on actual, perceived, or assumed economic conditions. Would as many of the possible economic discoveries justify as much exploration today as they did historically? That question cannot be answered, first, because it is not known how the perceived investment incentive of today and that of the past compare. However, we believe that differences in incentives would be modest in relation to the high levels of uncertainty inherent in exploration. A very high proportion of exploration expenditures is made for unsuccessful programs in the course of searching for possible economic deposits, the delineation of which may be influenced by the perceived incentive to invest. Therefore, we believe that the relationship between exploration expenditures and exploration results is not sensitive to the level of changes in investment incentive that occurred during the respective Australian and Canadian study periods, and that any bias in this regard would not necessarily lead to overestimation of expenditures relative to results.

Other shortcomings in the data base reflect a compromise between the ready availability of data and the research limitations imposed by time constraints. Although there is much room for improvement, we feel that the existing data base is adequate to support meaningful assessments of the economic productivity of exploration for broad government policy and corporate planning purposes.

Economic Productivity of Exploration

Base Case Conditions

The base case conditions to which other results are compared are as follows:

- All possible economic deposits discovered in Australia during the 1955–78 period and in Canada during the 1946–77 period;
- Expected metal price conditions;
- Minimum acceptable total revenue of $20 million for an economic deposit;
- Minimum acceptable rate of return of 10 percent;
- Annual exploration budget of $2.5 million;
- Utilization of regional milling opportunities.

Under these conditions, 12 of the 39 base metal discoveries in Australia and 100 of the 210 base metal discoveries in Canada are found to be economic. The cost and revenue characteristics of these economic deposits are used to assess the time distributions of average cash flows for an economic deposit, which are presented in figure 11-7. Average cash flow characteristics from these distributions are compared in table 11-6.[13]

Based on the cash flow results shown in figure 11-7, further assessments of the expected value of exploration have been made and are summarized in table 11-7. These base case results indicate that, while base metal exploration in Australia has been uneconomic overall, a decidedly positive economic return has been realized from base metal exploration in Canada.

---

[13]With reference to the time period for the production phase, the results do not mean that the average economic deposit produces for forty-three years in Australia and forty-eight years in Canada. To correctly account for time value in the assessment, the cash flows of the economic deposits in production each year are divided by (or averaged across) the total number of economic deposits. The average cash flow for an economic deposit generally declines over the production phase because the number of economic deposits remaining in production declines. Thus, forty-three years and forty-eight years are the productive lives of the longest-lived economic deposits in Australia and Canada, respectively. In fact, the average productive lives for these economic deposits are similar to the average lives of twelve years in Australia and thirteen years in Canada shown in table 11-5 for all possible economic discoveries.

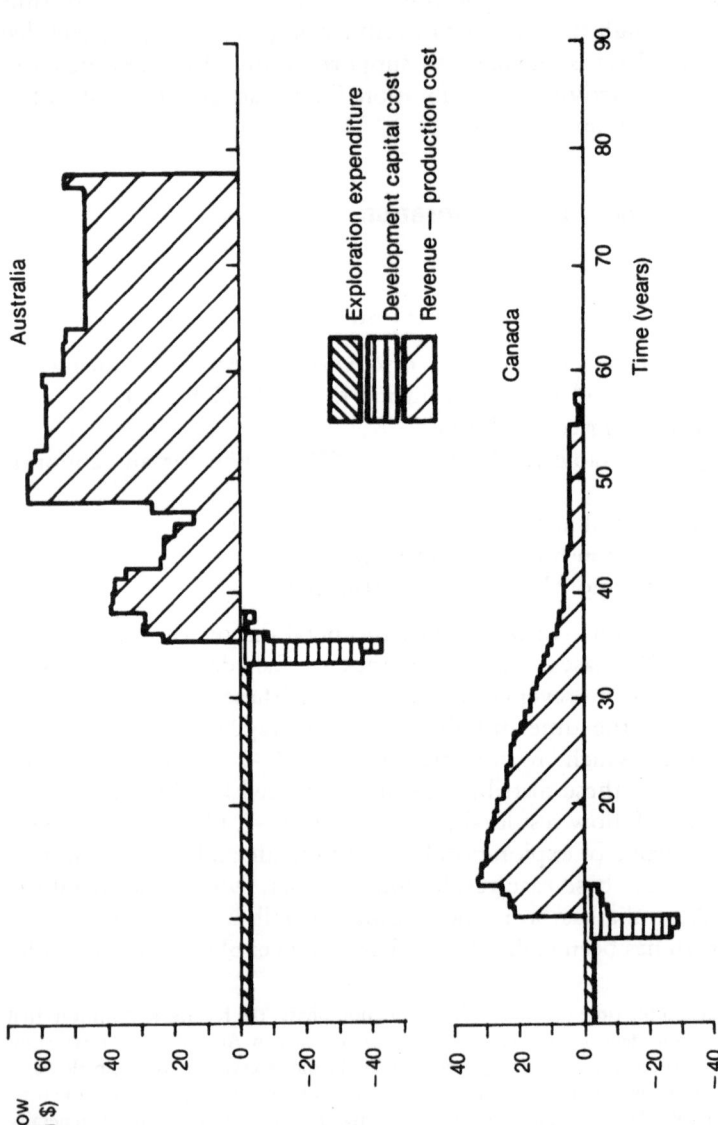

Figure 11-7. Time distributions of average cash flows for an economic deposit: base case conditions.
Source: Redrawn from B. W. Mackenzie and M. L. Bilodeau, *Comparison of the Economics of Base Metal Exploration in Australia and Canada.* Unpublished report prepared for Western Mining Corporation, Adelaide, Australia, 1983.

Table 11-6. Average Cash Flow Characteristics for an Economic Base Metal Deposit Under Base Case Conditions

Characteristic	Australia	Canada
Time period (years)		
Exploration phase	33	8
Development phase (maximum)	5	5
Production phase (maximum)	45	48
Undiscounted values (millions of constant 1980 A$)		
Revenue	3,431	1,137
Exploration expenditure	83	20
Development capital cost	97	71
Production cost	1,471	460
Cash flow	1,780	586

These base case results, as well as the results presented below for exploration environments and other variants, represent the average performance of all mining companies that have explored in Australia and Canada. There is no doubt a high degree of variability in the performance of individual companies, which is masked within the average results. For example, in Australia the exploration expenditures of Western Mining Corporation (WMC) represent about 4 percent of the total included in this study. That investment resulted in the discovery of 10 percent of the possible economic deposits made by all companies. WMC's exploration performance yields a 20 percent expected rate of return as compared with the 8 percent return evaluated above for all explorers. Thus, even in marginal or uneconomic exploration environments an acceptable economic payoff can be achieved by successful exploration organizations.

It is evident from the results presented in table 11-7 that there is a large difference in the economic productivity of base metal exploration of the two countries. The level of base metal exploration expenditures in Canada over the study period is twice that of Australia, but the resulting number of economic discoveries has been more than eight times greater. In other

Table 11-7. Assessments of Expected Value of Base Metal Exploration in Australia and Canada Under Base Case Conditions
(in millions of constant 1980 Australian dollars)

Category	Australia	Canada
Total exploration expenditures	1,001	2,001
Economic discoveries (number)	12	100
Average exploration expenditure (E)	24	13
Average return (R)	9	64
Expected value (EV)	− 15	51
Expected rate of return (percent)	8	20

words, these historical footprints suggest that an economic base metal deposit in Australia costs four times as much and takes four times as long to find and delineate as one in Canada. Given the presence of comparable geologic expertise, exploration technology, and organizational efficiency in the two countries, the results suggest either that base metal exploration in Australia is much more difficult than in Canada or that fewer base metal deposits actually occur in Australia. The results are certainly influenced by the Canadian advantage of lower capital and operating costs. Whereas almost half of the Canadian discoveries proved to be economic, less than a third of the Australian discoveries met this criterion.[14]

Base metal exploration has been more costly and time consuming in Australia than in Canada mainly because of the more difficult geologic conditions associated with concealed bedrock, but the economic deposits that have been discovered in Australia to date are typically about three times larger in terms of total revenue than Canadian economic deposits. This feature is masked in table 11-7 by the discounting effects associated with the long average exploration period in Australia. However, the superiority of Australian discoveries in this regard can be seen by inspecting the cash flow characteristics for the undiscounted development and production phases in figure 11-7 and table 11-6. To permit further examination of the return characteristics of economic discoveries in Australia and Canada, table 11-8 presents assessments of expected value for the development and production phases only. While the profitability of base metal deposits is somewhat higher in Canada than in Australia, as shown by the results for rate of return, the base metal deposits discovered in Australia tend to be significantly larger, reflected in part by the results for the average return per economic deposit.[15] Does this mean that there is a fundamental difference in the size characteristics of base metal deposits in the two countries, or is it a question of the relative effectiveness of exploration technology— the productivity of geophysical exploration in glaciated Canadian environments as compared to the problem of detecting medium and small

---

[14]Ten actual Australian deposits were evaluated under both Australian and Canadian conditions to examine real differences in the economics of developing mineral deposits in the two countries. In all but one of the cases, unit revenue was found to be marginally higher in Canada due to lower transportation costs for concentrates. In eight cases, preproduction capital costs would be lower in Canada than they are in Australia. For all ten cases, operating costs would be lower in Canada than in Australia. It was found that four of the ten deposits that are uneconomic to develop in Australia under today's conditions would be economic if located in Canada.

[15]The difference between the rates of return for each country in tables 11-7 and 11-8 is a consequence of the exploration time and cost necessary to find and delineate an economic deposit. For example, in Australia, given an economic deposit at the start of development, the average rate of return is 30 percent. However, at the beginning of the mineral supply cycle prior to the start of exploration, the rate of return for the assumed conditions is only 8 percent.

Table 11-8. Assessments of Expected Value of Development and Production Phases in Australia and Canada Under Base Case Conditions

Category	Australia	Canada
Return per economic deposit (millions of constant 1980 A$)	218	138
Rate of return (percent)	30	35

Note: The expected or average return per economic deposit is assessed as a net present value at 10 percent discounted back to the start of development.

deposits beneath the deeply weathered near-surface zone in Australia? These and other questions are addressed in the more detailed analysis of results that follows.

Exploration Environments

The assessments of expected value for the base metal environments of Australia and Canada presented in table 11-9 confirm the overall differences in the economic productivity of base metal exploration in the two countries shown in table 11-7. There is relatively little variability in expected values among exploration environments within Australia and Canada compared to the marked difference between the two countries.

The distribution of the number of economic discoveries in Australia and Canada is more or less in proportion to regional exploration expenditures. Thus, the average exploration expenditure associated with the discovery of an economic deposit is in each case close to the national average.

In Australia, only the Proterozoic base metal environment yields positive expected values for the assumed conditions. The Archean and Paleozoic regions give uneconomic overall results. While results for all three Canadian base metal environments evaluated show favorable expected values, there appear to be important regional differences in the economics of base metal exploration. In particular, the relatively large size of base metal deposits in the Cordilleran environment is reflected in the expected-value results.

Time Trends

Table 11-10 shows time trends in the economic characteristics of base metal exploration in Australia and Canada within the respective study periods. Results reflect the combined effects of advances in exploration

Table 11-9. Assessments of Expected Value for Base Metal Environments in Australia and Canada
(in millions of constant 1980 Australian dollars)

Country and environment	Total exploration expenditures	Economic discoveries (number)	Average exploration expenditure (E)	Average return (R)	Expected value (EV)	Expected rate of return (percent)
Australia						
Paleozoic	405	4	25	4	-21	5
Precambrian	596	8	23	14	-9	9
Archean	(184)	(1)	25	0	-25	-1
Proterozoic	(412)	(7)	22	28	6	11
Canada						
Appalachian	267	13	14	66	52	20
Shield	955	41	15	50	35	18
Cordilleran	726	38	13	81	68	21

Note: Figures in parentheses = breakdown of Precambrian. Environmental breakdowns for Canada do not include "other" regions.

Table 11-10. Time Trends in Assessments of Expected Value for Base Metal Exploration in Australia and Canada
(in millions of constant 1980 Australian dollars)

Country and period	Total exploration expenditures	Economic discoveries (number)	Average exploration expenditures (E)	Average return (R)	Expected value (EV)	Expected rate of return (percent)
Australia						
1955–62	80	2	20	27	7	11
1963–70	339	5	23	4	−19	3
1971–78	581	5	25	5	−20	7
Canada						
1946–54	188	17	9	78	69	25
1955–62	520	29	12	48	36	18
1963–70	764	29	16	90	74	21
1971–77	529	25	14	33	19	14

technology and of depletion.[16] There is evidence in both countries of some deterioration in exploration productivity with time. We would reserve judgment on the significance of these trends, because past experience has shown that there is an inherent bias in the data base that tends to undervalue discoveries made most recently.

The difficulty of base metal exploration in Australia, as discussed above, is reflected in the low rate of discovery. Seven of the twelve Australian economic discoveries since 1955 resulted from advances in exploration geochemistry—in particular, the recognition of the significance of anomalous metal values in leached, ironstone gossans. Airborne magnetic surveys were critical in several cases. These improvements in exploration technology were slow to have an effect before 1963, but since then success has been constant, if only at a very modest level. Electrical geophysics has been instrumental in discovery only since 1970, and then in only one instance. Significant advances in geologic science with respect to ore genesis began in the 1970s and, together with improvements in geochemical and geophysical technology, helped to maintain the discovery rate.

In Canada, the favorable results shown in table 11-10 for exploration expenditures and returns for the 1946–54 period reflect the application of new exploration technology, particularly airborne electromagnetic geophysical methods, on a relatively unexplored endowment base. The deterioration in base metal exploration characteristics evident in the 1955–62 period suggests that depletion effects—with regard to both increasing exploration costs and diminishing returns from the resulting economic discoveries—more than outweighed any advances in technology. Specifically, the poor results probably reflect application of the 1946–54 exploration approach to second- and third-priority targets.

While the average exploration expenditure per economic discovery in Canada increased further during the 1963–70 period, as shown in table 11-10, the average return per economic discovery almost doubled that of the preceding period. The application of conceptual geologic models for such purposes as area selection and the ranking of exploration targets was likely important to the improvement shown in the economics of base metal exploration. Average exploration costs declined for the most recent, 1971–77 period, but this improvement has been outweighed by the notable decline in the average return associated with an economic discovery. The sharp decrease in the expected values assessed may reflect the dominance of depletion forces relative to advances in exploration technology, or it

---

[16]These time trends do not take into account the real changes in mineral market conditions or the improvements in mining and mineral-processing technologies that have taken place during the respective study periods. In this evaluation, these factors are standardized on the basis of current outlook conditions. The intention is to isolate those aspects of the economic productivity of exploration that have resulted from the changing nature of the exploration process itself and from changes in mineral endowment characteristics as revealed through the base metal deposits discovered.

may be caused by the bias in the data base that tends to undervalue recent discoveries.

Metal Price Variants

Assessments of expected value are presented in table 11-11 as a function of lower-limit, expected-value, and upper-limit price conditions. These results indicate the high degree of sensitivity of the economic productivity of exploration to metal price uncertainties. As metal prices increase, there are sharp increases in the number of economic discoveries, in the average return associated with an economic discovery, and in the expected value per economic discovery and rate of return. At the same time, higher metal prices reduce the average exploration expenditure and the average time required to discover and delineate an economic deposit. (This is because more discoveries satisfy the definition of *economic deposit*; it is not due to integral change in the exploration process.) Thus, as indicated earlier, the uncertainty associated with metal price forecasts represents an important risk in exploration planning and government policy analysis. Flexibility is required in plans and policies to accommodate the resolution of price uncertainties with time.

As shown in table 11-11, differences between the Australian and Canadian results are a function of price. The expected values of base metal exploration in the two countries are unattractive and quite similar at lower-limit prices, but the gap widens dramatically as prices are increased. The economics of Canadian base metal exploration is more responsive to price movements because of the relatively short average exploration period in Canada (see table 11-6).

Variants with Respect to the Annual Exploration Budget

An annual exploration budget of $2.5 million—considered to be among the most efficient rates of expenditure, as discussed above—is incorporated in the base case conditions. This assumption has a critical bearing on study results, particularly because under this condition thirty-three years of exploration are required on average in Australia to find and delineate an economic deposit.

Assessments of the economic productivity of exploration have also been made for annual exploration budgets of $1.5 million and $4.0 million. The consequences of these variants in annual exploration budget on the expected-value measures are shown in table 11-12. As the annual exploration budget is increased, a significant improvement in results is apparent in both Australia and Canada. By reducing the average exploration time required for the discovery of an economic deposit, the time distribution of negative cash flows is shortened, and the time distribution of positive cash flows moves toward the present. Thus, although total exploration

Table 11-11. Assessments of Expected Value for Base Metal Exploration in Australia and Canada As a Function of Metal Prices

(in millions of constant 1980 Australian dollars)

Country and price variant	Economic discoveries (number)	Average exploration expenditure (E)	Average return (R)	Expected value (EV)	Expected rate of return (percent)
Australia					
Lower-limit prices	6	25	0	−25	4
Expected prices	12	24	9	−15	8
Upper-limit prices	24	20	57	37	15
Canada					
Lower-limit prices	31	23	9	−14	7
Expected prices	100	13	64	51	20
Upper-limit prices	144	10	135	125	32

Table 11-12. Assessments of Expected Value for Base Metal Exploration in Australia and Canada As a Function of Annual Exploration Budget (in millions of constant 1980 Australian dollars)

Country and budget variant	Average exploration expenditure (E)	Average return (R)	Expected value (EV)	Expected rate of return (percent)
Australia				
$1.5 million	15	1	−14	6
$2.5 million	24	9	−15	8
$4.0 million	34	30	−4	9
Canada				
$1.5 million	11	41	30	17
$2.5 million	13	64	51	20
$4.0 million	15	85	70	22

expenditures and the number of economic deposits remain constant, the expected-value measures are enhanced.

The economics of base metal exploration is more sensitive in Canada than in Australia to changes in the annual exploration budget, and hence the difference in results between the two countries widens as the exploration budget is increased. In Australia the benefits of a higher annual exploration budget tend to be dampened by the relatively long average exploration period. Also, as shown in table 11-8, economic discoveries in Australia are somewhat less profitable on average than are those in Canada.

Return Characteristics of Economic Deposits

Potential-value assessments of average return characteristics for economic deposits are summarized in table 11-13, subdivided by exploration environment. It should be noted that the results presented here make no allowance for exploration time and cost; the evaluation includes only the development and production phases of the mineral supply process. Discoveries made prior to 1955 in Australia and prior to 1946 in Canada are included.

Under the assumed conditions, 28 of the 77 base metal deposits discovered in Australia and 118 of the 275 in Canada are found to be economic, as indicated in table 11-13. The average profitability of economic base metal deposits in Australia, as reflected by the rate of return criterion, is shown to be somewhat higher than in Canada. This comparative result is the reverse of that previously shown for the economic discoveries made during the modern era of exploration for concealed deposits in the two countries (see table 11-8). Once again, however, the economic discoveries

**Table 11-13.   Assessments of Average Return Characteristics for
Economic Base Metal Deposits in Australia and Canada**
(in millions of constant 1980 Australian dollars)

Country and environment	Economic discoveries (number)	Total revenue	Return	Rate of return (percent)
Australia	28	3,054	324	41
Paleozoic	12	994	171	44
Precambrian	16	4,598	439	39
Archean	(1)	265	59	39
Proterozoic	(15)	4,887	465	39
Canada	118	1,206	156	34
Appalachian	17	1,068	131	36
Shield	50	894	144	37
Cordilleran	43	1,767	197	31

Notes: The assessments are based on expected metal prices. Total revenue is evaluated at the mine site. Return is assessed as a net present value at 10 percent discounted back to the start of development. Figures in parentheses = breakdown of Precambrian. Environmental breakdowns for Canada do not include "other" regions.

made in Australia tend to be very much larger,[17] and the question again arises as to whether this marked difference in the size of base metal deposits reflects a fundamental difference in mineral endowment or results from a difference in exploration difficulty and effectiveness.

It should be noted that, as shown in table 11-13, the average return characteristics of base metal discoveries in the Appalachian region of Canada are comparable to those of discoveries made in what are similar Paleozoic rocks of eastern Australia. Furthermore, it is only in the Paleozoic base metal environment of Australia that the weathered zone has been removed (see figures 11-5 and 11-6), so that in these regions in both countries exploration has probably been equally effective. This clue suggests that the overall difference in the size characteristics of base metal deposits in the two countries is the result of more effective exploration in Canada to date. The discovery of small and medium-size base metal deposits under the complex weathered zone that covers most of Australia awaits appropriate advances in exploration technology.

One of the important risks in mineral exploration previously discussed is the variability of the return, given the discovery of an economic mineral deposit. As a consequence of the variability in tonnage, grade, and other geologic characteristics among economic deposits, there is a typically wide range of possible returns distributed around the average return. To illustrate, we have examined the distribution of returns for the 28 economic

---

[17]Deposit size is measured here, as throughout the study, by average total revenue per economic discovery and is also incorporated, along with profitability, in the assessment of average return per economic discovery.

base metal deposits discovered in Australia and for the 118 economic discoveries in Canada around the average returns of $324 million and $156 million, respectively, presented in table 11-13. Key statistics from these distributions are summarized in table 11-14.

The variability of the return from economic discoveries is particularly pronounced in Australia, as illustrated by the 500-times difference in value between the lower-decile and upper-decile values (A$2 million and A$1,072 million, respectively). In Canada there is also a large variability in value, with a more than 60-times difference between the lower- and upper-decile returns (A$5 million and A$329 million, respectively).

These return distributions are also highly skewed. There are many deposits of low economic value and relatively few of very high value. Thus, in Canada, for example, the mean value of economic discoveries (A$156 million) is slightly higher than the upper-quartile value (A$151 million) and almost four times greater than the median or middle value of the distribution (A$44 million). Given the discovery of an economic deposit in Canada, there is more than a 75 percent chance that it will yield a lower-than-average return. On the other hand, there is almost a 25 percent chance that an economic discovery will provide an above-average return. There is even a small possibility of a "billion dollar" discovery. Although this is an extremely rare eventuality, the thought that any exploration program could lead to another discovery of the Kidd Creek variety (that deposit in Ontario, discovered in 1963, is the most important base metal deposit found in Canada to date) is a great motivator in mineral exploration.

With respect to the formulation of mineral policy in government, the important message that table 11-14 conveys is that there is no such thing as a "typical deposit." Thus, flexibility is required in government policies to equitably and efficiently accommodate the wide range of returns inherent in mining.

Figure 11-8 illustrates the average mix of metal products realized from the economic base metal deposits. Thus, there is a natural market hedge

Table 11-14. Variability of the Return from Economic Base Metal Discoveries in Australia and Canada: Key Statistics
(in millions of constant 1980 Australian dollars)

	Return from economic discoveries	
Statistic	Australia	Canada
Upper decile	1,072	329
Upper quartile	424	151
Mean (average)	324	156
Median	62	44
Lower quartile	11	19
Lower decile	2	5

Note: Returns are expressed as net present values at 10 percent discounted back to the start of development.

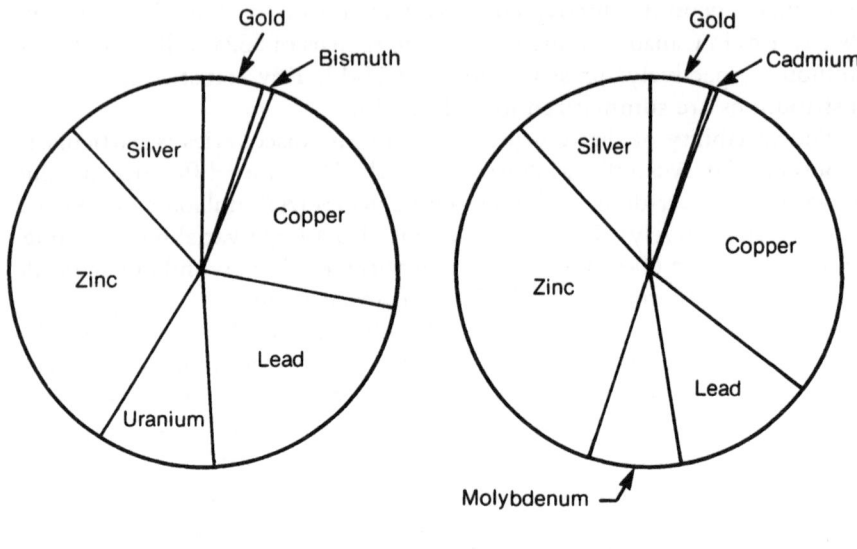

Australia   **Canada**

Figure 11-8.   The mix of metal products from economic deposits in Australia and Canada. *Source:* Redrawn from B. W. Mackenzie and M. L. Bilodeau, *Comparison of the Economics of Base Metal Exploration in Australia and Canada.* Unpublished report prepared for Western Mining Corporation, Adelaide, Australia, 1983.

associated with base metal exploration. Zinc, copper, and lead are relatively evenly balanced constituents in Australian deposits, together accounting for 72 percent of the value of production. In Canada, zinc and copper contribute 64 percent of the total revenue realized from base metal deposits, while lead contributes another 8 percent. Precious metal by-products account for 17 percent of the value of production in both countries. (The metal shares depend, of course, on relative metal prices. The evaluation just given assumes expected-value price conditions.)

The Olympic Dam deposit in Australia and the Kidd Creek deposit in Canada are the most important base metal discoveries made in the two countries during the respective study periods. The effect of these deposits on overall study results has been analyzed by removing them from the assessment of average return characteristics for economic deposits. Comparative results show that the two deposits represent a substantial part of the base metal mineral endowment in their respective countries. The Olympic Dam deposit has a greater impact on the average size of Australian deposits than the Kidd Creek deposit has in Canada. However, their relative effects on the average return of economic deposits are similar. Removal of the two deposits does not have such a significant effect on rate of return, the result of discounting relatively distant future values at high rates.

Endowment and Return Characteristics for Deposits Discovered Before and During the Study Periods

A previous section examined time trends in the economic characteristics of base metal exploration in Australia during the 1955–78 period and in Canada during the 1946–77 period. This section considers the complete time frame of historical experience, focusing only on the endowment and return characteristics of the mineral deposits.

Table 11-15 presents time trends in average endowment characteristics of possible economic deposits. Results are remarkably similar for the 38 base metal deposits discovered in Australia before 1955 and the 39 discoveries made during the 1955–78 period. In Canada the 210 base metal discoveries made during the 1946–77 period are significantly larger, on average, that the 65 deposits discovered before 1946. This difference primarily reflects the impact of large porphyry copper and copper-molybdenum deposits discovered during the study period. (Significant advances in large-scale open pit and mineral processing machinery, positive perceptions of future copper and molybdenum markets, and tax incentives in Canada all encouraged these porphyry developments.)

The time trends in average return characteristics of economic deposits are presented in table 11-16. In Australia the economic discoveries made during the 1955–78 period are typically larger, but less profitable, than those discovered before 1955. Economic deposits discovered in Canada

Table 11-15. Time Trends in Average Mineral Endowment Characteristics of Australian and Canadian Possible Economic Discoveries

Characteristic	Australian discoveries		Canadian discoveries	
	Pre-1955	1955–78	Pre-1946	1946–77
Possible economic discoveries (number)	38	39	65	210
Recoverable ore reserves (millions of metric tons)	24	29	13	38
Annual mill capacity (thousands of metric tons)	1,079	1,172	635	1,642
Productive mine life (years)	12	12	12	13
Development capital cost (millions of constant 1980 A$)	69	67	32	52
Unit operating cost ($/metric ton)	35	34	33	28
Unit revenue ($/metric ton)	61	58	42	41
Total revenue (millions of constant 1980 A$)	1,334	1,193	479	628

Note: Revenue estimates are based on expected metal prices and evaluated at the mine site.

**Table 11-16. Time Trends in Average Return Characteristics of
Australian and Canadian Economic Discoveries**
(in millions of constant 1980 Australian dollars)

Characteristic	Australian discoveries		Canadian discoveries	
	Pre-1955	1955–78	Pre-1946	1946–77
Economic discoveries (number)	16	12	18	100
Total revenue	2,771	3,431	1,468	1,137
Return	404	218	258	138
Rate of return (percent)	43	38	48	31

Notes: Revenue estimates are based on expected metal prices and evaluated at the mine site. Return is
assessed as a net present value at 10 percent discounted back to the start of development.

during the period 1946–77 are smaller and less profitable, on average, than
discoveries made before 1946.[18] In this case the influence of the porphyry
discoveries is more modest, because relatively few of these deposits would
be economic under the current outlook conditions assumed in the evaluation.

Several possible reasons account for the apparent decline in the return
characteristics of economic discoveries. First, this trend may be caused by
depletion; that is, the quality of deposits discovered decreases over time
as a consequence of systematic exploration. Second, advances in geologic
concepts and exploration technology alter the character of discoveries,
perhaps resulting in an increasing proportion of smaller or more-marginal
economic deposits. Finally, changing economic conditions beyond the
realm of exploration have no doubt influenced the type of exploration
targets sought by mining companies, contributing, for example, to the
focus on porphyry deposits in western Canada in the late 1960s. It is likely
that all three of these factors have played a part in determining the trends
shown in table 11-16.

Organizational Consequences of Discovery Risk

There is a very small chance of success each time an investment is made
in primary exploration. The high risk associated with the discovery of an
economic mineral deposit has important consequences for the exploration
organization. Even if the expected value of exploration is positive, funds

[18]Note that the total revenue and return results presented in table 11-16 for economic
discoveries in Australia during the 1955–78 period and in Canada for the 1946–77 period
are the same as presented earlier for base case conditions (see tables 11-6 and 11-8). However,
the results for rate of return differ. Since deposit size is incorporated in the total revenue
and return measures, the indicators are in both cases weighted averages and, thus, identical.
With respect to rate of return, the results shown in table 11-8 are derived from the time
distributions of average cash flows for an economic deposit (the development and production
phase estimates portrayed in figure 11-7) and, therefore, these, too, are weighted averages.
However, the results for rate of return presented in table 11-16 are not weighted for deposit
size; they are determined in each case by summing the rates of return for the individual
economic discoveries and dividing this total by the number of economic deposits.

Table 11-17. Exploration Expenditures Required for the Discovery of an Economic Base Metal Deposit in Australia and Canada (in millions of constant 1980 Australian dollars)

Country and environment	Total exploration expenditures	Economic deposits (number)	Average exploration funds required[a]	Exploration funds required to ensure success[b]
Australia	1,001	12	83	192
Paleozoic	405	4	101	233
Precambrian	596	8	75	171
Archean	(184)	(1)	184	423
Proterozoic	(412)	(7)	59	135
Canada	2,001	100	20	46
Appalachian	267	13	21	47
Shield	955	41	23	54
Cordilleran	726	38	19	44

Notes: Figures in parentheses = breakdown of Precambrian. Environmental breakdowns for Canada do not include "other" regions.

[a]The exploration expenditures required are expressed here on an *undiscounted* basis. Thus, the average exploration funds required for the discovery of an economic deposit are the undiscounted equivalents of the E values presented in tables 11-7 and 11-9 of this chapter.

[b]The exploration funds required for a 90 percent likelihood of making at least one economic discovery are approximately 2.3 times the average expenditure required.

may very well be expended without success. The important practical implication of discovery risk is the difference between the average exploration expenditure required to find and delineate an economic deposit and the exploration funds required to ensure that success is highly likely.

To illustrate the organizational consequences of discovery risk, assessments were made of the exploration funds required for a 90 percent level of confidence in making at least one economic discovery. This evaluation of funding requirements applies to the overall Australian and Canadian experiences in base metal exploration during the 1955–78 and 1946–77 study periods and is made as a function of exploration environments. Results are presented in table 11-17.

For typical Australian conditions, exploration funding of A$192 million would be required to ensure success, while only A$46 million would be necessary in Canada. These funding levels to ensure success compare with average expenditure requirements of A$83 million and A$20 million, respectively.

Conclusion

This chapter examines the economic relationship between real expenditures for base metal exploration and the potential returns from the resulting discoveries. It focuses on the modern era of base metal exploration in Australia and Canada, which began in Australia during the 1950s and in

Canada after World War II. The periods analyzed are 1955 through 1978 and 1946 through 1977 for the two countries, respectively. The purpose of the chapter is to assess and explain similarities and differences between the two countries in a way that is useful for corporate planning and the analysis of government policy. Its main conclusion is that the economic productivity of base metal exploration in Australia is quite different from the Canadian experience.

Overall results for a set of most-likely base case conditions indicate that, while base metal exploration in Australia has been uneconomic, decidedly positive expected values have been realized in Canada. The level of base metal exploration expenditures in Canada over the study period is found to be twice that in Australia, but the resulting number of economic discoveries assessed is more than eight times greater. An economic base metal deposit in Australia costs four times as much and takes four times as long to find and delineate as does one in Canada. On the other hand, the economic base metal deposits discovered in Australia are typically three times larger than those in Canada.

Mineral exploration in Australia and Canada utilizes comparable geologic expertise, exploration technology, and organizational efficiency. Thus, there are two possible explanations for the striking overall differences in exploration. First, there may be a fundamental difference in the base metal endowment characteristics of the two countries, with fewer and larger deposits actually occurring in Australia. Second, base metal exploration in Australia may be much more difficult and less effective than in Canada, reflecting the productivity of geophysical exploration in glaciated Canadian environments as compared to the problem of detecting medium and small deposits beneath the deeply weathered near-surface zone that blankets most of Australia.

More detailed study results show that the return characteristics of base metal discoveries in the Appalachian region of Canada are comparable to those of discoveries made in similar Paleozoic rocks of eastern Australia. Furthermore, it is only in the Paleozoic environment of Australia that the weathered zone has been removed, so that in these regions in both countries base metal exploration has probably been equally effective. This evidence indicates that the overall differences in exploration costs and the size of economic discoveries in the two countries are the result of more effective exploration in Canada to date. It further suggests that an undiscovered population of small and medium-size base metal deposits under the complex weathered zone covering most of Australia awaits discovery by more advanced exploration technology.

Time trends in the economic productivity of base metal exploration reflect the combined effects of advances in technology and depletion effects. While the time trend evidence in both Australia and Canada is inconclusive and incomplete, it does not, in general, support the hypothesis that the economics of mineral exploration has deteriorated with time.

The results of the study indicate that the economic productivity of exploration is highly sensitive to metal prices. Thus, the uncertainty associated with forecasts of metal prices represents an important risk in exploration planning and government policy analysis. Differences between the base metal exploration results in Australia and Canada are shown to be a function of price.

As a consequence of the variability in tonnage, grade, and other geologic characteristics among economic deposits, there is typically a wide range of possible returns distributed around the average return for an economic base metal deposit. For example, assuming expected metal prices, the results for Australia show a 500-times differences in value between the lower-decile and upper-decile economic deposits. Thus, there is really no such thing as a "typical deposit." Flexibility is required in government policies to accommodate the wide range of returns inherent in mining equitably and efficiently.

The high risk associated with the discovery of an economic mineral deposit results in a large difference between the average exploration expenditure required to find and delineate an economic deposit and the exploration funds required to ensure success. For typical Australian conditions, exploration funding of A$192 million would be required to provide a 90 percent certainty of making at least one economic discovery; only A$46 million would be necessary for base metal exploration in Canada. These funding levels to ensure success compare with average exploration expenditure requirements per economic discovery of A$83 million and A$20 million, respectively.

References and Bibliography

Bailly, P. A. 1977. "Changing Rates of Success in Metallic Exploration." Paper presented at Annual Meeting of the Geological Association of Canada, Vancouver, April.

Ball, Ray. 1981. "Northern Mining Corporation NL: Estimate of Discount Rate for Valuing Equity." Unpublished paper. Sydney, Australian Graduate School of Management, August.

Boldy, J. 1977. "(Un)Certain Exploration Facts and Figures," *CIM Bulletin* vol. 70, no. 781, pp. 86–95.

———. 1981. "Prospecting for Deep Volcanogenic Ore," *CIM Bulletin* vol. 74, no. 834, pp. 55–65.

Cranstone, D. A. 1982. "An Analysis of Ore Discovery Cost and Rates of Ore Discovery in Canada over the Period 1946 to 1977" (Ph.D. thesis, Harvard University).

Derry, D. R. 1970. "Exploration Expenditure, Discovery Rate and Methods" *CIM Bulletin* vol. 63, no. 5, pp. 362–366.

Derry, D. R., and J. K. B. Booth. 1978. "Mineral Discoveries and Exploration Expenditure—A Revised Review," *Mining Magazine* (May) pp. 430–433.

Fabricant, S. 1968. "Productivity," pp. 523–536 in *International Encyclopedia of the Social Sciences* vol. 12 (New York, Macmillan and The Free Press).

Harris, D. P., and B. J. Skinner. 1982. "The Assessment of Long-Term Supplies of Minerals," pp. 247–326 in V. K. Smith and J. V. Krutilla, eds., *Explorations in Natural Resource Economics* (Baltimore, Johns Hopkins University Press for Resources for the Future).

Kilburn, L. C. 1980. "Economic Viability of Canadian Massive Sulphide Base Metal Deposits As It Relates to Exploration in the 1980s." Paper presented at Annual Meeting of the Canadian Institute of Mining and Metallurgy, Toronto, April.

King, H. F. 1973. "A Look at Mineral Exploration 1934–1973." Paper presented at Australian Institute of Mining and Metallurgy Conference, Western Australia, May.

Lewis, P. J., and C. G. Streets. 1978. "An Analysis of Base-Metal Smelter Terms," in *Proceedings of the Eleventh Commonwealth Mining and Metallurgical Congress*, Hong Kong, 1978 (London, Institution of Mining and Metallurgy).

Mackenzie, B. W. 1973. "Corporate Exploration Strategies," pp. 1–8 in *Application of Computer Methods in the Mineral Industry* (Johannesburg, South African Institute of Mining and Metallurgy).

———. 1980. *Looking for the Improbable Needle in a Haystack: The Economics of Base Metal Exploration in Canada.* Working Paper No. 19 (Kingston, Ontario, Queen's University, Centre for Resource Studies).

———. 1985. *Geological Aspects of Mining Productivity: Canada's Base Metal Resources.* Technical Paper No. 6 (Kingston, Ontario, Queen's University, Centre for Resource Studies).

———. 1987. "Mineral Exploration Productivity: Focusing to Restore Profitability," in F. D. Anderson, ed., *Selected Readings in Mineral Economics* (Oxford and New York, Pergamon).

Mackenzie, B. W., and M. L. Bilodeau. 1979. *Effects of Taxation on Base Metal Mining in Canada* (Kingston, Ontario, Queen's University, Centre for Resource Studies).

Mackenzie, B. W., and M. L. Bilodeau. 1982. *Economic Effects of Smelter Controls on the Canadian Nonferrous Mining Industry.* Technical Paper No. 2 (Kingston, Ontario, Queen's University, Centre for Resource Studies).

Mackenzie, B. W., and M. L. Bilodeau. 1983. "Comparison of the Economics of Base Metal Exploration in Australia and Canada." Unpublished report prepared for Western Mining Corporation, Adelaide, Australia.

Mackenzie, B. W., and M. L. Bilodeau. 1984. *The Economics of Mineral Exploration in Australia* (Adelaide, Australian Mineral Foundation).

Pemberton, R. H. 1979. "Significant Technical Innovations in Exploration Technology." Paper presented at Annual Meeting of the Canadian Institute of Mining and Metallurgy, Montreal, April.

Preston, L. E. 1960. *Exploration for Non-Ferrous Metals: An Economic Analysis* (Washington, D.C., Resources for the Future).

Snow, G. G., and B. W. Mackenzie. 1981. "The Environment of Exploration: Economic, Organizational and Social Constraints," pp. 871–896 in B. J. Skinner, ed., *Economic Geology: Seventy-Fifth Anniversary Volume* (El Paso, Texas, Economic Geology Publishing Co.).

Woodall, R. 1984. "Success in Mineral Exploration: A Matter of Confidence," *Geoscience Canada* vol. 11, no. 1, pp. 41–46.

———. 1984. "Success in Mineral Exploration: Confidence in Prosperity," *Geoscience Canada* vol. 11, no. 2, pp. 83–90.

———. 1984. "Success in Mineral Exploration: Confidence in Science and Ore Deposit Models," *Geoscience Canada* vol. 11, no. 3, pp. 127–132.

Appendix 11-A
Australia:
Possible Economic
Discoveries for Evaluation

Table 11-A-1. Classification of Australian Possible Economic Discoveries by Exploration Environment and Discovery Date

Discoveries by exploration environment	Discovery Date Pre-1955	1955–62	1963–70	1971–78	Discoveries by exploration environment (continued)	Discovery Date Pre-1955	1955–62	1963–70	1971–78
Paleozoic region									
Beltana			X		Magnet	X			
Burraga	X				Mount Cannindah	X			
Cadia	X				Mount Chalmers	X			
Cape Horn		X			Mount Lyell	X			
Captain's Flat	X				Mount Morgan	X			
Chesney	X				New Cobar	X			
CSA East		X			North Lyell	X			
CSA West	X				Nymagee	X			
Currawong				X	Pillara Spring				X
Elura				X	QTS				X
Farrell	X				Que River				X
Galwadgere		X			Roseberry	X			
Great Cobar	X				Sorby Hills				X
Hercules	X				Thalanga				X
Kanmantoo		X			West Lyell	X			
Liontown		X			Wilga				X
Lyell Comstock	X				Woodlawn			X	

Table 11-A-1 (continued)

Discoveries by exploration environment	Pre-1955	1955–62	1963–70	1971–78	Discoveries by exploration environment (continued)	Pre-1955	1955–62	1963–70	1971–78
Precambrian regions									
Archean environment									
Angelo			X		Teutonic Bore				X
Gossan Hill				X	Whim Creek	X			
Mons Cupri			X		Whundo	X			
Scuddles				X					
Proterozoic environment									
Attutra		X			Lady Annie	X			
Broken Hill	X				Lady Loretta			X	
Brown's		X			Mammoth	X			
Burra	X				Martins			X	
Cattle Grid				X	Moonta	X			
Duchess Copper	X				Mount Elliott	X			
Dugald River	X				Mount Gunson	X			
Fitzpatrick				X	Mount Lyndhurst				X
Gecko			X		Olympic Dam				X
Hilton	X				Orlando		X		
HYC		X			Parabarana				X
Intermediate			X		Peko	X			
ISA Copper	X				Redbank			X	
ISA Silver-Lead-Zinc	X				Wallaroo	X			
Ivanhoe			X		Warrego		X		
Juno			X		Wee MacGregor	X			
Kapunda	X				White's	X			
Koongie Park				X	Woodcutters			X	

Appendix 11-B
Canada:
Possible Economic
Discoveries for Evaluation

Table 11-B-1. Classification of Canadian Possible Economic Discoveries by Exploration Environment and Discovery Date

Discoveries by exploration environment	Pre-1946	1946–54	1955–62	1963–70	1971–77	Discoveries by exploration environment (continued)	Pre-1946	1946–54	1955–62	1963–70	1971–77
Appalachian region											
Brunswick 6		X				Madeleine				X	
Brunswick 12		X				McLean			X		
Buchans	X					Ming				X	
Capelton	X					Murray Brook				X	
Cape Ray					X	Needle Mountain	X				
Caribou			X			Oriental No. 2			X		
Clearwater			X			Rambler Main	X				
Cupra			X			Rothermere			X		
Daniel's Harbour				X		Salmon River				X	
East				X		Solbec				X	
Eustis	X					Stirling	X				
Gay's River					X	Stratmat					X
Gaspé Copper Mountain	X					Suffield	X				
Great Burnt Lake			X			Third Portage Lake				X	
Gullbridge	X					Tilt Cove	X				
Halfmile Lake			X			Walton				X	
Huntington		X				Wedge				X	
Key Anacon		X				Weedon	X				
Little Bay	X					Whalesback				X	
Little River		X									

Table 11-B-1 (continued)

Discoveries by exploration environment	Pre-1946	1946–54	1955–62	1963–70	1971–77	Discoveries by exploration environment (continued)	Pre-1946	1946–54	1955–62	1963–70	1971–77
Shield region											
Agricola Lake					X	Copper-lode A				X	
Aldermac	X					Copper-lode E				X	
Amos	X					Copper-Man	X				
Amulet A	X					Copper Rand			X		
Amulet C	X					Corbet				X	
Amulet F	X					Coronation			X		
Anderson Lake				X		Cuprus	X				
Anglo American			X			Delbridge				X	
Barvallee			X			Detour A1					X
Barvue		X				Detour A2					X
Belfort		X				Detour B					X
Bell Allard			X			Dickstone	X				
Big Nama Creek			X			Dixie					X
Birch Lake		X				Dunraine	X				
Bob	X					East Sullivan	X				
Bouzan			X			East Waite			X		
Bruce	X					Embury Lake					X
Campbell Main	X					Errington	X				
Cedar Bay	X					F Group				X	
Centennial				X		Flexar			X		
Chibougamau Explorers		X				Flin Flon	X				
Chisel Lake			X			Fox				X	
Coldstream		X				Garon Lake				X	
Coniagas		X				Geco			X		
Cooke				X		Ghost Lake				X	
Copper Cliff			X			Golden Manitou	X				
Coppercorp		X				Grandroy			X		

(Continued)

Table 11-B-1 (continued)

Discoveries by exploration environment	Pre-1946	1946–54	1955–62	1963–70	1971–77	Discoveries by exploration environment (continued)	Pre-1946	1946–54	1955–62	1963–70	1971–77

Shield region (continued)

Discoveries by exploration environment	Pre-1946	1946–54	1955–62	1963–70	1971–77	Discoveries by exploration environment (continued)	Pre-1946	1946–54	1955–62	1963–70	1971–77
Grasset Lake					X	Louvem Zinc					X
Gwillam					X	Lyon Lake					X
Hackett			X			Magusi					X
Henderson		X				Mattabi				X	
High Lake		X				Mattagami			X		
Hood River					X	Merrill Island	X				
Hope Lake			X			Millenbach				X	
Horne	X					Munro Copper			X		
Hyers Island	X					Nanisivik			X		
Icon			X			New Calumet	X				
Indian Mountain		X				New Hosco			X		
Izok Lake				X		New Insco					X
Jaculet		X				Norbec			X		
Jameland			X			Norita			X		
Jamieson			X			Normetal	X				
Joliet	X					North Star			X		
Joutel		X				Old Waite	X				
June			X			Orchan			X		
Jungle		X				Osborne Lake			X		
Kam Kotia	X					Pater			X		
Kidd Creek			X			Perry			X		
Kokko Creek		X				Pinebay			X		
Lacorne	X					Poirier			X		
Lemoine					X	Portage			X		
Lessard					X	Preissac	X				
Lost Lake					X	Quebec Chibougamau			X		
Louvem				X		Quemont			X		

Table 11-B-1 (continued)

Discoveries by exploration environment	Pre-1946	1946–54	1955–62	1963–70	1971–77	Discoveries by exploration environment (continued)	Pre-1946	1946–54	1955–62	1963–70	1971–77

Shield region (continued)

Discoveries by exploration environment	Pre-1946	1946–54	1955–62	1963–70	1971–77	Discoveries by exploration environment (continued)	Pre-1946	1946–54	1955–62	1963–70	1971–77
Rail Lake				X		Temagami		X			
Reed Lake				X		Tribag			X		
Riocanex Obalski					X	Turnback	X				
Robataille			X			Vamp Lake	X				
Rod				X		Vauze			X		
Ruttan				X		Vendome		X			
Schist Lake		X				Vermillion	X				
Schumacher Copper			X			Waden Bay		X			
Scott					X	Waite Dufault	X				
Sherridon	X					Westarm					X
South Bay				X		West MacDonald	X				
Springer	X					White Lake				X	
Spruce Point					X	Willecho		X			
Stall			X			Willroy		X			
Sturgeon				X		Wim				X	
Sylvia					X						

Cordilleran region

Discoveries by exploration environment	Pre-1946	1946–54	1955–62	1963–70	1971–77	Discoveries by exploration environment (continued)	Pre-1946	1946–54	1955–62	1963–70	1971–77
Adanac				X		Bethlehem			X		
Afton					X	Big Bull and Tulsequah Chief	X				
Annex				X							
Anniv					X	Bluebell		X			
Bear Twit					X	Blue Grouse		X			
Bell			X			Bonanza	X				
Benson				X		Boss Mountain				X	
Berg				X		Brenda				X	
Best Chance	X					Britannia	X				

(Continued)

Table 11-B-1 (continued)

Discoveries by exploration environment	Pre-1946	1946-54	1955-62	1963-70	1971-77	Discoveries by exploration environment (continued)	Pre-1946	1946-54	1955-62	1963-70	1971-77

Cordilleran region (continued)

Discoveries by exploration environment	Pre-1946	1946-54	1955-62	1963-70	1971-77	Discoveries by exploration environment (continued)	Pre-1946	1946-54	1955-62	1963-70	1971-77
Cariboo Bell				X		Indian Chief	X				
Casino				X		Ingerbelle				X	
Catface			X			Island Copper				X	
Chuchua Creek					X	JA					X
Cirque					X	Jason					X
Copper Mountain	X					Jersey			X		
Cowley Park		X				Keewenaw	X				
Craigmont			X			Kitsault			X		
Davis Keays				X		Kutcho Creek					X
DY					X	Little Chief			X		
Endako			X			Lornex				X	
Faro			X			Lucky Jim	X				
Galore Creek			X			Lynx			X		
Gayna River					X	Maggie			X		
Gem			X			Magnum			X		
Gibraltar			X			Mel					X
Glacier Gulch			X			Mineral King		X			
Goldstream					X	Minto					X
Granduc		X				Morrison				X	
Granisle			X			Mother Lode and Sunset	X				
Grum					X	Myra Falls				X	
Harper Creek				X		Nadaleen River					X
HB		X				OK				X	
Hidden Creek	X					Oro Denoro	X				
Highmont			X			Phoenix	X				
Howard's Pass					X	Red					X
Huckleberry						Red Mountain				X	

Table 11-B-1 (continued)

Discoveries by exploration environment	Pre-1946	1946–54	1955–62	1963–70	1971–77	Discoveries by exploration environment (continued)	Pre-1946	1946–54	1955–62	1963–70	1971–77
	Discovery Date						Discovery Date				

Cordilleran region (continued)

Discoveries by exploration environment	Pre-1946	1946–54	1955–62	1963–70	1971–77	Discoveries	Pre-1946	1946–54	1955–62	1963–70	1971–77
Redstone Copper			X			Sustut					X
Reeves	X					Swim Lake				X	
Robb Lake				X		Tom			X		
Sam Goosly				X		Trojan	X				
Schaft Creek				X		Trout Lake				X	
Silver Giant	X					Valley Copper				X	
Snowshoe	X					Vangorda			X		
Sullivan	X					War Eagle	X				
Sunro			X								

Other regions

Discoveries	Pre-1946	1946–54	1955–62	1963–70	1971–77	Discoveries	Pre-1946	1946–54	1955–62	1963–70	1971–77
Coronet				X		Pine Point 3					X
Eclipse				X		Pine Point 4					X
Great Slave Reef					X	Polaris					X
Pine Point 1		X				Pyramid				X	
Pine Point 2				X							

12

Are Ore Grades Declining?
The Canadian Experience, 1939–89

HENRY L. MARTIN
LO-SUN JEN

Economists claim that it pays to mine high-grade mineral deposits first and lower-grade deposits later. Moreover, it is often assumed that relatively few ore bodies remain to be found that would be of higher quality than those already discovered. According to this line of reasoning, in the course of a nation's mining history the average grades of ore mined will decline as the higher-grade deposits are gradually depleted. The example most commonly cited as evidence of this sequence of events is the drop in the average grade of copper ore mined in the United States—from more than 3.0 percent around 1900 to about 0.5 percent today. It is often inferred from such evidence that countries with a significant history of mining have been forced to exploit lower-grade ores and are at a competitive disadvantage vis-à-vis countries with a shorter history of mining.

The logic of this theory—that resource depletion inevitably leads to exploitation of lower-grade resources—seems unassailable, but the length of time and the rate of exploitation necessary to bring about such change remain unclear. In Canada, the dramatic growth in mining during the past quarter-century has led some people to wonder whether resource depletion might not already have pushed that country past the peak of its golden age of mining. Challenging the idea that Canada may be running out of good ores at this stage in its mining history, this chapter assesses that concern by examining the evidence with regard to the decline of ore grades in Canadian mining for seven important metals: copper, zinc, lead, nickel, molybdenum, silver, and gold.

The next section explains how the assessment is carried out, after which trends in the number of operating mines and their output since

1939 are examined. A metal-by-metal analysis of grade changes since 1939 is then given, and the final section presents the conclusions of the chapter.

Methodology

Covering a half-century of Canadian metal mining, this study examines grades of mined ore of copper, zinc, lead, nickel, molybdenum, silver, and gold at approximately ten-year intervals from 1939 through 1979 and provides estimates for 1989. Instead of 1959, 1960 is used for its more reliable data. Estimates for 1989 are based on expected production from existing mines, deposits committed for production, and likely extensions to existing operations, given their current ore reserves and production capabilities.[1]

Each of the seven metals is mined in Canada not only as the sole or principal product of an ore body but also as a co-product or by-product; indeed, as many as five of the seven metals are being produced from a single ore body. The classification of a metal as a principal product, co-product, or by-product is normally based on that metal's relative value either in the mill feed or in the concentration product. This study uses average 1979 metal prices[2] for calculating a gross value of each metal in each mine's mill feed[3] for each of the years examined. Principal products, co-products, and by-products are then defined on the basis of these values. Specifically, the *principal product* is the metal with the greatest value in mill feed; any metal whose value is more than 50 percent of the value of the principal product is a *co-product*; and any metal whose value is less than 50 percent of the value of the principal product is a *by-product*.

In this chapter the terms *copper mine*, *zinc mine*, and so on, and *copper ore*, *zinc ore*, and so on refer to a mine or to ore containing the indicated metal as the principal product or as a co-product. The terms *copper-bearing ore*, *zinc-bearing ore*, and so on refer to ore containing the indicated metal as the principal product, as a co-product, or as a by-product.

---

[1] These estimates are from the mine-by-mine data and analysis that form the basis of Energy, Mines and Resources Canada (1983). In that annual publication on mines and mining, the Canadian government department Energy, Mines and Resources Canada presents the current perspective on reserves, supply capability, development, and exploration.

[2] The following 1979 prices (in Canadian dollars) are used—copper: C$1.07/lb (C$2,359/metric ton); zinc: C$0.44/lb (C$970/metric ton); lead: C$0.60/lb (C$1,323/metric ton); nickel: C$2.85/lb (C$6,283/metric ton); molybdenum: C$13.40/lb (C$29.50/kg); silver: C$11.90/troy oz (C$382.60/kg); gold: C$355/troy oz (C$11.40/g). After experimentation with current prices and various fixed-year prices, it was decided to use 1979 prices. Use of another price series would affect the classification of some mines as to their principal product, co-product, or by-product. The use of 1979 prices did not lead to a bias favoring the authors' conclusions.

[3] It was assumed that, for each mine, the tonnage and grade of ore concentrated in a year equaled the ore tonnage and grade mined during that year.

Only a few Canadian metal mines are single-metal operations. Since co-products and by-products are vital to the economics of many mining operations, all of the metals of interest in each mining operation need to be considered, and the combination of metals recovered needs to be treated as a whole in assessing changes in the quality of the deposits being mined. The procedure that this study uses to take account of all metals of interest is as follows: it assigns the 1979 average metal prices to all of the metals in ore concentrated at each operation for each of the years analyzed. Then, dividing the total value of production from each ore body by the price of the metal under discussion, it obtains a measure of the total tonnage of that metal contained in the ore, plus its equivalents (that is, the tonnage of the metal whose value just equals that of all the metals found in the ore. For instance, copper equivalents in a copper ore represent other metals present in terms of tonnages of copper of equal value.). This approach allows overall ore quality to be expressed in terms of a single grade that takes account of all associated metals.

Using this procedure, each of the seven metals was examined with respect to the average grade of all ores concentrated in which the metal was the principal product or a co-product. The average grade was assessed both including and excluding the equivalent tonnage of associated metals.

In the following sections on the individual metals, by-product metals are taken into account as equivalents of the principal product. However, in the determination of grades of, say, copper ores, ores with only by-product copper are *not* considered, although such by-product output is included in total copper production.

Ore Mined and Concentrated

As shown in figure 12-1, the ore mined and concentrated in Canada containing the seven metals of interest quadrupled from 1939 to 1979 and is expected to increase even more by 1989. Ore tonnage mined and concentrated rose slowly from 1939 to 1960. The rapid increase after 1960, expected to continue to 1989, is largely due to production from porphyry deposits.[4] In addition to the dramatic increase in porphyry ores as a source of metals, the figure also illustrates the historically waning importance of gold mining. The tonnage of nonporphyry metal ores was abnormally low in 1979, mostly as a result of a prolonged strike that year at Inco Limited's operations at Sudbury, Ontario.

---

[4]Because of their low grade, large size, and near-surface location, porphyry deposits of copper and molybdenum ores receive separate treatment in this chapter. While the technical definition of porphyry deposits is too complex to present here, it should be mentioned that not all large, low-grade, near-surface deposits are necessarily porphyries. All types of ore deposits other than porphyry deposits are grouped here into the category of nonporphyry ores, regardless of their grade, size, origin, or location.

The expansion of ore tonnage mined and concentrated does not simply represent an increase in production at existing mines. Figure 12-2 shows the number of mines (except gold mines) in operation in each of the years analyzed and also indicates the mine openings and closings that occurred between those years. The numbers of mines opened and closed are from one sample year to the next; they do not include openings and closings within a ten-year period—for example, a mine that opened in 1952 and

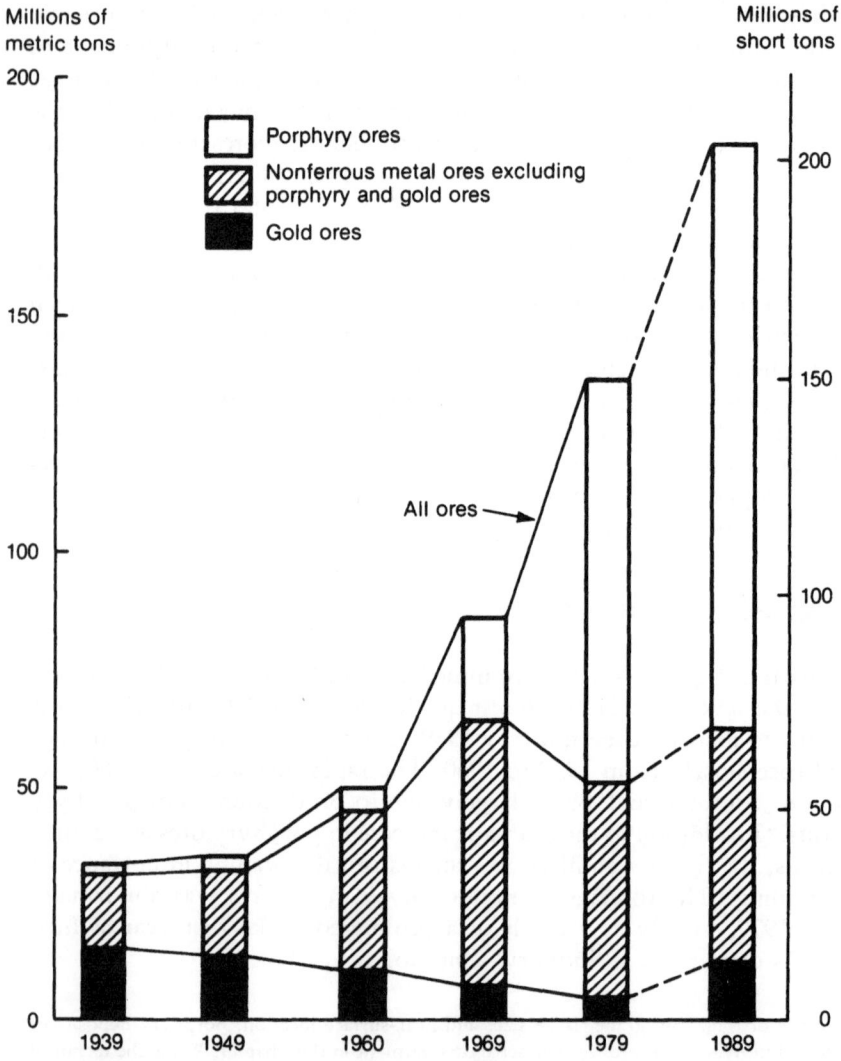

Figure 12-1.   Nonferrous metal ores mined and concentrated in Canada at ten-year intervals, 1939–79, and estimated for 1989 (in millions of metric and short tons). *Source:* Energy, Mines and Resources Canada.

closed in 1958 would not be counted. Excluding gold operations, there were 21 mines in 1939 and an estimated 84 mines in 1989, after 184 mine openings and 120 closings. Only 10 mines were in continuous operation between 1939 and 1979; 8 of these are likely to be in operation by 1989. Figure 12-2 also shows that there are relatively few porphyry mines in

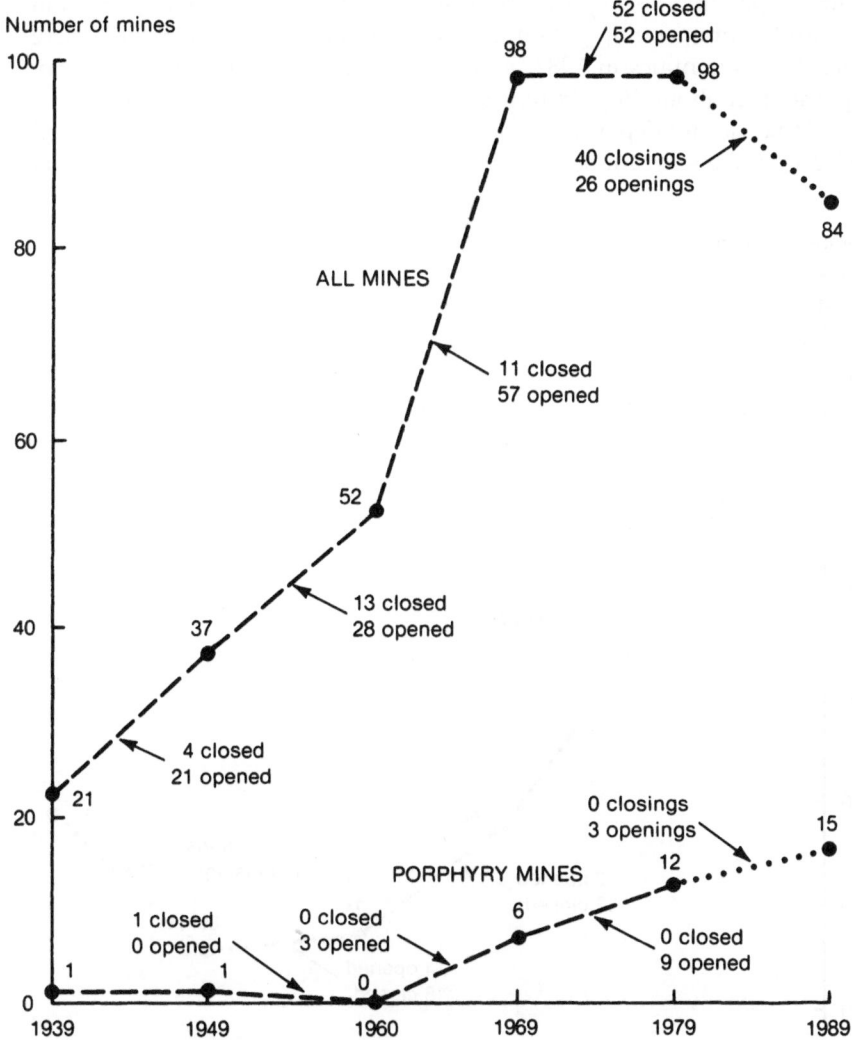

Figure 12-2. Number of operating mines (excluding gold mines) in Canada at ten-year intervals, 1939–79, and estimated for 1989. Openings and closings are from one sample year to the next. The seventy or more very small silver-mining operations in 1939 were grouped in three distinct mining camps. Each of these camps was counted as one mine. *Source:* Energy, Mines and Resources Canada.

Canada, although their contribution to ore mined is large, as indicated in figure 12-1.

The number of gold mines examined is shown in figure 12-3. Beginning with 154 mines in 1939, 52 mines came on-stream, while 176 closed, leaving 30 mines in operation in 1979. Only 10 gold mines were in continuous operation during these four decades. The long, declining trend in the number of operating gold mines appears to have bottomed out—there were 39 active mines in 1982, and production planning now under way suggests that about 56 gold mines could be in operation by 1989.

If Canadian ore deposits of the seven metals analyzed here, discovered and brought on-stream since 1939, had indeed declined in economic at-

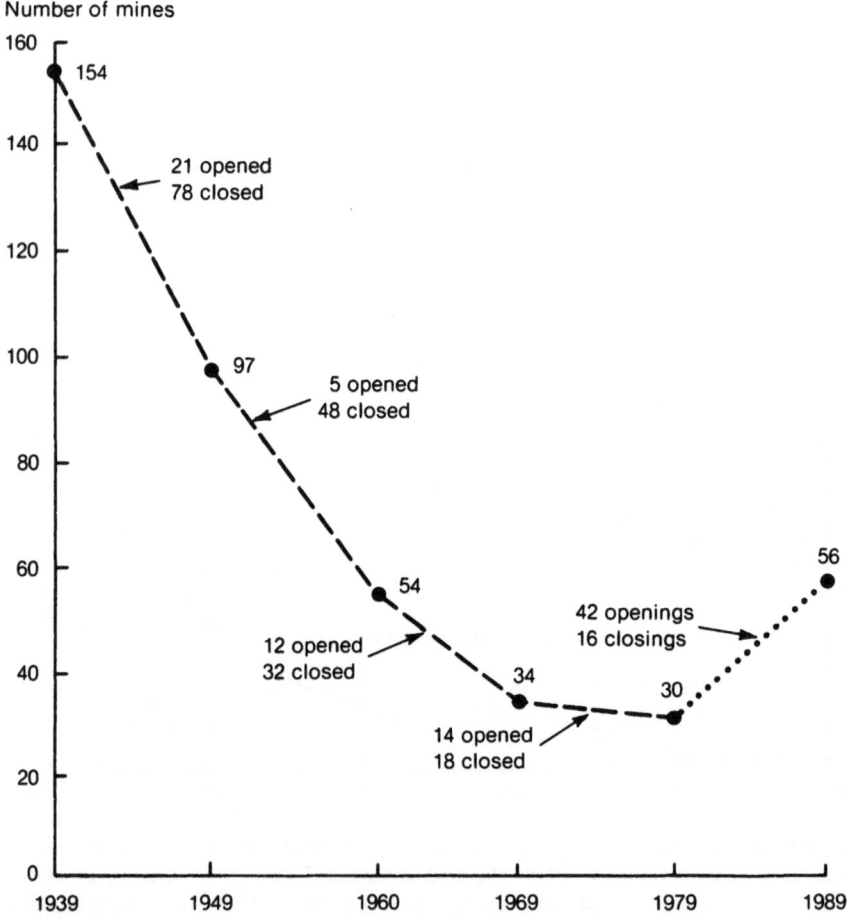

Figure 12-3.   Number of operating gold mines in Canada at ten-year intervals, 1939–79, and estimated for 1989. Openings and closings are from one sample year to the next. *Source:* Energy, Mines and Resources Canada.

tractiveness, the sizable sample of mines that contributed to the strongly rising production during this period would have exhibited a clearly diminishing grade. Discoveries of deposits containing these metals made between 1946 and 1977 alone, which were analyzed in detail by Cranstone (1982), accounted for at least 250 mines. Many additional mines were the result of earlier discoveries.

Since 1946 in Canada, the period required for progressing from discovery to initial production has averaged six years for metal mines, regardless of their size (Martin, Cranstone, and Zwartendyk, 1976; Cranstone, 1982). Therefore, the ore grades mined in each of the years analyzed reflect to some extent the discoveries of the preceding decade shifted back by a six-year lag—that is, some six to sixteen years before each year of analysis—although some operations resulting from even earlier discoveries were still active.

Copper

The number of mines with copper as a principal product or co-product increased from ten in 1949 to forty-four in 1969, following a high rate of copper discoveries in the 1951–56 and 1963–66 periods (Cranstone, 1982). Figure 12-4 illustrates the marked shift, starting in the 1960s, into the mining of porphyry copper ores in British Columbia. Although low-grade, large-tonnage porphyry copper deposits had been mined by open pit methods since the early 1900s in the western United States, their presence was largely unrecognized in Canada until the 1960s. Canadian base metal and precious metal mining focused on underground methods until opportunities arose for supplying the growing Japanese market for copper concentrates. This market provided the impetus for applying large-scale surface mining techniques to near-surface deposits while keeping unit costs competitive with higher-grade deposits that had to be mined with costlier underground methods.

Figure 12-4 also shows that the amounts of copper-bearing ores mined containing only by-product copper were substantial. In 1939 the total by-product copper content of such ores was almost 50 percent of total copper-in-ore produced. Almost all of the by-product copper came from the Sudbury nickel-copper deposits. The copper values in these deposits were still less than half of the nickel values, even though the grade of copper mined at Sudbury in 1939 averaged 1.64 percent, almost equaling the average grade of 1.72 percent of copper ores treated that year.

In the 1960s and 1970s the Sudbury by-product copper became relatively much less significant as production from porphyries rose strongly, especially after 1969 (see figure 12-4). Copper equivalents from all copper mines increased strongly as well after 1969 (figure 12-5). Copper production from nonporphyry sources increased 150 percent from 1939 to 1960

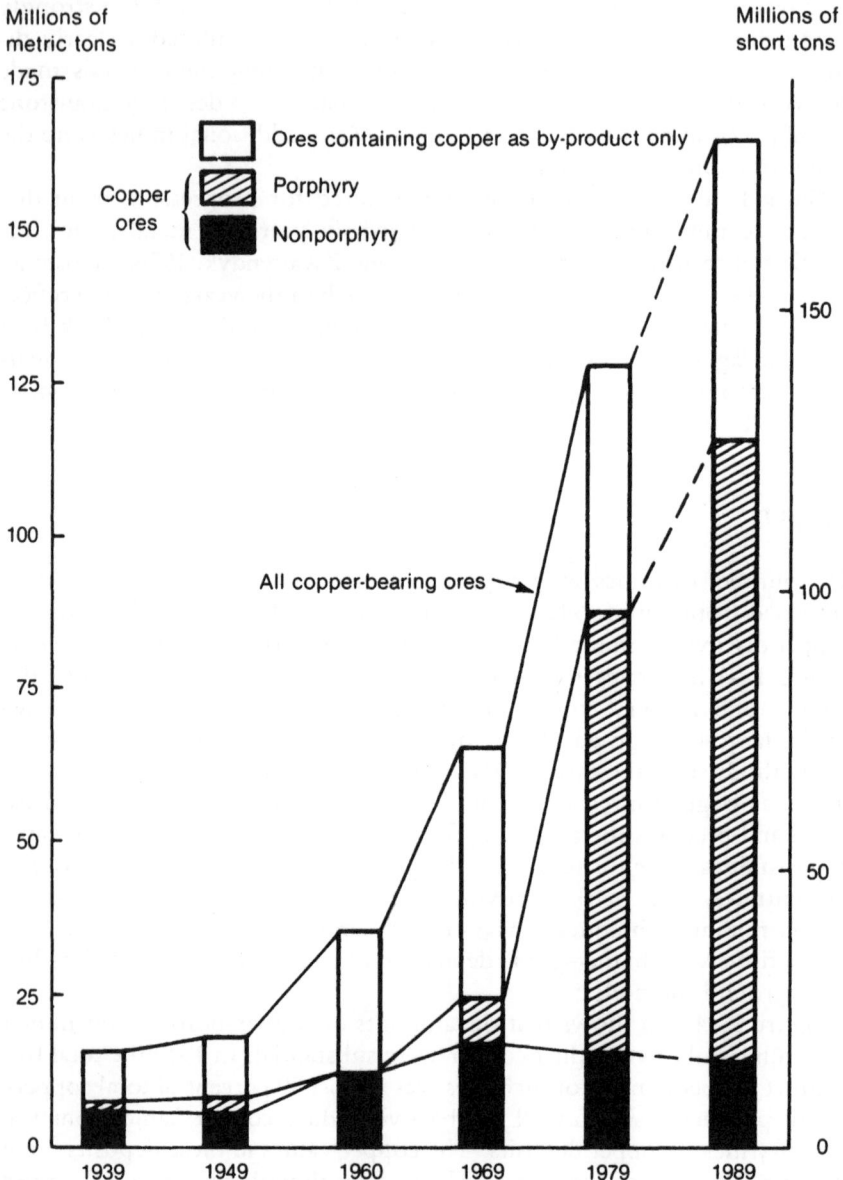

Figure 12-4. Copper-bearing ores mined and concentrated in Canada at ten-year intervals, 1939–79, and estimated for 1989 (in millions of metric and short tons). *Source:* Energy, Mines and Resources Canada.

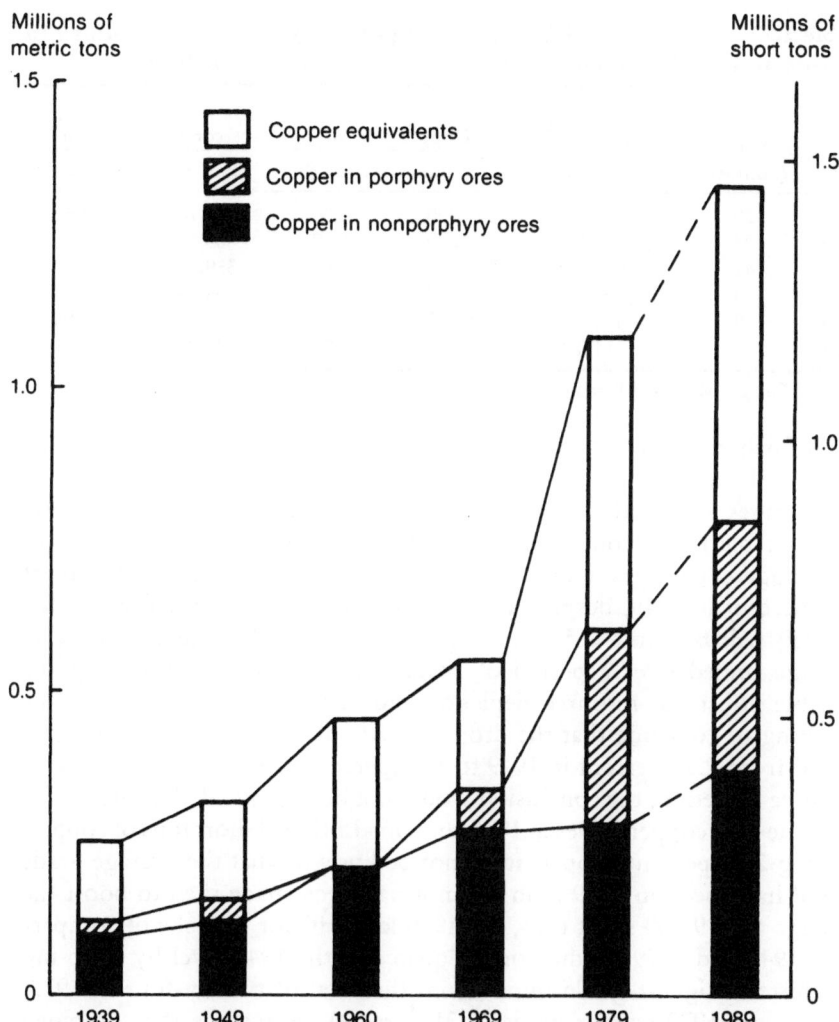

Figure 12-5. Copper and copper equivalents in copper ores mined and concentrated in Canada at ten-year intervals, 1939–79, and estimated for 1989 (in millions of metric and short tons). *Source:* Energy, Mines and Resources Canada.

Table 12-1. Tonnage and Grade of Nonporphyry Copper Ores Mined in
Canada at Ten-Year Intervals, 1939–79, and Estimated for 1989

Year	Mines (number)	Copper ores concentrated (short tons)	Copper content		Including copper equivalents	
			Short tons	Grade (%)	Short tons	Grade (%)
1989	25	17,116,200	392,700	2.30	771,700	4.51
1979	24	19,133,500	319,200	1.67	662,640	3.46
1969	41	19,889,623	325,273	1.64	559,513	2.82
1960	23	14,807,424	284,147	1.92	495,281	3.34
1949	9	6,376,200	141,900	2.23	315,600	4.95
1939	8	6,614,000	113,700	1.72	245,200	3.71

Source: Energy, Mines and Resources Canada.

and a further 12 percent by 1979; an additional 25 percent increase can be expected by 1989.

Tonnages and grades of nonporphyry copper ores mined are summarized in table 12-1. Nonporphyry ores mined in 1989 are expected to have an average copper grade of 2.30 percent, matching the historical peak of 2.23 percent in 1949, despite the fact that tonnage of copper ores treated in 1989 will be some 2.5 times higher than in 1949. The difference in tonnage mined over a period of years is important in considering the possible effects of resource depletion. For example, while it might be tempting to conclude that the drop in grade of nonporphyry copper ore mined from 2.23 percent in 1949 to 1.67 percent in 1979 is attributable to resource depletion, the conclusion would not be warranted. The ore mined from the nine copper mines in 1949 contained 141,900 short tons of copper, grading 2.23 percent copper. It cannot be known what the average grade would have been in 1949 if an attempt had been made then to boost the tonnage to 319,200 short tons, the 1979 level. If, for the sake of comparison, 1949 and 1979 production is equated at the 1949 level by choosing the thirteen highest-grade mines from the total of twenty-four in 1979, the average 1979 copper grade (2.31 percent), as well as the combined copper and copper equivalent grade (5.64 percent), would *exceed* the 1949 grades. Thus, the outlook for an average grade in 1989 comparable to that in 1949 is more significant in view of the much higher production level in 1989 than in 1949.

The decline in grades of porphyry copper ores (table 12-2) can be explained without reference to depletion. One relatively rich but small underground porphyry mine was in operation in British Columbia from before 1939 until 1957. The grade of this single, small, unusual mine represents all porphyry mining in 1939 and 1949 (table 12-2). Large-scale open pit mining of porphyry deposits did not begin in Canada until 1962 when Bethlehem Copper Corporation started its operation in the Highland Valley area of British Columbia. In 1966 a second open pit porphyry mine in British Columbia came on-stream, and in 1968 a third opened, this time in the Gaspé region of Quebec. Between 1969 and 1979 the production of copper-in-ore from porphyry mines increased by a factor of seven due to expansion at these three mines and to the opening of six additional, large, open pit mines in British Columbia. The grade decline from 0.57 percent copper in 1969 to 0.45 percent copper in 1979 is attributable mostly to the fact that the earlier operations came on-stream cautiously, initially mining higher-grade ores at gradually increasing rates of production. By 1979 the three pioneer mines accounted for about 20 percent of the copper-in-ore from porphyry mines, with an average grade of 0.43 percent copper. Grades of copper plus copper equivalent were about the same in 1969 and 1979 and are expected to remain so to 1989.

The large increase in copper-in-ore from the eleven porphyry operations that can be expected by 1989 will have little effect on the grade: the decline from 0.45 to 0.42 percent will be accounted for mostly by significant expansion of production from existing operations. The observed decline in the grade of porphyry copper thus represents the exhaustion of one small deposit that operated in the first two decades of the fifty-year interval studied and the subsequent discovery and opening of mines that, by 1989, will have expanded production of copper-in-ore by some twenty-two times over the 1939–49 levels.

Rather than suggesting depletion effects, porphyry operations show a history of growth based on discovery and the use of new exploitation technology. By 1979, economies of scale in large, open pit operations had led to ore production from prophyry copper deposits that was four times greater than the production from nonporphyry copper mines, and the production from the porphyry deposits was at competitive costs even though the nonporphyry mines had grades almost four times as high, together with much higher values in associated metals. The copper content was of the same order of magnitude in porphyry and nonporphyry ores.

Table 12-2. Tonnage and Grade of Porphyry Copper Ores Mined in Canada at Ten-Year Intervals, 1939–79, and Estimated for 1989

Year	Mines (number)	Copper ores concentrated (short tons)	Copper content		Including copper equivalents	
			Short tons	Grade (%)	Short tons	Grade (%)
1989	11	111,475,150	466,300	0.42	711,000	0.64
1979	9	78,234,800	352,100	0.45	528,660	0.68
1969	3	8,694,045	49,344	0.57	56,419	0.65
1960	0	0	0	0	0	0
1949	1	1,806,800	22,200	1.23	27,200	1.51
1939	1	1,450,400	20,000	1.38	24,000	1.66

Source: Energy, Mines and Resources Canada.

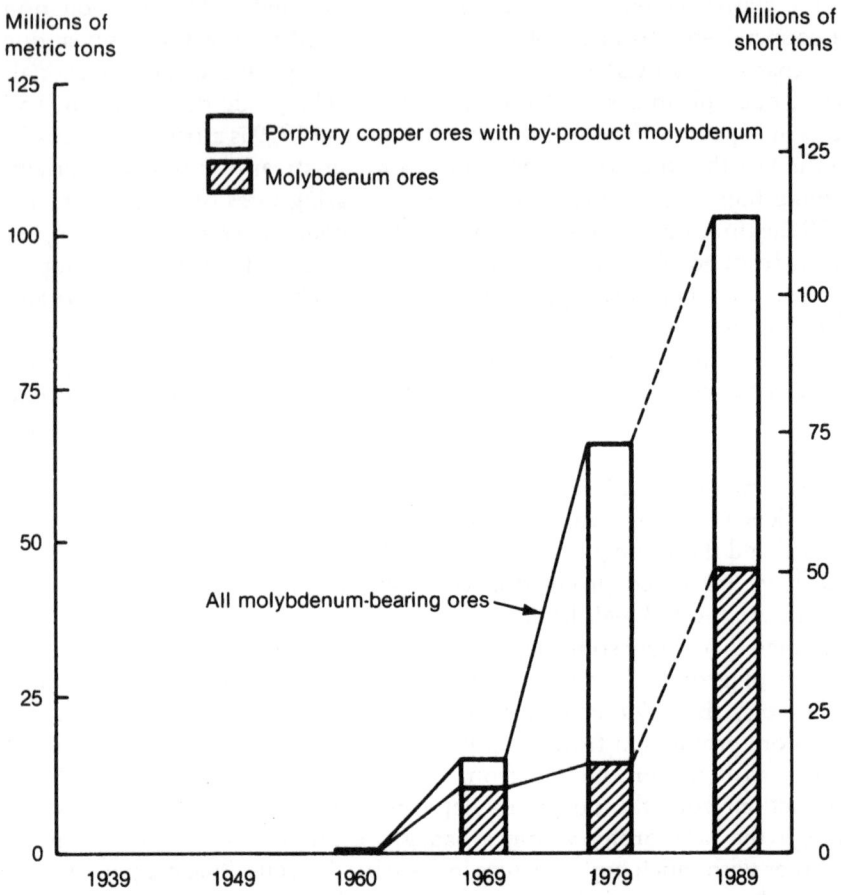

Figure 12-6. Molybdenum-bearing ores mined and concentrated in Canada at ten-year intervals, 1939–79, and estimated for 1989 (in millions of metric and short tons). *Source:* Energy, Mines and Resources Canada.

Molybdenum

As is the case with copper, molybdenum production in Canada comes from both porphyry and nonporphyry sources. One small nonporphyry mine accounted for all the molybdenum ore production in 1960. Although a few small, high-grade nonporphyry mines came on-stream, they were overshadowed by new porphyry mines.

By 1969 almost all molybdenum production in Canada was from porphyry sources. Most of the molybdenum-bearing ore now being mined is porphyry copper ore in which molybdenum is only a by-product (figure 12-6). In terms of contained molybdenum, these by-product ores accounted in 1982 for some 27 percent of all the molybdenum-in-ore mined.

Table 12-3. Tonnage and Grade of Porphyry Molybdenum Ores Mined in Canada in 1960, 1969, 1979, and Estimated for 1989

Year	Mines (number)	Molybdenum ores concentrated (short tons)	Molybdenum content		Including molybdenum equivalents	
			Short tons	Grade (%)	Short tons	Grade (%)
1989	4	26,473,100	17,270	0.065	18,750	0.071
1979[a]	3	22,800,094	14,792	0.065	16,194	0.071
1969	3	12,535,500	15,268	0.122	15,268	0.122
1960	0	0	0	0	0	0

Source: Energy, Mines and Resources Canada.
[a]Data for the Endako mine in British Columbia, one of the three mines included here, are for 1980, as that mine's operations were curtailed in 1979 by an eight-and-one-half month strike. The mine, owned by Placer Development Ltd, is Canada's largest producer of molybdenum.

Again, as in the case of copper, the shift to open pit porphyry production allowed economies of scale, so that lower grades could be mined without an increase in unit cost. Three molybdenum (porphyry) mines were in operation in Canada by 1969 (table 12-3), producing at grades above their average ore reserve grades. By 1979, two of these mines were mining molybdenum ores that were close to their average ore reserve grades, one had closed for lack of markets, and a copper-molybdenum co-producer had come on-stream. All four of these porphyry molybdenum mines are projected to be in operation by 1989, producing at grades equal to their average ore reserve grades. Thus, over their brief mining history the porphyry molybdenum mines produced at grades above the average of their ore reserves in their early years and then dropped back to those averages. This record is too limited to shed any light on possible depletion effects on Canada's porphyry molybdenum resources.

The numbers and sizes of nonporphyry molybdenum mines are too small to allow any conclusions with regard to grade trends (table 12-4). Only one mine was operating in 1960, at a grade of 0.201 percent molybdenum. This grade had dropped to 0.137 percent molybdenum by 1969,

Table 12-4. Tonnage and Grade of Nonporphyry Molybdenum Ores Mined in Canada, 1960, 1969, 1979, and Estimated for 1989

Year	Mines (number)	Molybdenum ores concentrated (short tons)	Molybdenum content		Including molybdenum equivalents	
			Short tons	Grade (%)	Short tons	Grade (%)
1989	1	750,000	900	0.120	2,450	0.327
1979	0	0	0	0	0	0
1969	4	848,966	1,459	0.172	1,513	0.178
1960	1	215,609	433	0.201	454	0.211

Source: Energy, Mines and Resources Canada.

when three additional operations resulted in an average molybdenum grade of ore treated of 0.172 percent. Molybdenum equivalents at all four mines were in the form of bismuth and had little effect on the overall grade of ore mined. No nonporphyry molybdenum mines were in operation in 1979. By 1989 a new tungsten mine with molybdenum co-product in New Brunswick is expected to produce at a grade of molybdenum and its equivalent significantly higher than that in 1960—0.327 percent compared with 0.211 percent.

Zinc

The tonnages of zinc ores mined increased sharply during the 1960s (figure 12-7), as did their total zinc content (figure 12-8). The increases resulted from the establishment of new mining camps during the 1950s and 1960s in various regions of Canada—in New Brunswick, northern Ontario and Quebec, the Yukon Territory, and the Northwest Territories (including the Arctic Islands). The tonnage of ores with zinc as a by-product has been gaining in significance (figure 12-7), though the zinc content of these ores is still well below 10 percent of the zinc contained in all ores concentrated.

Zinc production and the grades of ore concentrated are shown in table 12-5. Despite a tripling of the total amount of zinc produced from all zinc ores after 1960, the average grade mined was somewhat higher in 1979 and the early 1980s than it was in 1960. The average grade was still higher in 1949, at 6.16 percent, but not higher than that for the top four producers in 1979; these four produced a similar amount of zinc from ore, grading 9.66 percent zinc and 18.15 percent zinc and zinc equivalents.

Clearly, the trend in zinc ore grades gives no indication of resource depletion. As is the case with nonporphyry copper ores, the content of associated metals in Canadian zinc ores matches the zinc metal content in value, thus doubling the effective grade of zinc ores.

Table 12-5. Tonnage and Grade of Zinc Ores Mined in Canada at Ten-Year Intervals, 1939–79, and Estimated for 1989

Year	Mines (number)	Zinc ores concentrated (short tons)	Zinc content		Including zinc equivalents	
			Short tons	Grade (%)	Short tons	Grade (%)
1989	21	25,778,500	1,541,800	5.98	3,408,000	13.22
1979	18	23,528,300	1,418,900	6.03	3,009,300	12.79
1969	31	22,083,300	1,481,800	6.76	3,068,300	13.89
1960	15	9,670,800	472,100	4.88	1,229,000	12.71
1949	17	6,288,800	387,312	6.16	1,097,966	17.46
1939	5	6,571,692	301,203	4.58	1,043,523	15.88

Source: Energy, Mines and Resources Canada.

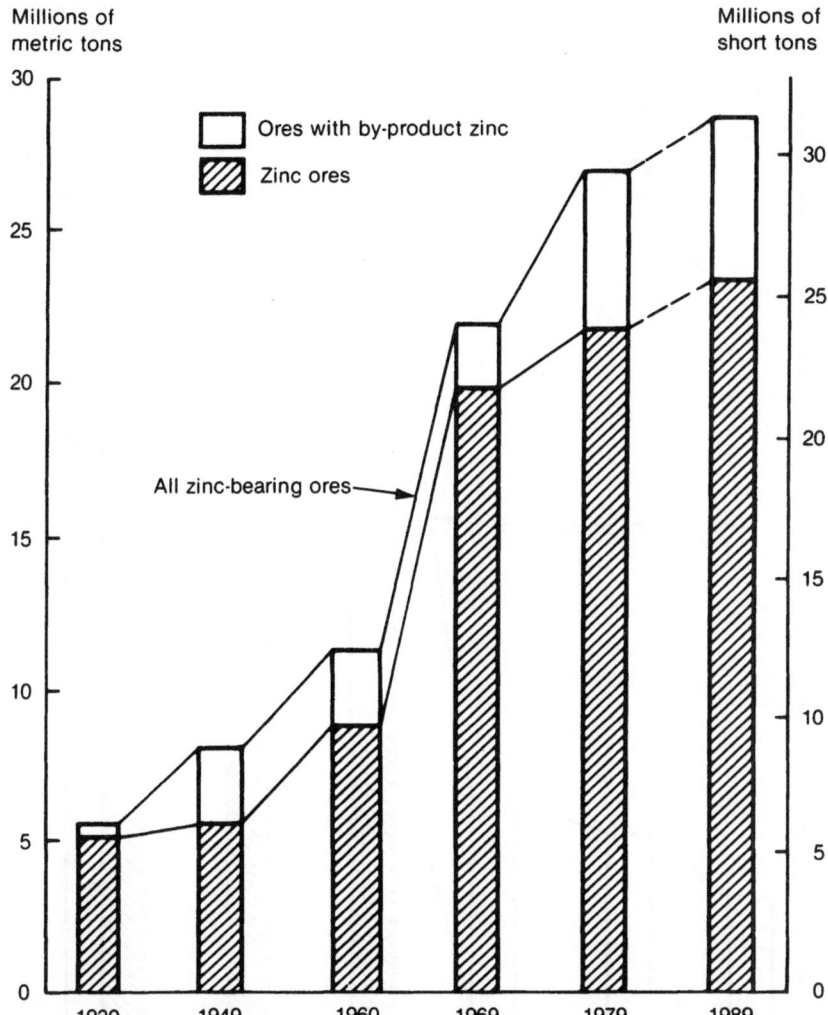

Figure 12-7. **Zinc-bearing ores mined and concentrated in Canada at ten-year intervals, 1939–79, and estimated for 1989 (in millions of metric and short tons).** *Source:* Energy, Mines and Resources Canada.

Lead

Mine production of ores containing lead as well as the tonnage of their contained lead and lead equivalents rose over the entire period, most sharply from 1960 to 1969 (figures 12-9 and 12-10). Although the tonnage of ore mined with by-product lead has become substantial (figure 12-9), the lead content of such ores is only about 10 percent of all the lead mined.

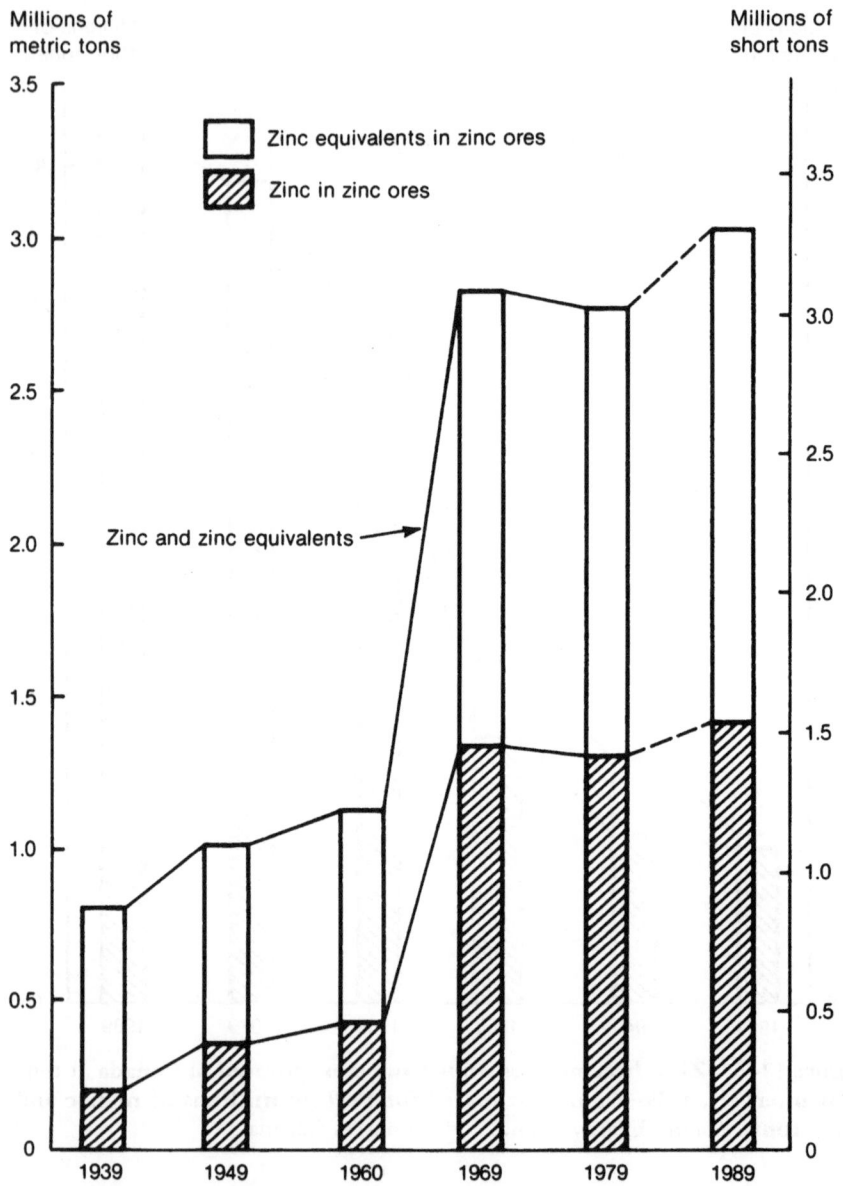

Figure 12-8. Zinc and zinc equivalents in zinc ores mined and concentrated in Canada at ten-year intervals, 1939–79, and estimated for 1989 (in millions of metric and short tons). *Source:* Energy, Mines and Resources Canada.

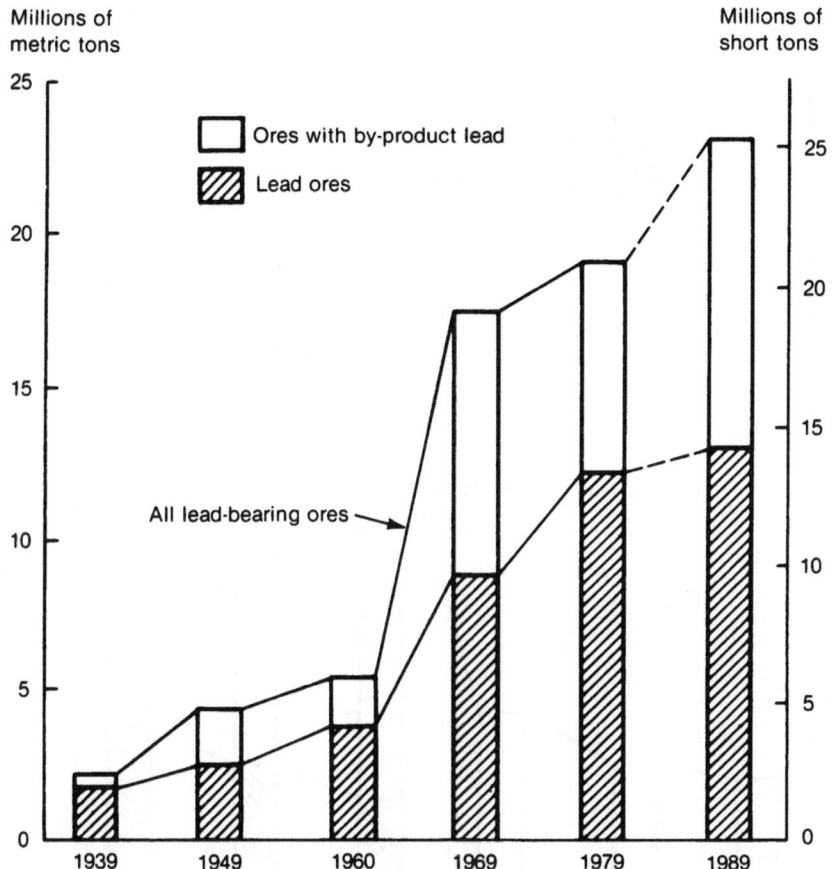

Figure 12-9. Lead-bearing ores mined and concentrated in Canada at ten-year intervals, 1939–79, and estimated for 1989 (in millions of metric and short tons). *Source:* Energy, Mines and Resources Canada.

Lead equivalents in lead ores more than doubled the effective lead tonnage during the 1939–60 period and almost tripled it in the 1969–89 period (figure 12-10). However, metal content rose at a lower rate than that of ore production. Lead tonnages, grades, and equivalents are summarized in table 12-6, which shows that the ore grade of lead and of lead equivalents has dropped steadily from 1939 to 1979 but is expected to recover somewhat by 1989. The decline in the grade of lead is a direct result of the dominant position in lead production held by the rich Sullivan mine of Cominco Ltd. in British Columbia: that mine accounted for more than four-fifths of the lead in lead ores in 1939 and 1949, dropping to about two-thirds in 1960 and less than one-third by 1969. Between 1939 and 1969, lead grades at the Sullivan mine dropped from about 9 percent to

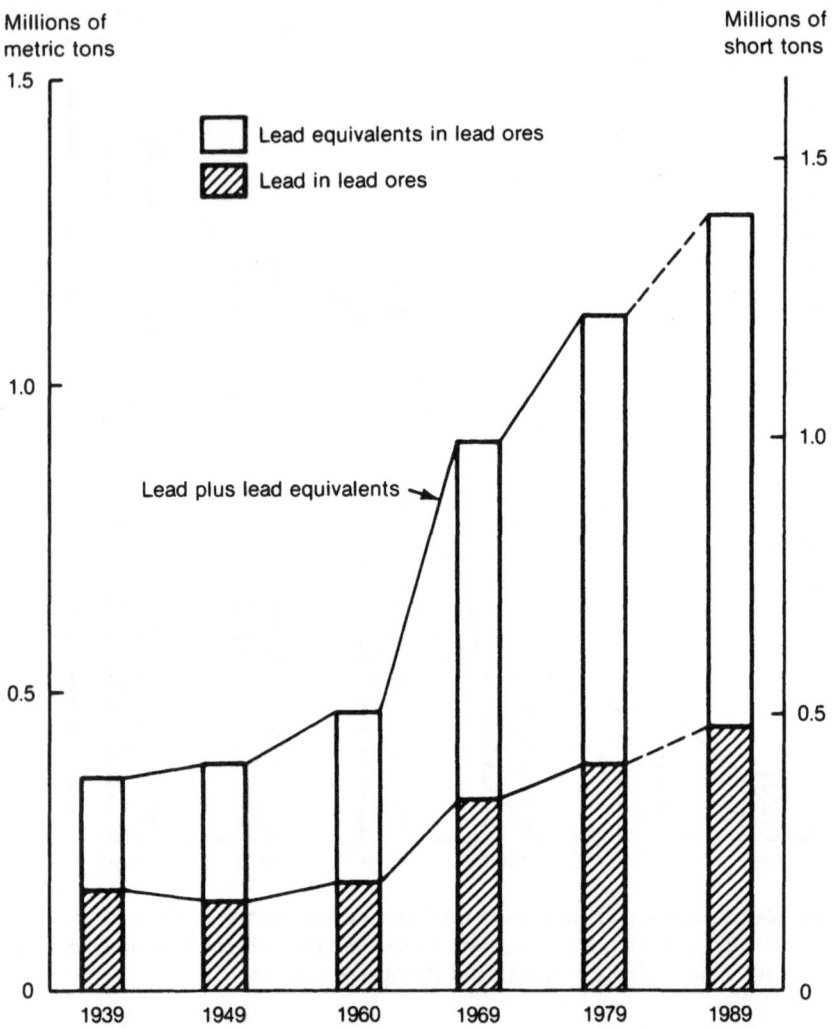

Figure 12-10.   Lead and lead equivalents in lead ores mined and concentrated in Canada at ten-year intervals, 1939–79, and estimated for 1989 (in millions of metric and short tons). *Source:* Energy, Mines and Resources Canada.

5 percent. Similarly, the high silver and zinc content of Sullivan ores were the main determinant of lead equivalents. Thus, the decline in the grade of lead reflects the predominance of one giant mine with a rich early history among a small number of lead mines.

No other ore body of a size or lead grade comparable to that of the Sullivan mine has been discovered and brought into production in Canada. The total amount of lead contained in lead ores mined in recent decades has doubled as a result of production from new mines in which lead is a

Table 12-6.   Tonnage and Grade of Lead Ores Mined in Canada at Ten-Year Intervals, 1939–79, and Estimated for 1989

Year	Mines (number)	Lead ores concentrated (short tons)	Lead content		Including lead equivalents	
			Short tons	Grade (%)	Short tons	Grade (%)
1989	5	14,364,600	496,800	3.53	1,409,500	10.02
1979	5	13,228,337	422,641	3.20	1,222,361	9.25
1969	8	9,633,387	352,134	3.66	990,639	10.28
1960	6	4,187,564	201,921	4.82	511,482	12.21
1949	5	2,774,984	173,676	6.26	421,758	15.20
1939	2	2,559,733	224,017	8.75	505,380	19.74

Source: Energy, Mines and Resources Canada.

co-product. Co-product status in this study is determined by the relatively high 1979 lead prices; more representative lead-zinc price differentials would shift much of the lead content in ores shown in table 12-6 for the years 1969, 1979, and 1989 into the by-product category, leaving the Sullivan mine as the sole lead mine for 1979 at an average grade of 5.33 percent lead (the five lead co-producers averaged only 2.77 percent lead).

Even the largest of mines eventually runs out of ore. In Canada, lead from the Sullivan mine—the only mine to have produced significant quantities between 1939 and 1979—is declining in grade. But the decline in ore grade at the mine of Canada's largest lead producer cannot be equated with the depletion of all lead resources in Canada. In new ore bodies that have been discovered and developed, lead is secondary to zinc, but the rate at which such ores are being mined is high enough to bring about a steady rise in the total lead content of lead-bearing ores mined in Canada.

Nickel

There are two major nickel-mining areas in Canada—one at Sudbury, Ontario, and the other at Thompson, Manitoba. Nickel production from these centers has eclipsed that from other regions in the country; the production from other regions has had no effect on average national ore grades mined. Production at Sudbury dates back to the nineteenth century, while mines at Thompson began to come on-stream in 1961 as a result of discoveries made between 1954 and 1962. Canadian production of nickel as a co-product or by-product has been insignificant, but nickel equivalents—mostly by-product copper—are of importance, especially in the Sudbury region.

The ore grades of nickel and associated by-product copper at Sudbury were at their peaks in 1939 and 1949 (table 12-7). By 1960 the number of mines had more than doubled, lowering the average grade of nickel and of nickel equivalents. The figures for 1969 record the highest production,

Table 12-7. Tonnage and Grade of Nickel Ores Mined in Canada at Ten-Year Intervals, 1939–79, and Estimated for 1989

Year	Mines (number)	Nickel ores concentrated (short tons)	Nickel content		Including nickel equivalents	
			Short tons	Grade (%)	Short tons	Grade (%)
1989	23	15,580,850	225,000	1.44	293,800	1.88
1979[a]	17	16,952,092	242,232	1.43	318,258	1.88
1969	19	23,352,419	297,482	1.27	375,505	1.61
1960	14	20,353,926	288,642	1.42	373,965	1.84
1949	6	9,660,784	154,573	1.60	213,756	2.21
1939	6	7,860,485	126,706	1.61	188,969	2.40

Source: Energy, Mines and Resources Canada.
[a]Data are from 1980 for Ontario operations of Inco Limited, which were curtailed by a strike from September 1978 to June 1979.

but the lowest grades, for the years analyzed; 1969 was a year of high nickel prices and unfilled world demand, exacerbated by a strike at Inco Limited's Sudbury operations that led to mill throughput of all available ore.

The integrated operational capabilities of Inco Limited at Sudbury are now constrained by sulfur dioxide emission controls imposed by the government of Ontario on Inco's smelter. Canada's other major nickel producer, Falconbridge Limited, also located at Sudbury, produces about one-quarter as much nickel-in-ore as do Inco's Sudbury operations, with slightly higher grades. The grade of ore from Inco's Thompson operations is about twice that of its Ontario operation, but annual production of nickel-in-ore from Thompson is only about one-third of Inco's Sudbury operations. Thus, grade decline at Sudbury is offset somewhat by Thompson's superior grade. If markets warranted, production facilities could be greatly expanded at Thompson.

Canada's nickel ore reserves are huge, and the full extent of the Thompson and Sudbury ore complexes is yet to be determined. The average nickel grade mined has not changed much since 1960 and is not expected to change in the 1980s.

Silver

There is some silver in the ores of almost all mines in Canada that produce one of the metals discussed in this chapter. Although silver ores constitute only a small fraction of all silver-bearing ores treated, they account for a high percentage of the total silver mined from 1939 to 1979. (Data limitations preclude estimates for 1989.)

It is particularly difficult to determine grade trends for silver ores for several reasons. First, this study uses 1979 metal prices, and, at $11.90 per

Table 12-8.   Tonnage and Grade of Silver Ores Mined in Canada at Ten-Year Intervals, 1939–79

Year	Type[a]	Mines (number)	Silver ores treated (short tons)	Silver content Troy ounces	Silver content Grade (oz/short ton)	Including silver equivalents Troy ounces	Including silver equivalents Grade (oz/short ton)
1979	P	13	456,260	7,393,980	16.17	8,309,322	18.17
	C	7	10,982,334	24,211,119	2.20	155,720,677	14.18
	Total	20	11,439,594	31,605,099	2.76	164,029,999	14.34
1969	P	14	1,206,736	11,726,174	9.72	15,568,776	12.90
	C	6	7,919,588	27,273,950	3.44	99,131,603	12.51
	Total	20	9,126,324	39,000,124	4.27	114,700,379	12.57
1960	P	10	657,931	13,789,676	20.97	17,688,688	26.84
	C	3	4,382,503	9,872,180	2.25	44,036,103	10.05
	Total	13	5,040,434	23,670,856	4.70	61,724,791	12.25
1949	P	12	551,242	6,423,529	11.65	8,623,059	15.66
	C	3	2,659,931	7,891,583	2.97	34,621,581	13.02
	Total	15	3,211,173	14,315,112	4.46	43,253,640	13.47
1939	P	3[b]	142,524	6,174,514	43.32	6,748,714	47.35
	C	1	2,091,064	8,406,077	4.02	37,596,121	17.98
	Total	4	2,233,588	14,580,571	6.52	44,344,835	19.85

Source: Energy, Mines and Resources Canada.

[a]P = ores in which silver was a principal product. C = ores in which silver was a co-product.

[b]The seventy or more very small silver mining operations in 1939 were grouped in three distinct mining camps, each of which was counted as one mine in this study.

troy ounce in 1979, silver happened to be at its peak price for the entire 1939–79 period, which relatively depresses the value of silver equivalents in the base metal ores containing co-product silver. To take this effect into account, table 12-8 indicates tonnage and grade both for ores mined in which silver was a primary product and for those in which silver was a co-product.

Another reason for the difficulty in determining grade trends for silver ores is that most of the mines classified here as co-producers of silver are close to the borderline separating co-product from by-product producers by the definitions adopted in this study. As a result, some mines may have produced silver as a co-product in some of the years analyzed and as a by-product in others. To provide consistent tonnage figures, mines such as Cominco's Sullivan mine that produced silver as a co-product for at least two of the years analyzed and as a by-product for the other years were counted as co-product silver mines in each year analyzed.

In 1939 some 70 "mines"—mostly very small operations—with silver as the principal product produced a total of only 142,500 tons of silver ores with a very high average grade of 43.3 ounces per ton. In contrast, the sole co-product silver producer (Sullivan) averaged only 4 ounces per ton, for a combined total average grade of 6.5 ounces per ton. The high (9 percent) lead grade at Sullivan in 1939 is largely responsible for the peak silver-equivalent grade in 1939. In the subsequent, 1949–79 period, the grades of silver ore mined by producers of silver as a principal product fluctuated considerably, although the grades of combined silver and silver equivalents from both principal producers and co-producers of silver remained fairly stable. It is noteworthy that during the 1960s the new copper-zinc-silver mines brought into production with silver as a co-product introduced a new, large source of silver that more than doubled the annual mine production of silver in silver ores.

In this complex record of silver mining, it would be difficult indeed to discern any evidence of depletion effects on the basis of changes in grades of silver ores mined in Canada.

Gold

Before World War II, gold was the focus of Canadian metal mining. Of the 175 metal-mining operations in 1939, 154 were gold mines,[5] which accounted for one-half of the annual ore tonnage mined (see figures 12-1, 12-2, and 12-3). By 1979 only 30 of the 129 nonferrous mining operations were gold mines, accounting for only 4 percent of the ore tonnage produced. The total gold content of gold ores mined in 1979 was only one-quarter of that recorded in 1939 (figure 12-11). That decline was offset to a small degree by the production of by-product gold from other metal ores (figures 12-11 and 12-12). By 1979 the total gold content of all gold-bearing ores mined was about 40 percent of the 1939 total (figure 12-11).

From 1939 to 1969 the average grade of gold ores mined showed little change (table 12-9). It had, however, declined appreciably by 1979 in response to the steep rise in the gold price during the 1970s, although this decline in grade seems to have bottomed out.

Gold mining is now experiencing a revival, as demonstrated by the increase in the number of gold mines from a low of thirty in 1979 to thirty-nine by the end of 1982. Production planning is either under way or is being seriously considered for at least thirty gold deposits throughout Canada. As many as fifty-six gold mines are expected to be in production by 1989. Estimated gold grades for that year, as shown in table 12-9, range from 0.21 to 0.24 ounces per short ton. The recent discoveries of large and relatively high grade gold deposits at Hemlo in northern Ontario will add at least three major gold mines with planned total production of some 1 million troy ounces per year by the early 1990s. As estimates of ore tonnages in the Hemlo area are reported to be in excess of 75 million short tons, at an average grade of 0.25 ounces per short ton, the coming on-

---

[5]The discussion of gold mining in this section excludes placer gold operations.

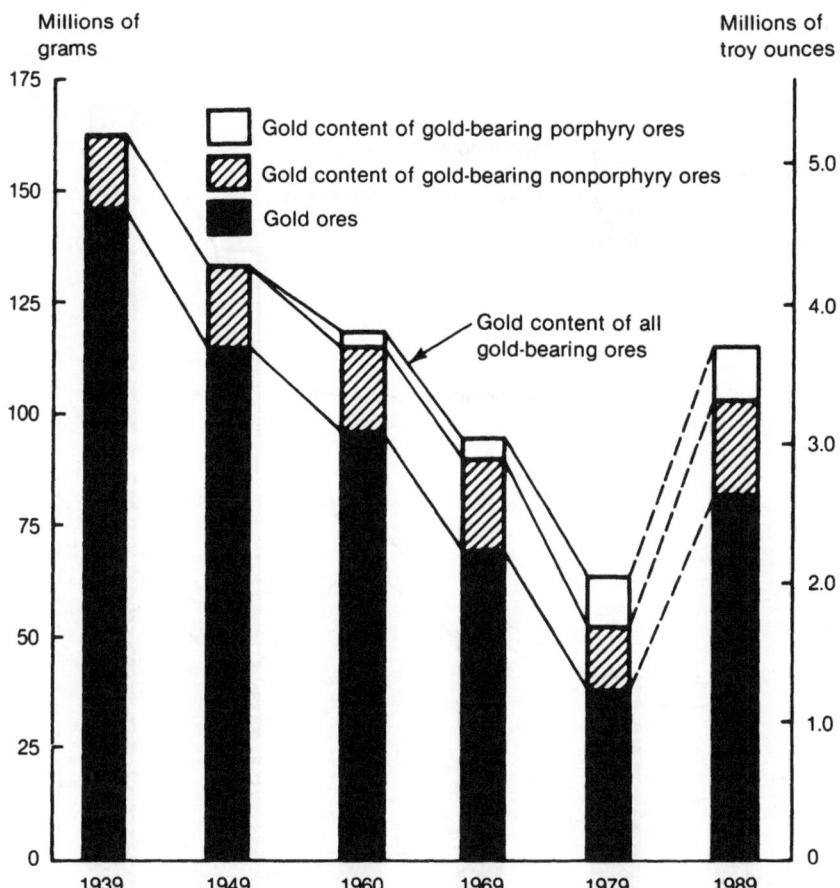

Figure 12-11. **Gold content of gold-bearing ores mined and milled in Canada at ten-year intervals, 1939–79, and estimated for 1989 (in millions of grams and troy ounces).** *Source:* Energy, Mines and Resources Canada.

stream of these three major gold mines may raise the average production grade to a level approaching the average of the 1939–69 era.

Higher gold prices have made lower grades profitable to mine. Therefore, a rising average grade after 1979 means that the effect of producing greater amounts of lower-grade ore is expected to be more than offset by the rising grades of other ores.

The discoveries at Hemlo represent a new deposit type for Canada. In these deposits gold distribution is not related to quartz veins as is typically the case with gold ore in the Canadian Shield. The Hemlo discoveries demonstrate that new types of mineral deposits of world–class size remain to be found in Canada. However, it will take more such major discoveries to return Canada to its peak level of lode gold production—which approached 5 million troy ounces a year from 1939 through 1941.

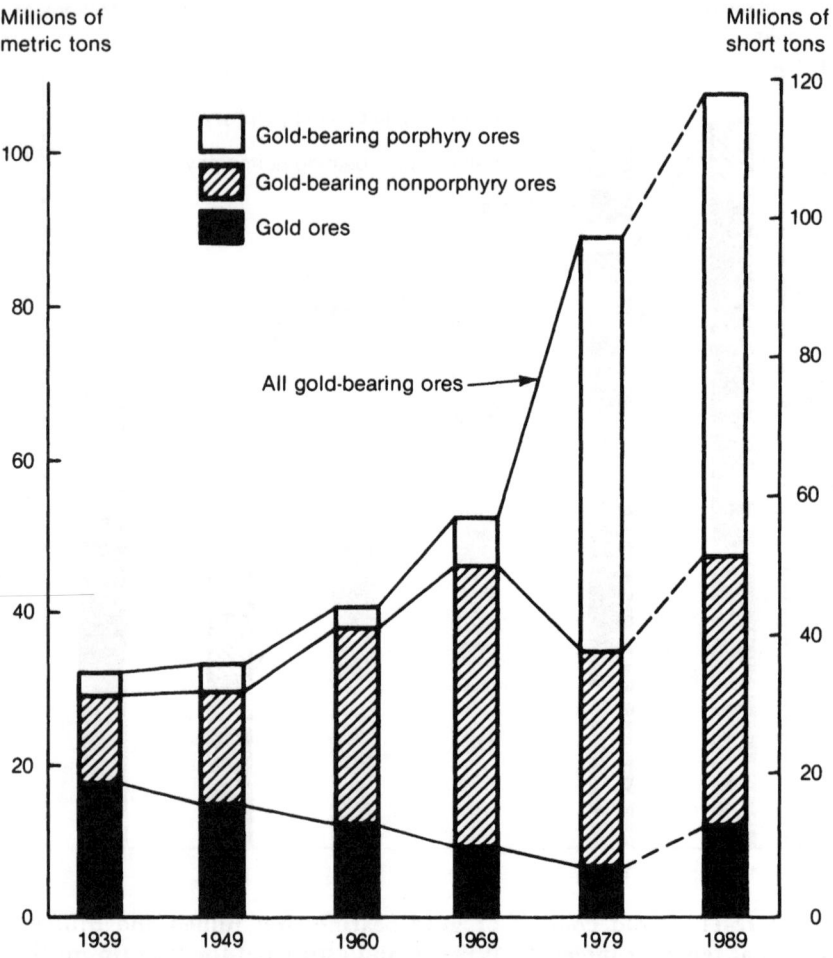

Figure 12-12. Gold-bearing ores mined and milled in Canada at ten-year intervals, 1939–79, and estimated for 1989 (in millions of metric and short tons). *Source:* Energy, Mines and Resources Canada.

Conclusions

The Canadian mining industry has grown significantly over the past fifty years as new deposit types have been found, economic relationships have changed, and technology has evolved. While such major changes preclude simple answers to the question of whether ore grades have declined in Canada, on the whole a consistent decline over time in grades of ore mined is not found for most of the metals analyzed in this study. As seen from the individual analyses of the different types of deposits, no clear decline

Table 12-9. Tonnage and Grade of Gold Ores Mined in Canada at Ten-Year Intervals, 1939–79, and Estimated for 1989

Year	Mines (number)	Gold ores treated (short tons)	Gold content	
			Troy ounces	Grade (oz/short ton)
1989	56	12,873,700	2,729,200	0.21–0.24[a]
1979	30	6,253,775	1,181,143	0.1889
1969	34	9,047,315	2,238,793	0.2475
1960	54	12,076,129	3,039,645	0.2517
1949	97	15,782,249	3,675,824	0.2329
1939	154	18,302,009	4,667,012	0.2550

Source: Energy, Mines and Resources Canada.
[a]The low end of the range includes a number of relatively low-grade deposits with doubtful economic prospects below U.S.$500 per ounce. The high end of the range excludes some of these deposits and is weighted by the large-tonnage, relatively high-grade deposits at Hemlo, Ontario.

in ore grades is evident in the case of copper, molybdenum, zinc, nickel, or silver. The grades of lead and gold have declined, but these declines can be attributed to directly observable reasons more plausible than that of the depletion of all higher-grade resources in Canada.

Although in theory it no doubt pays to mine high-grade deposits first, in reality discovery does not follow a path that leads so systematically from all of the high-grade deposits to all of the low-grade deposits. Moreover, even if depletion ultimately forces society to exploit lower-grade ores, the question remains as to how rapidly the decline will occur. The Canadian experience suggests that the decline is proceeding a good deal more slowly than is generally assumed. Depletion so far has had no clearly identifiable effect on grades of ore mined.

This finding is all the more surprising in view of the fact that there has been no pronounced move toward the less accessible North in Canadian mining. Indeed, Canada's most important gold discovery since the early 1900s—at Hemlo in Ontario in 1982—underlies both the Trans-Canada Highway and the main line of the transcontinental Canadian Pacific Railway. The grade of this deposit equals the average grade of gold mined in Canada prior to the decline in the ore grade mined in the 1970s in response to rising gold prices.

Complicating the analysis of the interaction between depletion and technological advances is the tremendous upswing in total mine production during the 1960s. The thrust of technology was directed less toward counteracting the adverse cost effects of depletion than toward finding ways of exploiting additional types of deposits that would allow the growth of profitable production at a much greater rate but still at comparable costs, notably by the large-scale, open pit mining of large, low-grade porphyry copper deposits. This had the incidental effect of slowing down the depletion rate of massive sulfide copper resources minable by underground methods, a type of deposit that continues to be mined today at grades comparable to those of several decades ago.

Although the preceding observations apply directly only to the Canadian experience, they merit consideration in any economic models of long-term mineral supply. For instance, the absence of a historical rise in metal prices has invariably led to the conclusion that technology must have offset the never-questioned adverse effects of depletion. It may be, however, that the effects of depletion have been so slow as to be hardly noticeable, and had little to do with technological advances that brought lower grades in different deposit types within economic reach.

Thus, the theory that the best deposits are the first to be mined loses much of its practical interest when considered in the light of a discovery record showing that the best deposits are not necessarily found first and in view of the possibility that the dynamics of technology may over time shift our judgment about what deposits are the "best."

For Canada, the record of mining since 1939 rounded off with a look at what can be expected in 1989 forms a fifty-year overview which suggests that exploration geologists have managed to continue discovering deposits comparable in quality to previous discoveries. Thus, the question of "when Canada will run out of good ores" seems premature and can safely be set aside until some time in the twenty-first century.[6]

■

The authors gratefully acknowledge the substantive editorial comments and many helpful suggestions contributed by Jan Zwartendyk of the Mineral Policy Sector at Energy, Mines and Resources Canada. They are also indebted to W. H. Laughlin of the Mineral Policy Sector for providing the information on possible gold mine openings by 1989.

References

Cranstone, D. A. 1982. "An Analysis of Ore Discovery Costs and Rates of Ore Discovery in Canada over the Period 1946 to 1977" (Ph.D. thesis, Harvard University, Cambridge, Mass.).

Energy, Mines and Resources Canada. 1983. *Canadian Mines: Perspective from 1982—Reserves, Production Capability, Exploration, Development.* MR 197 (Ottawa).

Martin, H. L., D. A. Cranstone, and J. Zwartendyk. 1976. *Metal Mining in Canada, 1976–2000.* Mineral Bulletin MR 167 (Ottawa, Energy, Mines and Resources Canada).

---

[6] The authors of this chapter, with more recent data at hand, confirmed their conclusions in 1987.

Index

Index

447